Master Math: Trigonometry

By

Debra Anne Ross

Course Technology PTR

A part of Cengage Learning

COURSE TECHNOLOGY
CENGAGE Learning™

Australia • Brazil • Japan • Korea • Mexico • Singapore • Spain • United Kingdom • United States

COURSE TECHNOLOGY
CENGAGE Learning™

Publisher and General Manager, Course Technology PTR:
Stacy L. Hiquet

Associate Director of Marketing:
Sarah Panella

Manager of Editorial Services:
Heather Talbot

Marketing Manager: Jordan Casey

Senior Acquisitions Editor:
Emi Smith

Interior Layout Tech:
Judith Littlefield

Illustrations and Equations:
Judith Littlefield and
Mike Tanamachi

Cover Designer: Jeff Cooper

Indexer: Larry Sweazy

Proofreader: Jenny Davidson

For product information and technology assistance, contact us at **Cengage Learning Customer & Sales Support, 1-800-354-9706**
For permission to use material from this text or product, submit all requests online at **cengage.com/permissions**
Further permissions questions can be emailed to **permissionrequest@cengage.com**

All trademarks are the property of their respective owners.

Library of Congress Control Number: 2009924622

ISBN-10: 1-59863-985-4
ISBN-13: 978-1-59863-985-8

Course Technology, a part of Cengage Learning
20 Channel Center Street
Boston, MA 02210
USA

Cengage Learning is a leading provider of customized learning solutions with office locations around the globe, including Singapore, the United Kingdom, Australia, Mexico, Brazil, and Japan. Locate your local office at: **international.cengage.com/region**

Cengage Learning products are represented in Canada by Nelson Education, Ltd. For your lifelong learning solutions, visit **courseptr.com**
Visit our corporate website at **cengage.com**

Printed in Canada
2 3 4 5 6 7 12 11 10 09

Table of Contents

Acknowledgments

I am indebted to Dr. Cyndy Lakowske for reading this book for accuracy and for her helpful comments. I am also indebted to Dr. Melanie McNeil, Professor of Chemical Engineering at San Jose State University, for reading the *Master Math* books and for her helpful comments. I am grateful to Dr. Channing Robertson, Professor of Chemical Engineering at Stanford University, for reviewing this book and, in general, for all his guidance. I especially thank my mother, Maggie Ross, for her editorial help.

Without my wonderful agent, Sidney B. Kramer, and the staff of Mews Books, the *Master Math* series would not have been published. Thank you, Sidney! I am also thankful to Ron Fry and the staff of Career Press for their work in publishing and launching the original *Master Math* books as a successful series.

I am grateful to Emi Smith, Senior Acquisitions Editor, and Course Technology, a part of Cengage Learning, for invigorating the *Master Math* series and improving the presentation. I particularly appreciate Judith Littlefield for her tireless and expert work on the illustrations, equations, and layout. I also thank Mike Tanamachi for his work on illustrations. Much thanks to Jenny Davidson for proofreading, Jeff Cooper for cover design, Larry Sweazy for indexing, as well as Stacy L. Hiquet, Sarah Panella, Heather Talbot, and Jordan Casey.

Finally, I deeply appreciate my beautiful and brilliant husband, David A. Lawrence, who worked side-by-side with me as we meticulously edited text and figures.

About the Author

Debra Anne Ross Lawrence is the author of six books of the *Master Math* series: *Basic Math and Pre-Algebra, Algebra, Pre-Calculus, Calculus, Trigonometry*, and *Geometry*. She earned a double Bachelor of Arts degree in biology and chemistry with honors from the University of California at Santa Cruz and a Master of Science degree in chemical engineering from Stanford University.

Her research experience encompasses investigating the photosynthetic light reactions using a dye laser, studying the eye lens of diabetic patients, creating a computer simulation program of physiological responses to sensory and chemical disturbances, genetically engineering bacteria cells for over-expression of a protein, and designing and fabricating biological reactors for in-vivo study of microbial metabolism using nuclear magnetic resonance spectroscopy.

Debra was a member of a small team of scientists and engineers who developed and brought to market the first commercial biosensor system. She managed an engineering group responsible for scale-up of combinatorial synthesis for pharmaceutical development. She also managed intellectual property for a scientific research and development company. Debra's work has been published in scientific journals and/or patented.

Debra is also the author of *The 3:00 PM Secret: Live Slim and Strong Live Your Dreams* and *The 3:00 PM Secret 10-Day Dream Diet*. She is the coauthor with her husband, David A. Lawrence, of *Arrows Through Time: A Time Travel Tale of Adventure, Courage, and Faith*. Debra is President of GlacierDog Publishing and Founder of GlacierDog.com. When Debra is not engaged in all-season mountaineering near her Alaska home, she is endeavoring to understand the incomprehensible workings of the universe.

Introduction

Master Math: Trigonometry is part of the *Master Math* series, which includes *Master Math: Basic Math and Pre-Algebra, Master Math: Algebra, Master Math: Pre-Calculus, Master Math Geometry,* and *Master Math: Calculus. Master Math: Trigonometry* and the *Master Math* series as a whole are clear, concise, and comprehensive reference sources providing easy-to-find, easy-to-understand explanations of concepts and principles, definitions, examples, and applications. *Master Math: Trigonometry* is written for students, tutors, parents, and teachers, as well as for scientists and engineers who need to look up principles, definitions, explanations of concepts, and examples pertaining to the field of trigonometry.

Trigonometry is a visual and application-oriented field of mathematics that was developed by early astronomers and scientists to understand, model, measure, and navigate the physical world around them. Today, trigonometry has applications in numerous fields, including mathematics, astronomy, engineering, physics, chemistry, geography, navigation, surveying, architecture, and the study of electricity, light, sound, and phenomena with periodic and wave properties. Trigonometry is one of the more interesting and useful areas of mathematics for the non-mathematician. This book provides detailed, comprehensive explanations of the fundamentals of trigonometry and also provides applications and examples, which will hopefully provide motivation for students to learn and become familiar with this truly interesting field of mathematics.

Trigonometry involves measurements of angles, distances, triangles, arc lengths, circles, planes, spheres, and phenomena that exhibit a periodic nature. The six trigonometric functions, sine, cosine, tangent, cotangent, secant, and cosecant, can be defined using three different approaches: as ratios of the sides of a right triangle (Chapter 3), in a coordinate system using angles in standard position (Chapter 4), and as arc lengths on a unit circle, called circular functions (Chapter 4). Trigonometric functions are found, described, and illustrated in numerous venues including graphs, equations, vectors, polar coordinates, complex numbers, exponential functions, series expansions, and spherical surfaces.

The contents of this book include a review of basic geometry, definitions pertaining to triangles, definitions of the trigonometric functions and circular functions, graphs and the periodic nature of trigonometric functions, inverse trigonometric functions, trigonometric identities, trigonometric equations and inequalities, vectors, polar coordinates, complex numbers, relationships between trigonometric functions and exponential functions, hyperbolic functions, and series expansions, as well as spherical trigonometry.

A note on calculators and computers: A widely used tool in trigonometry is the graphing calculator. More than 60 of the graphs in this book are drawn using a graphing utility that is a computer-software version of a graphing calculator. Graphing calculators and graphing utility software packages provide a means to quickly graph both simple and complicated equations containing trigonometric functions that would be tedious to do by hand. Many graphing calculators and graphing software packages will graph a range of equation types, including rectangular, polar, and parametric equations. There are also on-line graphing resources. In this book, I used the TI Interactive! software package by Texas Instruments. Computers, in general, are extensively used in mathematical calculations and in recent years have provided a means to model and measure extremely complex aspects of the physical world that would be virtually impossible to do by hand. Computers also allow us to accurately analyze complex measurements and to design intricate structures that expand from strict geometric shapes created by hand to geometrically elaborate architecture.

Chapter 1

Review of Numbers and Coordinate Systems

1.1 Review of Numbers, Including Natural, Whole, Integers, Zero, Rational, Irrational, Real, Complex, and Imaginary Numbers

• The following is a hierarchy of numbers, in which groups above encompass groups below. Each group of numbers is described after the hierarchy.

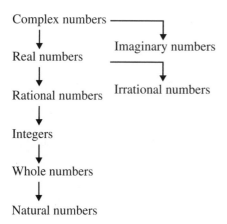

• *Natural numbers* are the original counting numbers beginning with number 1. The set of natural numbers written in the form of a set is:

Natural numbers = {1, 2, 3, 4, 5, 6, 7, ...}

• *Whole numbers* include zero and the natural numbers greater than zero. Whole numbers are depicted on the number line and include zero and numbers to the right of zero.

Whole numbers written in the form of a set is:

Whole Numbers = {0, 1, 2, 3, 4, 5, 6, 7, 8, 9, 10, 11, ...}

• Negative numbers and numbers in the form of fractions, decimals, percents, or exponents are *not* whole numbers.

• *Integers* include positive numbers and zero (whole numbers) and also *negative numbers*. Integers can be depicted on the number line:

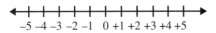

The set of all integers represented as a set is:

Integers = {... –6, –5, –4, –3, –2, –1, 0, 1, 2, 3, 4, 5, 6, 7, ...}

• *Consecutive integers* are integers that are arranged in an increasing order according to their size from the smallest to the largest without any integers missing in between. The following are examples of consecutive integers:

{–10, –9, –8, –7}
{–2, –1, 0, 1, 2, 3, 4, 5}
{99, 100, 101, 102, 103, 104, 105)

• Numbers that are *not* integers include numbers in the form of fractions, decimals, percents, or exponents.

• To *add or subtract negative and positive integers*, remember the number line:

When numbers are added and subtracted, think of moving along the number line. Begin with the first number and move to the right for positive numbers and addition or left for negative numbers and subtraction, depending on the sign of the second number and whether it is being added or subtracted to the first number.

• *Multiplication and division of negative and positive integers*: If a negative and a positive number are multiplied or divided, the result will be a negative number. If two negative numbers are multiplied or divided, the result will be a positive number. Summarizing, where $(+)$ denotes a positive number and $(-)$ a negative number:

$$(+)(+) = (+), (-)(-) = (+), (+)(-) = (-)$$
$$(+)/(+) = (+), (-)/(-) = (+), (+)/(-) = (-), (-)/(+) = (-)$$

• ***Zero*** is both an integer and a whole number, and is also an even number.

If zero is multiplied by any number n, the result is zero: $n \times 0 = 0$

If zero is divided by any number n, the result is zero: $0 / n = 0$

Dividing any number n by zero is undefined: $n / 0 = $ undefined

• A number is a ***rational number*** if it can be expressed in the form of a fraction, x/y, and the denominator is not zero. Every *integer* can be expressed as a fraction and is therefore a rational number. *Whole numbers* are included in the set of integers, and whole numbers are also rational numbers. The result of dividing two integers (with a nonzero divisor) is a rational number. Rational numbers can be represented in the form of decimals that either terminate or end, or as a decimal that repeats one or more digits over and over. For example, $1/3 = 0.33333...$ and $1/4 = 0.25$ are both rational numbers. Remember: The notation for repeating decimals is a bar above the digit or digits that repeat:

$$1/3 = 0.\overline{3}$$

• A number is an ***irrational number*** if it is not a rational number and therefore cannot be expressed in the form of a fraction. Examples of irrational numbers are numbers that possess endless non-repeating digits to the right of the decimal point, such as, $\pi = 3.1415...$, and

$$\sqrt{2} = 1.414213562..., \quad \sqrt{3} = 1.732050807..., \quad \sqrt{5} = 2.236067977...$$

where the ... represents that endless non-repeating digits follow. Irrational numbers not only include square roots but also cubed roots and other roots. Remember: Roots are represented by $\sqrt{x}$ for square roots, $\sqrt[3]{x}$ for cubed roots, and $\sqrt[4]{x}$ for fourth roots. An alternative notation is $x^{1/2}$, $x^{1/3}$, and $x^{1/4}$, respectively. (See *Master Math: Basic Math and Pre-Algebra*, Chapter 9 Roots and Radicals, for an in-depth explanation of roots.)

• The ***real number system*** is comprised of rational and irrational numbers. *Real numbers* also include natural numbers, whole numbers, integers, fractions, and decimals. Real numbers can be expressed as the sum of a decimal and an integer. All real numbers except zero are either positive or negative. All real numbers correspond to points on the real number line, and all points on the number line correspond to real numbers. The real number line reaches from negative infinity ($-\infty$) to positive infinity ($+\infty$).

$$\xleftarrow{\hspace{1cm}}\underset{-4 \quad\quad -3 \quad\quad -2 \quad\quad -1 \;\; -.5 \;\; 0 \quad\quad\; 1 \;\; \sqrt{2} \;\; 2 \;\; 5/2 \;\; 3 \;\; \pi \quad\; 4}{\hspace{8cm}}\xrightarrow{\hspace{1cm}}$$

Real numbers include $-0.5, -2, 5/2$, and π (where $\pi = 3.14159...$).

All numbers to the left of zero are negative.

All numbers to the right of zero are positive.

• ***Complex and imaginary numbers***: Every *real number* corresponds to a point on a number line. There is, however, no real number equal to $\sqrt{-1}$ and no point on a number line corresponding to $\sqrt{-1}$. This means that the equation $x^2 = -1$ has no *real* solutions. Because there is no number that when squared equals -1, the symbol i was introduced, such that $\sqrt{-x} = i\sqrt{x}$, where x is a positive number and $(i)^2 = -1$. For example, $(\sqrt{-4})^2 = (i\sqrt{4})(i\sqrt{4}) = i^2\sqrt{(4)(4)} = -4$. Numbers involving $\sqrt{-1}$ are called ***complex numbers***.

• Complex numbers involve i and are generally in the form $(x + iy)$, where x and y are *real numbers* and i is *imaginary*. In the expression, $(x + iy)$, the x term is referred to as the *real part* and the iy term is referred to as the *imaginary part*. A real number multiplied by i forms an ***imaginary number***, such that:

(real number) $\times$ i = (imaginary number)

A real number added to an imaginary number forms a complex number, such that:

(real number) + (real number)(i) = (complex number)

- *Complex numbers are added or subtracted* by adding or subtracting the real terms and imaginary terms separately. The result is in the form $(x + iy)$. For example:

$$(1 + 2i) + (3 + 4i) = (1 + 3) + (2i + 4i) = 4 + 6i$$

- *Complex numbers are multiplied* as ordinary binomials, and $(i)^2$ is replaced by -1. For example:

$$(1 + 2i)(3 + 4i) = (1)(3) + (1)(4i) + (2i)(3) + (2i)(4i)$$
$$= 3 + 4i + 6i + 8(i)^2 = 3 - 8 + 10i = -5 + 10i$$

- *Complex numbers are divided* by first multiplying the numerator and denominator by what is called the **complex conjugate** of the denominator. Then the numerator and denominator are divided and combined as with multiplication. For example, the complex conjugate of $(3 + 2i)$ is $(3 - 2i)$. The product of a complex number and its conjugate is a real number. Remember to replace $(i)^2$ by -1 during calculations. For example, divide the following:

$$(1 + 2i) \div (3 + 4i) = (1 + 2i)(3 - 4i) \div (3 + 4i)(3 - 4i)$$
$$= (3 - 4i + 6i - 8i^2) \div (9 - 12i + 12i - 16i^2)$$
$$= (3 + 2i - 8(-1)) \div (9 - 16(-1)) = (11 + 2i)/25 = 11/25 + 2i/25$$

1.2 Absolute Value

- The distance between zero and a number on the number line is called the **absolute value** or the **magnitude** of the number. The absolute value is always positive or zero, never negative. The symbol for absolute value of a number represented by n is $|n|$. For example, positive 4 and negative 4 have the same absolute value: $|4| = 4$ and $|-4| = 4$

- Properties of absolute value include (x and y represent numbers):

$$|x| \geq 0, |x - y| = |y - x|, |x|\,|y| = |xy|, |x + y| \leq |x| + |y|.$$

1.3 Significant Digits and Rounding Numbers and Decimals

- When solving complex mathematical or engineering problems, it is important to retain the same number of *significant digits* in the intermediate and final results. The significant digits of a number indicate the accuracy of the number. The significant digits in a number are determined from actual measurements. For example, if the length of a rod is measured in meters to be 22 meters, the number of significant digits is 2. If the rod is measured to the nearest tenth of a meter, and the result is 22.0 meters, the number of significant digits is 3.

• In general, the number of significant digits in a number with no decimal point includes each of the digits from left to right up to the last nonzero digit. The number of significant digits in a number with a decimal point includes each of the digits from left to right beginning with the first nonzero digit and up to the last digit which may be zero. For example, 22,022 has 5 significant digits; 22,000 has 2 significant digits; and 22.200 has 5 significant digits.

• It is often easier to determine the exact number of significant digits when a number is written using *scientific notation*. Remember: Scientific notation uses powers of 10 to represent a number. For example, 2,200,000 = 2.2×10^6. The exponent is +6 because the decimal point was moved from the far right six places to the left. Consider a very small number written using scientific notation, $0.00000023 = 2.3 \times 10^{-7}$. The exponent is –7 because the decimal point was moved from the far left seven places to the right. Positive exponents describe very large numbers and negative exponents describe very small numbers.

• When two or more numbers are combined, the number of decimal places in the resulting numbers should not exceed the number of decimal places in the initial numbers, because the resulting numbers cannot be known with greater accuracy than the original numbers. The number of decimal places is equal to the number of significant digits to the right of the decimal point. Therefore, depending on the least number of significant digits in the initial numbers, rounding intermediate and final results will be required to maintain the decimal places to ones, tenths, hundredths, thousandths, etc. For example, if 45.689 and 1.9654 are added, the accuracy of the result cannot be greater than three decimal places. 45.689 + 1.9654 = 47.6544 = 47.654

• Numbers can be *rounded* to the nearest ten, hundred, thousand, etc. or in decimals to the nearest tenth, hundredth, thousandth, etc. To round numbers and decimals, the last retained digit should either be increased by one or left unchanged according to the following rules:

1. If the left most digit to be dropped is less than 5, leave the last retained digit unchanged.

2. If the left most digit to be dropped is greater than 5, increase the last retained digit by one.

3. If the left most digit to be dropped is equal to 5, leave the last retained digit unchanged if it is even, or increase the last retained digit by one if it is odd. In other words, when the left most digit to be dropped is 5, if the digit preceding the 5 is even, do not change the last retained digit, or if the digit preceding the 5 is odd, round the last retained digit up. For example, 8.45 rounds to 8.4 and 8.55 rounds to 8.6. (Note that other rules for rounding suggest that if the left most digit to be dropped is 5 or greater, *always* increase the last, retained digit by one.) Rounding up and down for odd and even may be particularly beneficial in reducing rounding errors when many digits are being rounded off.

• Decimals may be rounded to the nearest tenth, hundredth, thousandth, etc., depending on how many decimal places there are and the accuracy required. For example:

45.64 rounded to the nearest ones is 46.

45.689 rounded to the nearest tenth is 45.7.

1.545454 rounded to the nearest ten-thousandth is 1.5454 using even/odd strategy.

1.545454 rounded to the nearest ten-thousandth is 1.5455 using the always round 5 up strategy.

• Remember when *comparing the size or value of decimals*, such as 0.0076 and 0.00076, in order to determine which decimal is larger or smaller, the following procedure can be applied:

1. Place the decimals in a column.
2. Align the decimal points.
3. Fill in zeros to the right so that both decimals have the same number of digits to the right of the decimal point.
4. The larger decimal will have the largest digit in the greatest column (farthest to the left).

1.4 Review of Coordinate Systems, Including Two- and Three-Dimensional Rectangular Coordinates, Polar Coordinates, Cylindrical Coordinates, and Spherical Coordinates

• Real numbers can be identified with points on a number line and depicted on the *real number line*. A number line is a one-dimensional coordinate system. If a number is represented by a point on the number line, the number is called the *coordinate* of that point. Pairs of real numbers that define a point on a plane can be depicted by identifying them with the two axes of a ***two-dimensional coordinate system***. Points in three-dimensional space can be depicted by identifying them with the three axes of a ***three-dimensional coordinate system***. Graphing on a coordinate system is often used to visualize quantitative data in a manner that will provide insight into trends, patterns, relationships, and so on. Graphing an equation on a coordinate system provides a depiction of the slope (in the case of a linear equation) or the shape of a curve in the case of a nonlinear equation. Geometrical figures defined by equations are also depicted on a coordinate system.

• Two-dimensional, planar ***rectangular coordinate systems*** consist of two axes, generally denoted x and y, for the horizontal and vertical axes that are at right angles to each other. A three-dimensional rectangular, or Cartesian, coordinate system consists of three axes, generally denoted x, y, and z, which are all three at right angles to each other. Types of coordinate systems include *rectangular*, *cylindrical*, and *spherical*.

• A *point* on the number line is represented in one dimension. On a number line, the numbers to the right are greater than the numbers to the left. Therefore, a number to the left of another number is less than a number to the right. To define the position of a point on a number line, the number the point corresponds to is identified.

The point is at the –3 coordinate position on the number line.

• A *point on a planar* ***rectangular coordinate system*** with two intersecting axes that are perpendicular to each other is represented in two dimensions. X represents the horizontal axis and Y the vertical axis. X is often called the abscissa and Y the ordinate. The axes intersect at zero on the X-axis and zero on the Y-axis. To define the position of a point on a two-axis planar coordinate system, the numbers on each axis that the point corresponds to are identified.

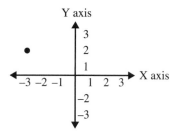

The point corresponds to –3 on the X-axis and +2 on the Y-axis.

• Each point on an X-Y coordinate system corresponds to a unique ordered pair of real numbers (x_1, y_1) where x_1 and y_1 are the coordinates of the point. In the preceding coordinate system, $(-3, 2)$ are the coordinates of the point depicted. The convention for writing numbers that a point corresponds to on each axis is: (X-axis-number, Y-axis-number).

• To identify the point (x, y) on an X-Y coordinate system:

   A positive x value is on the right of the Y-axis.

   A negative x value is on the left of the Y-axis.

   A positive y value is above the X-axis.

   A negative y value is below the X-axis.

• A *point on a spatial **three-dimensional rectangular coordinate system*** (see figure below) with three intersecting axes that are all perpendicular to each other is represented in three dimensions. X represents the horizontal axis, Y the vertical axis, and Z the axis that comes out of the page and is perpendicular to the page with the positive side of the axis above the page and the negative side of the axis below the page. The point is represented in the 3-dimensional space. The axes intersect at zero on the X-axis, Y-axis, and Z-axis. To define the position of a point on a three-axis (three-dimensional) coordinate system the number on each axis that the point corresponds to is identified.

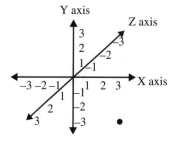

The point corresponds to x = 3, y = –3, and z = 3, or +3 on the
X-axis, –3 on the Y-axis, and +3 on the Z-axis, and can be identified by
(x, y, z) = (3, –3, 3). (The position of the point with respect to the Z-axis
cannot be accurately visualized in two dimensions.)

• A *line* can be defined by either a two-axis or a three-axis coordinate
system, such that each point on the line corresponds to a position on each
axis.

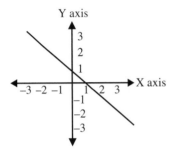

• *Coordinate systems* include rectangular coordinates, polar coordinates,
cylindrical coordinates, and spherical coordinates. Rectangular, or Cartesian,
coordinates described in the preceding paragraphs represent points in a
plane or in space. **Polar coordinates** describe points in a plane or in space,
similar to rectangular Cartesian coordinates. The difference is that in polar
coordinates, there is an r-coordinate that maps the distance of a point from
the origin of the coordinate system, and there is a θ-coordinate that measures
the angle the r-ray makes from the horizontal positive X-axis.

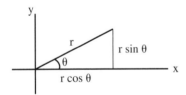

• The **relationship between polar and rectangular coordinates** can be
visualized in the preceding figure using the **Pythagorean Theorem** for
a right triangle, $r^2 = x^2 + y^2$. The r-coordinate is the *hypotenuse* and
measures the distance from the origin to a point of interest. The angle θ,
between r and the positive part of the X-axis, can be described by
$\tan \theta = y/x$. The relationships between these two coordinate systems are:

$x = r \cos \theta, y = r \sin \theta, r = \sqrt{x^2 + y^2}$, and

$\tan \theta = y/x$ or $\theta = \tan^{-1}(y/x)$

• In three dimensions, polar coordinates become *cylindrical coordinates* and are given in terms of r, q, and z, where:

$x = r \cos \theta, y = r \sin \theta, z = z, r = \sqrt{x^2 + y^2}$

• When comparing the Cartesian and cylindrical coordinate systems, the x- and y-components of the Cartesian coordinate system are expressed in terms of polar coordinates, and the z-component is the same component as in the Cartesian system. The r-component is measured from the Z-axis, the θ-component measures the distance around the Z-axis, and the z-component measures along the Z-axis.

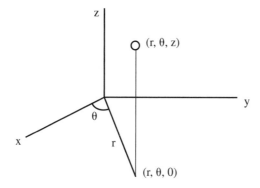

• It is possible to transform data from one coordinate system to another. Consider polar coordinates. To find x and y if r and θ are given:

$r = 5$ and $\theta = \pi/2$, simply calculate $x = r \cos \theta$ and $y = r \sin \theta$:

$x = (5) \cos(\pi/2) = 0$ and $y = (5) \sin(\pi/2) = 5$

Alternatively, to find r and θ if x and y are given:

$x = 2$ and $y = 3$, calculate $r = \sqrt{x^2 + y^2}$ and $\theta = \tan^{-1}(y/x)$:

$r = \sqrt{2^2 + 3^2} \approx 3.6$ and $\theta = \tan^{-1}(3/2) \approx 56° \approx 0.98$ rad

• Different shapes can be depicted on coordinate systems. For example, a *circle on a coordinate system* can be represented by the polar equations $r = \cos \theta$ or $r = \sin \theta$, where substituting values of θ around the coordinate system will produce points on the circle. Note that polar coordinates are more suitable and less complicated than Cartesian, or rectangular, coordinates for representing certain shapes, such as circles.

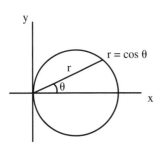

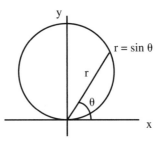

- Another coordinate system that is related to Cartesian coordinates is *spherical coordinates*. In three dimensions, spherical coordinates are expressed in terms of ρ (rho), θ (theta), and ϕ (phi), where ρ can range from 0 to ∞, θ can range from 0 to 2π, and ϕ can range from 0 to π. In spherical coordinates, the ρ component is measured from the origin, the θ component measures the distance around the Z-axis, and the ϕ component measures down from the Z-axis and is referred to as the *polar angle*. Note that ρ is measured from the origin rather than the Z-axis as is the case with r in cylindrical coordinates. Also, θ and ϕ are similar to longitude and latitude on a globe. Spherical coordinates can be defined in terms of Cartesian coordinates, x, y, and z as:

$$x = \rho \cos \theta \sin \phi, y = \rho \sin \theta \sin \phi, z = \rho \cos \phi, \rho = \sqrt{x^2 + y^2 + z^2}$$

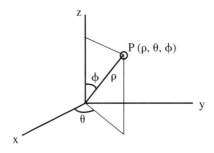

1.5 Chapter 1 Summary and Highlights

- This chapter provides a brief review of selected mathematics definitions that are pertinent to learning trigonometry. Included in this chapter is the basic hierarchy of types of number groups, as well as definitions and operations. In summary, complex numbers encompass both real and imaginary numbers, real numbers encompass both rational and irrational numbers, rational numbers include integers, which include whole numbers, which include natural numbers.

• Included also is a brief introduction to complex and imaginary numbers and their operations (which are described in detail in Chapter 11). Complex numbers involve *i* and are generally in the form $(x + iy)$, where x and y are *real numbers* and *i* is *imaginary*. In the expression $(x + iy)$, the x term is referred to as the real part and the iy term is referred to as the imaginary part. A real number multiplied by *i* forms an *imaginary number*, such that: (real number) $\times$ *i* = (imaginary number). A real number added to an imaginary number forms a complex number, such that: (real number) + (real number)(*i*) = (complex number), or equivalently, (real number) + (imaginary number) = (complex number).

• This chapter also provides basic definitions of different types of coordinate systems, including two- and three-dimensional rectangular coordinates, polar coordinates, cylindrical coordinates, and spherical coordinates.

Chapter 2

Review of Geometry

2.1 Introduction

• Most ancient peoples thought the Earth was flat, but as time went by there were signs that the Earth is in fact spherical. For instance, as a ship approaches from over the horizon, its mast becomes visible. As the ship comes closer, more of it comes into view. More recent evidence that the Earth is round was provided by circumnavigating the globe and by photographs taken from space. In order to measure and navigate the Earth, as well as the heavens, we need to understand geometry and trigonometry. Have you ever wondered how the radius and circumference of the Earth and Sun were determined?

• The Greek geographer and astronomer Eratosthenes is thought to be the first person (about 240 B.C.) to successfully measure the *circumference of the Earth*. The basis for his calculations was the measurement of the elevation of the Sun from two different locations. Two simultaneous observations were made, one from Alexandria, Egypt, and the other 5,000 *stadia* away from a site on the Nile near the present Aswan Dam. At noon in Syene (now Aswan) in Egypt when the Sun was directly overhead on the day of the summer solstice, which is the longest day of the year, the Sun's rays beamed down to the bottom of a deep well. North of Syene in Alexandria, at the same time, the Sun's rays shown at a 7.2° (which is 1/50th of a 360° circle) angle from the zenith when measured by the shadow of a pole sticking straight up out of the ground. Eratosthenes imagined that if the Earth was round, the noonday Sun could not appear in the same position in the sky as seen by two widely separated observers. He therefore compared the angular displacement of the Sun with the distance between the two ground locations. Because the Sun is so far away, it could be assumed that the Sun's rays at the two locations were parallel, and that the difference in the Sun's rays at the two locations was due to the spherical shape of the Earth.

Although the observer at Syene saw the Sun directly overhead at noon, the observer in Alexandria found the Sun was inclined at an angle of 7.2° to the vertical. Because a measure of 7.2° corresponds to one-fiftieth of a full circle (360°), Eratosthenes reasoned that the measured ground distance of 5,000 stadia must represent one-fiftieth of the Earth's circumference. (See the following figure.) Therefore, using the distance of 5,000 stadia between the two locations and the angle differing by 7.2° (or 1/50th of a 360° circle), the circumference of the Earth could be determined to be 50 times 5,000 stadia, or 250,000 stadia. A stadium is estimated to be equivalent to somewhere between 607 and 738 feet, therefore Eratosthenes' determination was between 29,000 and 35,000 miles. Today's measurements are 24,902 miles at the equator and 24,818 miles at the poles (as the Earth is an oblate spheroid flatter at the poles).

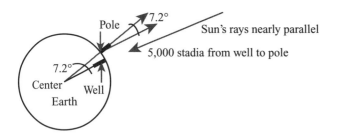

From geometry (discussed in Section 2.2) we know that if lines s_1 and s_2 are parallel, then angles α and β have the same measure.

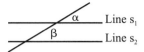

Line s_1

Line s_2

If we know the **circumference** C of the Earth, the **diameter** *of the Earth* d and the **radius** r can be found using the formulas:

$C = 2\pi r$ and $d = 2r$. Remember: $\pi = C/d$ *for all circles.*

• Data from orbiting Earth satellites have confirmed that the Earth is actually slightly flattened at the poles. It is an oblate spheroid, the polar circumference being 27 miles less than at the equator. The following measurements are currently accepted: average diameter 7,918 miles, average radius 3,959 miles, and average circumference 24,900 miles.

• It is possible to estimate the **diameter of the Sun** using the known value of the distance from the Earth to the Sun as 93,000,000 miles and given the Sun subtends an angle of 0°31'55" on the surface of the Earth.

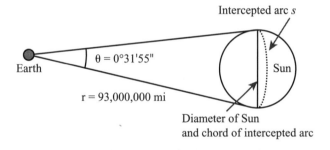

Intercepted arc *s*

$\theta = 0°31'55"$

Earth

Sun

r = 93,000,000 mi

Diameter of Sun
and chord of intercepted arc

Because the central angle θ is so small relative to the radius r, the arc opposite the central angle (or intercepted arc *s*) and the chord of the inter- cepted arc (in this case the diameter of the Sun) can be approximated as the same length. Using the relationship that the arc *s* over the circumfer- ence $2\pi r$ is equal to the angle of the arc θ over the circumference 360°, $s / 2\pi r = \theta / 360°$, and given that $\theta = 0°31'55" = 0.532°$, we can estimate the diameter of the Sun as:

$s = 2\pi r\theta / 360 = 2\pi(93,000,000 \text{ mi})(0.532) / 360 = 864,000 \text{ mi}$

2.2 Lines and Angles

• It is important to be familiar with the basic definitions of lines and angles. Different lines and angles have specific names according to their environment, position, size, measurement, and so on. This section includes common definitions for lines and angles. Additional definitions can be found in the discussion on circles in Section 2.5.

• The shortest **distance between two points** is a straight line. The distance between points (x_1, y_1) and (x_2, y_2) is given by:

$$d = [(x_2 - x_1)^2 + (y_2 - y_1)^2]^{1/2}$$

The coordinates of the midpoint between the two points are:

$(x_1 + x_2)/2$ and $(y_1 + y_2)/2$

• A **line segment** is a section of a line between two points. There are three line segments in the line below.

• Two lines are **perpendicular** if they intersect at right (90°) angles, and **parallel** if they do not intersect each other.

• A **ray** has an end point at one end and extends indefinitely in the other direction. If two straight lines meet or cross each other at a point, an **angle** is formed. The point where the lines meet is called the **vertex** of the angle (A) and the sides are called the **rays** (AB and AC) of the angle.

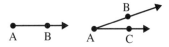

• The symbol for an angle is L. The angle above can be called $LBAC$ (where the middle letter names the vertex) or simply LA. Angles are also denoted by Greek letters, such as α, β, δ, ϕ, and θ.

• An angle can be formed by rotating a ray around its endpoint. If an angle is formed by rotating a ray counterclockwise, a positive angle results. If an angle is formed by rotating a ray clockwise, a negative angle results.

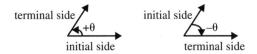

- Angles are measured in **degrees** or **radians**. The symbol for degrees is °, and radian is often shortened to *rad*. (A full circle is 360° or 2π rad.) Degrees can be divided into **minutes** (denoted by ') and **seconds** (denoted by ").

> 1 minute = 1/60th of a degree
>
> 1 second = 1/60th of a minute = 1/3600th of a degree

Examples:

> 30°15'22" = 30° + 15/60 + 22/3600 = 30.26°
>
> 63.23° = 63° + 0.23(60') = 63°13.8' = 63° + 13' + 0.8(60") = 63°13'48"

- A 180° angle is a straight line. In the following, $\angle ABC = 180°$.

- A **right angle** measures 90° and is drawn with a square at the vertex.

- Angles smaller than 90° are called **acute angles**, and angles larger than 90° are called **obtuse angles**.

- If two angles have the same initial and terminal sides, they are called **coterminal angles**. In the drawing there are two positive angles and one negative angle that are coterminal.

Coterminal angles can be formed by beginning at the same initial side and circling in a positive or negative direction more than 360° and ending with the same terminal side. Coterminal angles can therefore contain multiples of 360° if the angle circles more than one time.

• If two angles have the same vertex and are adjacent to each other, they are called *adjacent angles*. L a and L b are adjacent angles.

• If the sum of any two angles equals 180°, the two angles are called *supplementary angles*. The following examples are of angles that are adjacent and supplementary. Note that supplementary angles do not have to be adjacent. In the drawing, a + b = 180°.

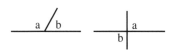

• If the sum of any two angles equals 90°, the two angles are called *complementary angles*. For example, if L ABC is a right angle, then angles L ABD and L DBC are adjacent and complementary angles. Note that complementary angles do not have to be adjacent.

• If two *lines intersect* each other, there are four angles formed. In the following diagram the sum of the four angles a, b, c, and d is 360°. The sum of the *adjacent angles*, a and b, c and d, a and c, and b and d are each 180°. The angles opposite to each other, a and d and b and c, are called *vertical angles* and they are equivalent.

• If two lines are *perpendicular* to each other, four *right angles*, each measuring 90°, are formed. In the drawing, angles a, b, c, and d each measure 90°, and the sum of angles, a + b + c + d, is 360°.

$$\frac{a \mid b}{c \mid d}$$

- A *transversal* is a line that intersects two other lines.

$$\frac{a / b}{c / d} \quad \text{Line A}$$
$$\frac{e / f}{g / h} \quad \text{Line B}$$

If lines A and B are parallel to each other, the following is true:

Angles $a = d = e = h$
Angles $b = c = f = g$
Angle c + Angle e $= 180°$ (Supplementary)
Angle d + Angle f $= 180°$ (Supplementary)

- As described in Section 2.5 on circles, there are 360° in a circle, 180° in a semi-circle, 90° in a quarter-circle, and 45° in an eighth-circle.

2.3 Triangles (Planar)

- Trigonometry involves triangles, and it is necessary to be familiar with the properties of triangles. *Triangles* are three-sided polygons and contain three angles. The symbol for a triangle is Δ. The three points where the three line segments that make up the triangle intersect or meet are called *vertices*.

- The *sum of the angles* in a planar triangle is always 180°. This is found by using the formula for sum of angles in a polygon: $(n - 2)180°$. Using this formula we find that the sum of the angles of every triangle is 180°. For a triangle, $n = 3$, $(3 - 2)180° = (1)180° = 180°$.

- *Properties of planar triangles* include:

1. If the value of two angles in a triangle is known, the third angle can be calculated by subtracting the sum of the two known angles from 180°.
2. The length of one side of a triangle is always less than the sum of the lengths of the other two sides.
3. In a triangle, the largest side is opposite the largest angle, the smallest side is opposite the smallest angle, and the middle-length side is opposite the middle-size angle.

- Types of triangles include:

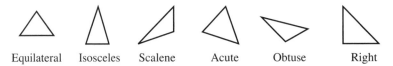

Equilateral   Isosceles   Scalene   Acute   Obtuse   Right

1. In an **equilateral triangle**, all three sides have equal lengths and all three angles have equal measurements of 60°.
2. In an **isosceles triangle**, two sides have equal lengths and the angles opposite the two equal sides have equal measurements.
3. In a **scalene triangle**, all three sides have different lengths and all three angles have different measurements.
4. In an **acute triangle**, all of the angles in the triangle are smaller than 90°.
5. In an **obtuse triangle**, one of the angles in the triangle is larger than 90°.
6. In a **right triangle**, one of the angles in the triangle is a right angle measuring 90°.

- In a **right triangle**, the side opposite to the right angle is called the **hypotenuse** and the two sides that meet to form the right angle are called **legs**. The hypotenuse is always the longest side.

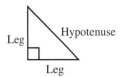

In a right triangle, the square of the length of the hypotenuse is equal to the sum of the squares of the lengths of the legs.

$$(Leg)^2 + (Leg)^2 = (Hypotenuse)^2$$

This is called the **Pythagorean Theorem** and it only applies to right triangles. If the lengths of the legs are x and y and the length of the hypotenuse is z, the Pythagorean Theorem can be written: $x^2 + y^2 = z^2$.

- The acute angles of a right triangle are *complementary* and therefore sum to 90°.

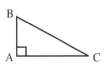

Angle B + Angle C = 90°

• Two noteworthy right triangles are the **30°:60°:90°** and the **45°:45°:90°** (right isosceles triangle). For a 30°:60°:90° triangle, the length of the hypotenuse equals two times the length of the leg opposite the 30° angle, and the length of the leg opposite the 60° angle equals the square root of 3 times the length of the leg opposite the 30° angle. For the 45°:45°:90° triangle, the length of the hypotenuse equals the square root of 2 times x, where x is the length of a leg.

30°:60°:90° triangles are often drawn as:

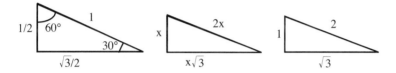

45°:45°:90° triangles are often drawn as:

• Other noteworthy right triangles are called ***triplet right triangles***. The most common triplets are 3:4:5, 5:12:13, and 7:24:25.

Any multiple of the ratios of these triangles is also a triplet, such as 6:8:10 and 10:24:26.

• If two corresponding sides of two *right triangles* are equal, the third corresponding sides are also equal, and the triangles are ***congruent***. This can be proven using the Pythagorean Theorem. Any two triangles, whether they are right triangles or not, are called ***congruent triangles*** if:

1. All three corresponding sides are equal; this is called *side-side-side*.
2. Two corresponding sides with their vertex angles are equal; this is called *side-angle-side*.

3. Two corresponding angles with the side in between are equal; this is
   called *angle-side-angle*.
4. Two angles and a non included side are equal; called *angle-angle-side*.

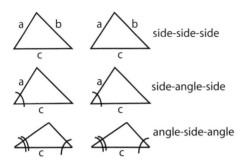

side-side-side

side-angle-side

angle-side-angle

• If all three pairs of corresponding angles in two triangles are equal to
each other, the two triangles are called **similar triangles**. Two similar
triangles can be created by drawing a line parallel to one of the sides of
a triangle. In the figure below, triangle ADC is *similar* to triangle AEB
because the three corresponding angles are equal.

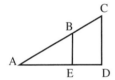

• The corresponding sides of **similar triangles** have the same proportion;
therefore, if one side is 3 times the length of its corresponding side in a
similar triangle, then the other two sides of the similar triangle will be
3 times the lengths of their corresponding sides. In the following triangles,
angles $A_1 = A_2$, $B_1 = B_2$, and $C_1 = C_2$:

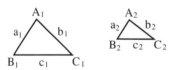

and corresponding sides are in proportion: $\dfrac{a_2}{a_1} = \dfrac{b_2}{b_1} = \dfrac{c_2}{c_1}$.

• **Example:** Determine the height of a tree using *similar triangles*. To
find the height of a tree, set a post of a known height next to the tree and
measure the length of the shadow. Then measure the length of the tree's

shadow. The post and the tree and their shadows form similar triangles as they are both right triangles and have an acute angle of the same measure (assuming the Sun's rays are parallel).

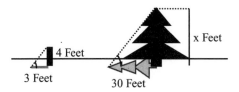

The height of the tree, x feet, is proportional to the height of the post, 4 feet, and the length of the shadow of the tree, 30 feet, is proportional to the length of the shadow of the post, 3 feet. Therefore, the height of the tree represented by x is:

$$\frac{x}{4} = \frac{30}{3}, \text{ or } x = (30)(4)/3 = 40 \text{ feet.}$$

- If two corresponding angles of two triangles are equal, then the third angles are also equal. Remember, the sum of the angles in a triangle is 180°.

- The following two triangles are *similar* providing *AB* and *DE* are parallel:

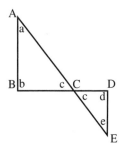

Angles c and c are vertical angles and therefore equal.

Angles a and e are equal because if parallel lines are cut by a transversal, then the alternate interior angles measure the same.

Angles b and d are equal because if two corresponding angles of two triangles are equal, then the third angles are also equal.

- Note the following angle properties for triangles:

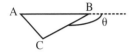

angle θ = angle A + angle C, angle A + angle B + angle C = 180°,
angle B + angle θ = 180°

When the dashed line (following) is parallel to the *AB* base of the triangle

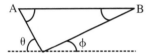

angle A = angle θ and angle B = angle φ.

- The *perimeter of a triangle* is the sum of the sides, and the *area of a triangle* is (1/2)(base)(height)

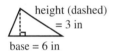

height (dashed)
= 3 in
base = 6 in

For example, in this triangle area is:

$(1/2)(6 \text{ inches})(3 \text{ inches}) = 9 \text{ inches}^2$ or 9 square inches

where the height is a perpendicular line from the base to the opposite angle. Note that the area of an equilateral triangle is $(1/4)(\text{side}^2)\sqrt{3}$. See Chapter 3 for a detailed discussion of triangles.

2.4 Polygons and Quadrilaterals

- A *polygon* is a closed planar figure that is formed by three or more line segments that all meet at their end points. There are no end points that are not met by another end point in a polygon. The line segments that make up a polygon only intersect at their end points. Examples of polygons include:

Triangle, Square, Rectangle, Octagon, Hexagon, Trapezoid, Heptagon, Pentagon

Note that rectangles, squares, and trapezoids are quadrilaterals (having four sides).

• Polygons are named according to the number of sides they contain. For example 3 sides is a triangle, 4 sides is a quadrilateral, 5 sides is a pentagon, 6 sides is a hexagon, 7 sides is a heptagon, 8 sides is an octagon, 9 sides is a nonagon, and 10 sides is a decagon. In a polygon, the number of sides equal the number of angles.

• If the lengths of the sides are equal and the angle measurements are equal, the polygon is called a ***regular polygon***. A *square* is a *regular quadrilateral*. If two polygons have the same size and shape, they are called ***congruent polygons***. If two polygons have the same shape such that their angle measurements are equal and their sides are proportional, however one is larger than the other, they are called ***similar polygons***.

• A useful equation for polygons is the equation that gives the ***sum of the angles*** *in a polygon*: $(n - 2)180° =$ Sum of all angles in n-gon, where n is the number of angles (or sides) in the polygon.

For example, because a pentagon has five sides and five angles, what is the sum of all angle measurements?

$(n - 2)180° =$ sum of all angles in n-gon

If n = 5, then $(5 - 2)180° = (3)180° = 540°$.

Therefore, the sum of the angles in a pentagon is 540°.

• The ***perimeter*** *of polygons and planar figures* is the sum of the lengths of its sides or the distance around. The units for perimeter are always singular because of the one dimension described. Remember to convert all measurements to the same units before adding. To find the ***area*** *of polygons and planar objects* that are not triangles, squares, rectangles, parallelograms, or trapezoids, find the area of sections of the polygon that form one of these figures, then add the areas of the sections. Units for area are always squared because of the two dimensions described.

• ***Quadrilaterals*** are four-sided polygons. The sum of the angles in a quadrilateral is: $(n - 2)180° = (4 - 2)180° = (2)180° = 360°$.

Therefore, the sum of the angles in all quadrilaterals is 360°.

• A *parallelogram* is a quadrilateral in which both of the two opposite sides are parallel to each other.

AB is parallel to *CD*. *AC* is parallel to *BD*. Both pairs of opposite sides are the same length. Both pairs of opposite angles are the same size. The diagonals of a parallelogram bisect each other. Consecutive interior angles are supplementary (A + B, B + D, etc.). The *area of a parallelogram* is base times height. In the figure below, area is (base)(height) = (5 ft)(3 ft) = 15 feet^2.

height (dashed)
= 3 ft

base = 5 ft

• A *rectangle* is a parallelogram with all four angles having equal measurements of 90°.

Opposite sides of a rectangle are also equal to each other. Diagonal lines have the same length, and their length can be determined using the Pythagorean Theorem if the side lengths of the rectangle are known.

(long-side length)^2 + (short-side length)^2 = (diagonal length)^2

The *area of a rectangle* is length times height, or area = (length)(height).

• A *square* is a parallelogram in which the four angles and four sides are equal. A square is also described as a rectangle in which two adjacent sides are equal. A square is a *regular polygon*.

The angles in a square each measure 90°. Diagonal lines have the same length, and their length can be determined using the Pythagorean Theorem if the side lengths of the square are known:

(side length)^2 + (side length)^2 = (diagonal length)^2

For example, if the side length is x and the diagonal length is d:

$$x^2 + x^2 = d^2 \text{, or } d^2 = 2x^2 \text{, or } d = \sqrt{2x^2} = x\sqrt{2}$$

The *area of a square* is the length of a side-squared or $(\text{side})^2$. Area for a square can also be given by:

$$(1/2)(\text{diagonal})^2$$

We can see this by applying to diagonal formula $d^2 = 2x^2$:

$$\text{Area} = (1/2)d^2 = (1/2)(2x^2) = x^2 \text{, where x is the side length.}$$

- A **rhombus** is a parallelogram with all four sides of equal length, or equivalently, a parallelogram in which two adjacent sides are equal (because if adjacent sides are equal and opposite sides are parallel, then all four sides are equal). Opposite angles of a rhombus are equal to each other. Diagonal lines in a rhombus are perpendicular to each other. The *area of a rhombus* is (side)(height). In the diagram below, angle A and angle B are supplementary.

- A **trapezoid** is a quadrilateral having only one pair of opposite sides parallel to each other. (In an *isosceles trapezoid*, the non-parallel sides have equal lengths.)

AB is parallel to *CD*. *AC* is not parallel to *BD*. *AB* and *CD* are called *bases*. *AC* and *BD* are called *legs*. The *area of a trapezoid* is the average of the bases times height, or area = (average of bases)(height).

2.5 Conic Sections, Including Circles, Arcs and Angles, Ellipses, Parabolas, and Hyperbolas

- Parabolas, circles, ellipses, and hyperbolas are important curves in mathematics. They are often referred to as **conic sections**, because each of these curves can be represented as the intersection of a plane with right circular cones.

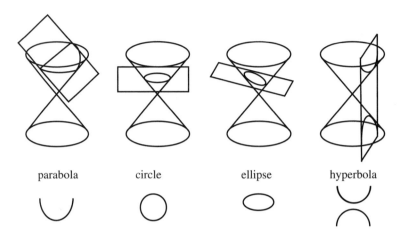

| parabola | circle | ellipse | hyperbola |

- Examples of applications where conic sections are used for modeling include ellipses for orbits of planets, parabolas for the path of a projectile, and hyperbolas for the reflection of sound.

Circles, Arcs, and Angles

- A *circle* is a planar shape consisting of a closed curve in which each point on the curve is the same distance from the center of the circle.

- The *radius* of a circle is the distance between the center and any point on the circle. All radii drawn for a given circle have the same length. The radius is one-half of the diameter. A line segment drawn through the center point with its end points on the circle is called the *diameter* of the circle. The diameter is twice the radius, or (2)(radius) = diameter. A diameter line divides a circle into two equal semicircles. Any line segment whose ends are on the circle is called a *chord* (including the diameter line segment). A *semicircle* is an arc joining the endpoints of a diameter of a circle.

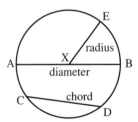

X depicts the point at the center of the circle. *AB* is the diameter chord. *CD* is a chord. *XE, AX*, and *XB* are radii. If two chords are equal, then they are of equal distance from the center of the circle.

• *Pi*, or π, defines the ratio between the circumference and the diameter of a circle. More specifically, Pi is equivalent to the circumference divided by the diameter of a circle. The value of Pi is approximately 3.141592654.

• A circle always measures 360° around, equivalent to 2π radians. Half of the circle measures 180°, which is equivalent to π radians. A quarter of a circle measures 90°, which is equivalent to $\pi/2$ radians. The *degrees of a circle* are:

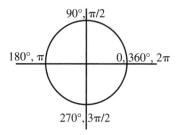

2π ***radians*** = 360 degrees

1 radian = $360°/2\pi$ = $180°/\pi$

1 degree = 2π radians/360° = π radians/180°

Because there are 360° in a circle, 1° = 1/360th of a circle

1 *Minute*, denoted by ', is defined as (1/60) of 1°, or 0.0167°

1 *Second*, denoted by " is (1/60) of 1 Minute or (1/3600) of a degree or 0.00027778°

• A *tangent line* passes through only one point on a circle. If a radius line segment is drawn from the center of the circle to the point of tangency, the tangent line and the radius line segment are perpendicular to each other.

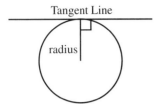

Tangent Line

radius

• A *secant* is a line that intersects a circle in two points.

• If two or more circles have the same center point, they are called *concentric circles*.

concentric circles

• If angles or polygons are drawn inside circles, any angle whose vertex is at the circle's center point is called a *central angle*, and any angle whose vertex is on the circle is called an *inscribed angle*. The sides of a central angle are radii of the circle.

central angle   inscribed angle

• *Arc length*: A section of a circle defined by two or more points is called an *arc*.

Arc length = (radius)(central angle measured in radians)

$\quad\quad\quad = r\theta$, with θ measured in radians

$\quad\quad\quad = (\pi/180°)r\theta°$, with θ the central angle measured in degrees

For example, if r = 10 and θ = 90° = π/2 radians:

Arc length = (r)(θ) = (10)(π/2) ≈ 15.7

= (π/180°)rθ° = (π/180°)(10)(90°) ≈ 15.7

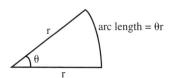

arc length = θr

• The measure of a ***central angle*** is proportional to the measure of the arc it intercepts. Note the following relationship for central angles and arcs:

∠ABC = arc AC

$$\frac{\text{measure of angle B}}{360} = \frac{\text{length of arc AC}}{\text{circumference}} = \frac{\text{area of sector BAC}}{\text{area of circle}}$$

• A central angle *subtending* an arc equal in length to the radius of a circle is defined as a ***radian***. In other words, a radian is the measure of the central angle subtending an arc of a circle that is equal to the radius of the circle. Using the definition, 2π radians = 360°, then:

1 radian = 360°/2π = 180°/π ≈ 57.296°

1 degree = 2π/360° = π/180° ≈ 0.017453 radians

• The measure of an ***inscribed angle*** is equal in measure to *half of the arc it intercepts* (measured in radians). In the following drawing, ∠ABC = (1/2)(Arc AC).

• If an inscribed angle has its rays ending at the end points of a diameter chord, the vertex of the angle will be a right angle (90°), which is one-half of the 180° measurement of the arc. In the diagram below, $\angle ABC = 90° = (1/2)(\text{Arc AC}) = (1/2)180°$.

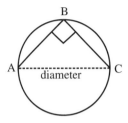

• Inscribed angles with the same endpoints defined by the same arc have the same measure. In the diagram below, $\angle ADC$ has the same measure as $\angle ABC$.

• An inscribed angle is equal to one-half of the central angle formed from the endpoints of the same arc.

In both cases, angle $\angle ABC$ is an inscribed angle with endpoints on arc AC. Angle ADC is a central angle with endpoints on arc AC. Because they are formed by the same arc, $\angle ABC$ is one-half the measure of $\angle ADC$ or, alternatively, angle $\angle ADC = 2\angle ABC$.

• The *perimeter of a circle* is called the **circumference**.

Perimeter of a circle = circumference = $2\pi r = \pi d$

where r = radius, d = diameter, and $\pi \approx 3.14$.

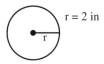

r = 2 in

In this example, perimeter = circumference = $2\pi r$
= $2\pi 2$ inches = 4π inches ≈ 12.56 inches

Because circumference = $2\pi r = \pi d$, we can see that **Pi** is:

π = circumference/diameter

- The **area of a circle** is given by:

Area of a circle = $\pi r^2 = \pi(d/2)^2$

where r = radius, d = diameter, and $\pi \approx 3.14$.

r = 2 in

In this example, area = $\pi r^2 = \pi(d/2)^2 = \pi(2 \text{ inches})^2$
= $\pi 4$ inches ≈ 12.56 inches^2

- The **area of a sector** of a circle is a fraction of the area of the whole circle.

For a circle with its central angle = ABC, the following is true:

$$\frac{\text{measure of angle B}}{360} = \frac{\text{length of arc AC}}{\text{circumference}} = \frac{\text{area of sector BAC}}{\text{area of circle}}$$

If $\angle B = 60°$, then the central angle is 60°/360°, or 1/6th of the circle. Also, the length of arc AC is 1/6th of the circumference of the circle. Finally, the area of sector BAC is 1/6th of the area of the circle. The area of a sector is also given by: (1/2)(radius)^2(central angle in radians).

For example, use two different equations from the above relationship to calculate the area of sector ABC.

If $LB = 90° = \pi/2$ and r = 10, find area of sector ABC.

Area of sector ABC = $(LB/360°)$(area of circle) = $(\pi 10^2)90°/360° \approx 78.54$

Area of sector ABC = (length of arc AC/circumference)(area of circle)

$\quad = (\pi r^2)(r\theta/2\pi r) = (r^2)(\theta/2) = (1/2)(\text{radius})^2(\text{central angle in radians})$

$\quad = (1/2)(10^2)(\pi/2) \approx 78.54$

• Equations for a *circle* located at the origin of a coordinate system can be written in the form: $x^2 + y^2 = r^2$, where r is the radius and r > 0.

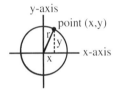

For a circle whose center is located at a point (x = p, y = q) other than the origin, the equation becomes: $(x - p)^2 + (y - q)^2 = r^2$.

• To plot an equation of a circle, choose values for x and solve for the corresponding y values. Alternatively, set x = 0 and solve for the corresponding y value, and set y = 0 and solve for the corresponding x value. If a circle has its origin at (0,0), it is possible to choose x values in one quadrant of the coordinate system and use symmetry to complete the circle.

Ellipses

• *Ellipses* are flattened circles.

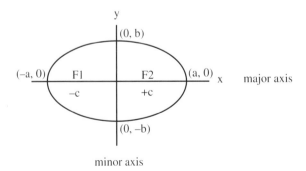

A circle has one *focus* at the center. An ellipse has two *foci*, designated F1 and F2, along the *major axis* on either side of the center. The sum of the distances of the foci to any point on an ellipse is 2a. Therefore at any point on the ellipse: (the distance to F1) + (the distance to F2) = 2a.

The ellipse can also be described by:

$$[(x - c)^2 + y^2]^{1/2} + [(x + c)^2 + y^2]^{1/2} = 2a$$
$$= \text{distance from F1} + \text{F2 to a point } (x, y),$$

where +c and –c represent the location of F1 and F2 on the major axis.

• Because the distance from F1 to F2 is 2a, as sound waves, etc. are reflected off the ellipse, a sound generated at F1 will be concentrated at F2. (Note that there is also a direct path between F1 and F2 where sound is not reflected.)

• The equation for an ***ellipse*** at the origin can be written:

$$(x^2/a^2) + (y^2/b^2) = 1, \text{ where } a \neq b, a > 0, b > 0, \text{ origin at } (0, 0).$$

If $a = b$, the ellipse becomes a circle. The equation for an ellipse with $a = b = r$ is the equation for a circle: $x^2 + y^2 = r^2$.

For an ellipse located at a point $(x = p, y = q)$ other than the origin, the equation becomes: $((x - p)^2/a^2) + ((y - q)^2/b^2) = 1$ where $a \neq b, a > 0$, $b > 0$, origin at point (p, q).

• The *equations for ellipses and circles* having the form $(x^2/a^2) + (y^2/b^2) = 1$, with origin at $(0, 0)$ and $a \neq b$ for ellipses and $a = b$ for circles, can be solved for y as follows:

$$y/b = \pm[1 - (x^2/a^2)]^{1/2} = \pm[(a^2/a^2) - (x^2/a^2)]^{1/2} = \pm[(a^2 - x^2)/a^2)]^{1/2}$$
$$= \pm(1/a)[a^2 - x^2]^{1/2}$$

Therefore, $y = \pm(b/a)[a^2 - x^2]^{1/2}$

where the (+) values represent the top half of an ellipse or circle and the (–) values represent the bottom half of an ellipse or circle. The curve crosses from (+) to (–) or (–) to (+) at $y = 0, x = a$, and $y = 0, x = -a$, respectively. The maximum and minimum of the curves are at $y = b$ and $y = -b$.

• The shape of an ellipse (rounder or flatter) is designated by its eccentricity, which is given by e = $\sqrt{a^2 - b^2}/a$, where a is the length of the semimajor (longer) axis (from center to point a on X-axis) and b is the length of the semiminor axis (from center to point b on Y-axis). Eccentricity ranges from 0 to 1 with higher numbers indicating flatter ellipse. If a = b, eccentricity = 0, which is a circle.

• To plot an equation of an ellipse, choose values for x and solve for the corresponding y values. Alternatively, set x = 0 and solve for the corresponding y value and set y = 0 and solve for the corresponding x value.

Parabolas

• *Parabolas* are sets of points in a plane that form a curve with symmetry. Parabolas can point in any direction. Equations for parabolas are quadratic equations.

• The equations for a parabola with a vertical and a horizontal axis are:

$y = ax^2 + bx + c$ with a vertical axis
$x = ay^2 + by + c$ with a horizontal axis

Following are vertical-axis parabolas:

vertex at bottom   vertex at top

• In the vertical form of the equation, $y = ax^2 + bx + c$, if a is positive, the parabola is open at the top with the *vertex* at the bottom. Conversely, if a is negative, the parabola is open at the bottom with the vertex at the top. For $y = ax^2 + bx + c$, the graph crosses the X-axis at y = 0, and the *vertex point* is a minimum or maximum point (where dy/dx = 0).

- For a vertical-axis parabola, if the parabola lies above its vertex, then the y-coordinate (x, y) of its vertex is the smallest y value of the parabola that satisfies the equation for that parabola. Conversely, if the parabola lies below its vertex, then the y-coordinate (x, y) of its vertex is the largest y value of the parabola that satisfies the equation for that parabola.

- The *axis of symmetry* can be drawn through the center of a parabola to divide it in half. The equation for the axis of symmetry in a vertical parabola is $x_v = -b/2a$. For example, if the solution to this equation is $x_v = 2$, then a vertical line through the point 2 on the X-axis can be drawn to represent the axis of symmetry.

- The *vertex point of a parabola* can be found by substituting x_v into the equation $y = ax^2 + bx + c$ (if x_v is known) and solving for the corresponding y or y_v value resulting in (x_v, y_v). The vertex point of a parabola with a vertical axis can be found using the equation for the parabola as follows:

1. Separate the terms that contain x from the terms not containing x.
2. Complete the square on the terms that contain x.
3. Set each side of the equation equal to zero.
4. Solve the resulting equations for x and y resulting in (x_v, y_v).

For example, if $y = x^2 - 2x - 3$, what is the vertex point?

   Rearrange: $x^2 - 2x = y + 3$

Complete the square by finding 1/2 of the coefficient b ($b = 2$). Square 1/2 of the coefficient b, $(b/2)^2$, and add the result to each side of the equation. First, square 1/2 of the coefficient b:

   $(b/2)^2 = (-2/2)^2 = (-1)^2 = 1.$

Add the result to each side of the equation:

   $x^2 - 2x + 1 = y + 3 + 1$
   $x^2 - 2x + 1 = y + 4$

Factor the resulting perfect square, and set each side of the equation equal to zero and solve:

   $(x - 1)(x - 1) = y + 4$
   $(x - 1)^2 = 0$, therefore, $x = 1$
   $y + 4 = 0$, therefore, $y = -4$
   The vertex point is $(1, -4)$.

cus of a parabola is a point on the axis of symmetry on to which
(for example, light, sound, etc.) coming toward the bottom of the
ola, parallel with the axis of symmetry is reflected. An example of
ase of the *focus point* is a receiver of radio waves or TV signals where
ε rays are concentrated at the focus. This principle applies in reverse as
well. When light energy is emitted from a focus point and reflected off the
inner surface of the parabola, it will point out of the parabola parallel to
the axis of symmetry.

• The **directrix** is a line that exists perpendicular to the axis of symmetry
such that every point on the parabola is the same distance from the focus
point as it is from the directrix line. Therefore, the distance from the vertex
to the focus (d2 on graph) is equal to the distance from the vertex to the
directrix (d1 on graph).

The following is a *horizontal-axis parabola* with its vertex at the left:

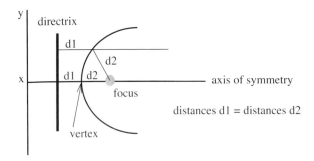

• For a *vertical-axis parabola* with its vertex at the bottom, if the focus
is at $(0, a)$, the directrix is a line at $y = -a$, and the vertex is at $(0, 0)$, then
the equation for the parabola can be written: $y = x^2/4a$. If the focus is at
$y = 1/4$ and the directrix is at $y = -1/4$, then $y = x^2$.

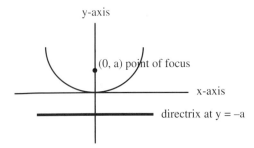

- To *graph a parabola*, which is a *quadratic equation*, one method involves finding the x component of the vertex point, $x_v = -b/2a$, and substituting x_v into the equation, $y = ax^2 + bx + c$, and solving for the corresponding y or y_v value, resulting in (x_v, y_v). Then, choose other values for x on both sides of x_v and solve for their corresponding y values using the original equation. Finally, the points can be plotted and the parabola sketched.

- It is possible to *solve a quadratic equation graphically* using the fact that it forms a parabola as follows:

1. Write the equation in the standard form $y = ax^2 + bx + c$.
2. Graph the parabola by identifying $x_v = -b/2a$, substitute x_v into the equation $y = ax^2 + bx + c$ and solve for y_v resulting in the vertex point (x_v, y_v). Then choose other values for x on both sides of x_v and solve for their corresponding y values using the original equation to plot the parabola.
3. Determine the solutions for x (called the roots of the equation because of the x^2 term) by estimating the two points where the parabola crosses the X-axis (at $y = 0$).

Hyperbolas

- **Hyperbolas** are a set of points in a plane that form two parabola-like curves that are mirror images of each other. The *equations for hyperbolas* can be written in the form:

$$(x^2/a^2) - (y^2/b^2) = 1 \text{ or } -(x^2/a^2) + (y^2/b^2) = 1$$

where a and b have opposite signs and the center of the hyperbola is at the origin, $(0, 0)$.

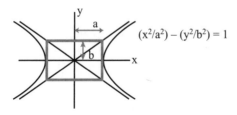

- For a *hyperbola* located at a point $(x = p, y = q)$ other than the origin, the equation becomes: $((x-p)^2/a^2) - ((y-q)^2/b^2) = 1$

- The equation for a hyperbola can be solved for y:

$$y/b = \pm[1 + (x^2/a^2)]^{1/2} = \pm[(a^2/a^2) + (x^2/a^2)]^{1/2}$$
$$= \pm[(a^2 + x^2)/a^2]^{1/2} = \pm(1/a)[a^2 + x^2]^{1/2}$$

Therefore, $y = \pm(b/a)[a^2 + x^2]^{1/2}$ where the (+) expression represents the side of the hyperbola with $y \geq b$, and the (−) expression represents the side of the hyperbola with $y < b$.

In the following figure, v1 and v2 are **vertexes** at $(0, b)$ and $(0, -b)$, and F1 and F2 are *foci*. The ray drawn coming toward one focus, F2, and contacting the outside of that side of the hyperbola will be reflected to the other focus, F1. The foci of the hyperbola are inside the curve of each side such that for the points on the hyperbola, the difference between the distances to the foci is 2b.

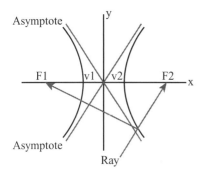

- A hyperbola can be drawn along the X-axis or Y-axis and is symmetric with respect to its axis. The diagonal lines are called **asymptotes**, and each hyperbola has two asymptotes such that the curve of a hyperbola approaches its asymptotes. The *equations for the asymptotes* have zero replaced for the constant terms, and for a hyperbola centered at the $(0, 0)$ in a coordinate system, the equations for the asymptotes are $y = \pm(b/a)x$, and the slopes are $+b/a$ and $-b/a$.

- Another equation for a hyperbola is $xy = k$. If k is positive, the hyperbola will graph in the upper right and lower left quadrants. Conversely, if k is negative, the hyperbola will graph in the upper left and lower right quadrants.

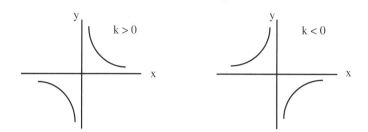

- To plot an equation of a hyperbola, choose values for x and solve for the corresponding y values. Because the graph of a hyperbola is symmetric with respect to both axes, the points plotted in one quadrant (for example, $x \geq 0, y \geq 0$) will mirror the points in the opposite quadrant.

2.6 Three-Dimensional Objects, Including Cubes, Rectangular Solids, Cylinders, Spheres, Cones, and Pyramids

- Three-dimensional objects take up space in three dimensions. Measurements of three-dimensional objects include volume and surface area. *Volume* is a measure of the three-dimensional space that an object occupies. The units for volume are always cubed because of the three dimensions described, $(x)(x)(x) = x^3$. Remember to convert all measurements to the same units before calculating. The *surface area* of three-dimensional objects such as cubes, rectangular solids, cylinders, and spheres is a sum of the areas of the surfaces. The units for surface area are always squared because of the two dimensions described, $(x)(x) = x^2$. Remember to convert all measurements to the same units before calculating. Following are sample formulas of *volume*, *surface area,* and *main diagonal measurements.*

- *Cubes* have six surfaces that are each squares and have the same measurements. The cube below has a side length of 5 inches:

5 in

Volume of a cube = (edge)^3 = (5 inches)^3 = 125 inches^3

Surface area of a cube = (6 sides)(area of each side)

= (6 sides)(25 inches^2) = 150 inches^2 or 150 square inches

The *main diagonal* d of a cube is given by: $d^2 = l^2 + w^2 + h^2$, or $s^2 + s^2 + s^2 = 3s^2 = d^2$, where s = the length of the edge.

$$d^2 = 3s^2 = 3(5)^2 = 75, d \approx 8.7 \text{ inches}$$

• **Rectangular solids** have six rectangular surfaces with three pairs of opposite surfaces that have the same measurements. The rectangular solid below has length 8 inches, width 2 inches, and height 3 inches:

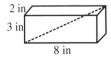

Volume of a rectangular solid = (length)(width)(height)
= (8 in.)(2 in.)(3 in.) = 48 inches³ or 48 cubic inches

Surface area of a rectangular solid = sum of area of 6 faces where opposite sides are identical

= (2)(length)(width) + (2)(length)(height) + (2)(width)(height)
= (32 in.²) + (48 in.²) + (12 in.²) = 92 in.² or 92 square inches

The *main diagonal* of a rectangular solid is given by: $d^2 = l^2 + w^2 + h^2$.

$$d^2 = (8 \text{ in.})^2 + (2 \text{ in.})^2 + (3 \text{ in.})^2 = 77 \text{ in.}^2, d \approx 8.8 \text{ inches}$$

• **Cylinders,** or *circular solids*, are three-dimensional objects that have two identical circles connected by a tube. In the cylinder below, r = 2 inches and h = 10 inches.

Volume of a cylinder = (area of circle)(height) = $(\pi r^2)(h)$
$(\pi)(2 \text{ in})^2(10 \text{ in}) \approx 125.6$ inches³ or 125.6 cubic inches

Surface area of a cylinder or circular solid
= (area of both circles) + (area of tube)
= $2\pi r^2 + 2\pi rh$ (where $2\pi r$ = circumference)
= $(2)(\pi)(2 \text{ in})^2 + (2)(\pi)(2 \text{ in})(10 \text{ in}) \approx 150.72 \text{ in.}^2$

• *Spheres* or *spherical solids* are three-dimensional objects consisting of points that are all the same distance from the center.

> *Volume* of a sphere = $(4/3)\pi r^3$
>
> If the radius is 2 feet, volume is: $(4/3)(\pi)(2 \text{ ft})^3 \approx 33.49 \text{ feet}^3$
>
> *Surface area* of a sphere = $4\pi r^2 = 4\pi(2 \text{ ft})^2 \approx 50.24 \text{ feet}^2$

• *Cones* are three-dimensional objects that have a circle connected to a point. The depth of a cone forms a triangular solid.

> *Volume* of a cone = $(1/3)$(area of circle)(height) = $(1/3)\pi r^2 d$

Note that the volume of a cone is one-third the volume of a cylinder of the same radius and height.

• *Pyramids* are three-dimensional objects that have a square, rectangle, triangle, or other polygon base connected to a point.

> *Volume* of a pyramid
> = $(1/3)$(area of base)(height) = $(1/3)$(area of base)(d)

Remember: The area of a triangle is $(1/2)$(base)(height), the area of a square is $(\text{side})^2$, and the area of a rectangle is (length)(height).

2.7 Chapter 2 Summary and Highlights

• This chapter provides a brief review of basic principles of geometry that are pertinent to learning trigonometry. Geometry and trigonometry are important for modeling the world around us. Early scientists used simple calculations to determine such measurements as the circumference and radius of the Earth. There are fundamental principles, relationships, and formulas from geometry that should be known when learning trigonometry. These include definitions pertaining to lines and angles, especially degrees, radians, right angles (90°), acute angles (<90°), and obtuse angles (>90°), as well as supplementary angles (sum to 180°), complementary angles (sum to 90°), and coterminal angles (same initial and terminal sides).

• Trigonometry involves *triangles*, and it is necessary to be familiar with their properties. One particularly important property of planar triangles is that the angles always sum to 180°. *Right triangles* are also especially useful in trigonometry. Properties and relationships of right triangles should be known, including the *Pythagorean Theorem*, which is $x^2 + y^2 = r^2$, and special right triangles, such as the 30:60:90 and 45:45:90 triangles, and the 3:4:5, 5:12:13, and 7:24:25 triplet triangles.

• Conic sections are curves that can be represented as the intersection of a plane with right circular cones. Important conic sections include parabolas, circles, ellipses, and hyperbolas. Especially relevant to trigonometry are circles. It is important to know basic definitions relating to circles, such as π (circumference/diameter), degrees and radians (2π radians = 360°), minutes (1/60th of a degree), seconds (1/60th of a minute), central angles (vertex at center), inscribed angles (vertex on interior), and arc length (radius × central angle measured in radians).

• It is also important to be familiar with the shapes of common polygons and the equation for the sum of the angles in a polygon which is: $(n - 2)180°$ = sum of the angles in a polygon, where n is the number of angles in the polygon. This chapter concludes with a brief review of three-dimensional objects and the formulas for volume, surface area, and main diagonals.

Chapter 3

Triangles and Trigonometric Funtions

• Why are triangles and trigonometry so interesting and important? What can the use of triangles help us figure out? There are many questions that can be answered by setting up a model involving a right triangle or an oblique triangle. For example, how do we measure the distance to a star, the distance across a canyon, the angle of elevation of the Sun, the distance of a ship from a lighthouse, the height of a mountain, the distance of a UFO from radar towers using bearing, or the distance across a lake? We can set up triangles and determine unknown distances and angles as well as finding the area of a model triangle. See sections 3.3 and 3.6 for examples of applications of triangles.

3.1 Right Triangles and the Trigonometric Functions

• **_Right triangles_** consist of one right (90°) angle and two acute (<90°) angles that sum to 90°, so that the total sum of the angles is 180°. Solving right triangles involves measuring distances and angles.

• **Trigonometric functions** can be defined using ratios of sides of a right triangle. *Sine, cosine, tangent, cotangent, secant,* and *cosecant* are the trigonometric functions. As we will see in Chapter 4, trigonometric functions can also be described using the coordinates of points on a circle of radius one. Trigonometric functions have a periodic nature that can be depicted on a graph, as described in Chapter 5.

• **Right triangle relationships**: The **six trigonometric functions** are defined according to the ratios of the three sides of a right triangle. A right triangle can be drawn alone or at the origin of a coordinate system. Consider the right triangle with sides x, y, and r:

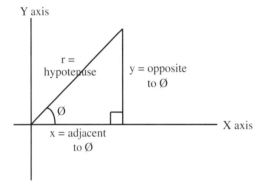

If r is the hypotenuse and the terminal side of the angle, y is the side opposite Ø, and x is the side adjacent to Ø, then the **six trigonometric functions** are:

    sine Ø = sin Ø = opposite/hypotenuse = y/r
    cosecant Ø = csc Ø = hypotenuse/opposite = r/y = 1 / sin Ø
    cosine Ø = cos Ø = adjacent/hypotenuse = x/r
    secant Ø = sec Ø = hypotenuse/adjacent = r/x = 1/ cos Ø
    tangent Ø = tan Ø = opposite/adjacent = y/x = sin Ø / cos Ø
    cotangent Ø = cot Ø = adjacent/opposite = x/y = 1/ tan Ø

Also note that $(y/r)^2 + (x/r)^2 = 1$ or $y^2 + x^2 = r^2$ and that the **Pythagorean Theorem** is $r^2 = x^2 + y^2$

(To remember sin Ø = y/r, cos Ø = x/r, and tan Ø = y/x, think of the word SohCahToa or $S^o{}_hC^a{}_hT^o{}_a$ or $S^o/{}_hC^a/{}_hT^o/{}_a$.)

• In a right triangle the trigonometric functions can be written with respect to either of the acute angles. Consider the following triangle with trigonometric functions for angle α rather than angle Ø.

$$\sin \alpha = \text{opposite/hypotenuse} = x/r$$
$$\csc \alpha = \text{hypotenuse/opposite} = r/x$$
$$\cos \alpha = \text{adjacent/hypotenuse} = y/r$$
$$\sec \alpha = \text{hypotenuse/adjacent} = r/y$$
$$\tan \alpha = \text{opposite/adjacent} = x/y$$
$$\cot \alpha = \text{adjacent/opposite} = y/x$$

Notice how these functions differ from the functions written for angle Ø listed above.

• There are a few common right triangles that it is a good idea to be able to recognize. These include the **30°:60°:90°** triangle and the **45°:45°:90°** triangle (right isosceles triangle). For the 30°:60°:90° triangle, the length of the hypotenuse equals two times the length of the leg opposite the 30° angle, and the length of the leg opposite the 60° angle equals the square root of 3 times the length of the leg opposite the 30° angle. For the 45°:45°:90° triangle, the length of the hypotenuse equals the square root of 2 times x, where x is the length of a leg.

45:45:90 triangles are often drawn as:

where 45° = π/4 and 90° = π/2.

30:60:90 triangles are often drawn as:

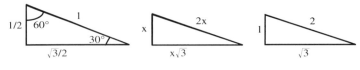

where $30° = \pi/6$, $60° = \pi/3$, $90° = \pi/2$.

The 30°:60°:90° triangle is half of an equilateral triangle:

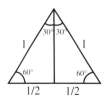

• Other noteworthy right triangles include ***triplet right triangles***. The most common triplets are 3:4:5, 5:12:13, and 7:24:25.

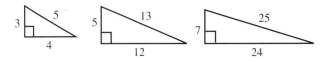

Note that any multiple of the ratios of these triangles is also a triplet, such as 6:8:10 and 10:24:26.

• The two acute angles in a right triangle are always *complementary angles* (sum to 90°) because the angles in all triangles sum to 180°.

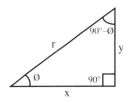

There are **cofunction identities** that describe the complementary nature of the acute angles in a right triangle. The trigonometric ratios of cosine, cotangent, and cosecant are the *cofunctions* of sine, tangent, and secant, respectively:

$$\sin \varnothing = \cos (90° - \varnothing); \quad \cos \varnothing = \sin (90° - \varnothing)$$
$$\tan \varnothing = \cot (90° - \varnothing); \quad \cot \varnothing = \tan (90° - \varnothing)$$
$$\sec \varnothing = \csc (90° - \varnothing); \quad \csc \varnothing = \sec (90° - \varnothing)$$

• Other relationships for right triangles include the following ***reciprocal relationships***:

$\csc Ø = 1 / \sin Ø$ or $\sin Ø = 1 / \csc Ø$

$\sec Ø = 1 / \cos Ø$ or $\cos Ø = 1 / \sec Ø$

$\cot Ø = 1 / \tan Ø$ or $\tan Ø = 1 / \cot Ø$

These can be verified by rearranging:

$\csc Ø \sin Ø = 1 = (r/y)(y/r) = 1$

$\sec Ø \cos Ø = 1 = (r/x)(x/r) = 1$

$\cot Ø \tan Ø = 1 = (x/y)(y/x) = 1$

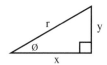

• In a right triangle, the *distance between points* can be calculated using the Pythagorean Theorem. These points can be defined by X and Y axes of a coordinate system. The distance d between the points is represented using: $d^2 = (x_2 - x_1)^2 + (y_2 - y_1)^2$, and depicted by:

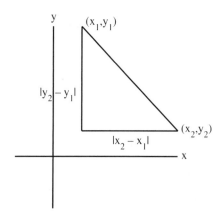

Note that the distance d between two points in three-dimensional space is represented using: $d^2 = (x_2 - x_1)^2 + (y_2 - y_1)^2 + (z_2 - z_1)^2$.

3.2 Solving Right Triangles

• *Solving right triangles* involves finding the measurements of the unknown sides and angles. In a right triangle we know that the measure of one of the angles is 90°; therefore, if we know the measures of 1 acute angle and 1 side, or if we know the measures of 2 sides, we can find all measures of all sides and angles. To solve right triangles, trigonometric ratios, complementary angles (sum to 90°), reciprocal relationships, and the Pythagorean Theorem can be used. There are usually several approaches that can be used to solve a right triangle, such as solving sides and angles in different orders or using different trigonometric ratios.

• In general, when solving a right triangle:

1. If you know one of the acute angles, you can find the other angle because acute angles of a right triangle are complementary and sum to 90° (the sum of all angles in a planar triangle is always 180°).
2. To find the first acute angle, use one of the trigonometric functions for sin Ø, cos Ø, tan Ø, csc Ø, sec Ø, or cot Ø, depending on which sides are known.
3. To find the third side, use the Pythagorean Theorem or one of the six trigonometric functions in combination with a known angle.

• When solving triangles and other problems involving degree measurements, the general guidelines for the accuracy of the results are:

1. Angles expressed to nearest degree, 1°, correspond to length measures expressed to 2 significant digits.
2. Angles expressed to nearest degree, 0.1° (10'), correspond to length measures expressed to 3 significant digits.
3. Angles expressed to nearest degree, 0.01° (1'), correspond to length measures expressed to 4 significant digits.
4. Angles expressed to nearest degree, 0.001° (0.1'), correspond to length measures expressed to 5 significant digits.

Remember: ' denotes *minute*.

• When using a *calculator* to solve trigonometric functions for the value of the function or for an angle, use the indicated buttons on the face of your calculator, and also make sure the calculator is in *degree mode* for calculations involving degrees or in *radian mode* for calculations using radian angle measurements. For example, to calculate the cosine, sine, or tangent of an angle, enter the angle and press the cosine, sine, or tangent key. To calculate the secant, cosecant, or cotangent of an angle, enter the angle and press the secant, cosecant, or cotangent key if it is present on the calculator. If the secant, cosecant, and cotangent keys are not present on the calculator, then use the ***reciprocal relationships***: csc Ø = 1 / sin Ø, sec Ø = 1 / cos Ø, and cot Ø = 1 / tan Ø to calculate the values. If the cosine, sine, or tangent of an angle is known, to find the value of the angle Ø, use the appropriate ***inverse function*** *key* on your calculator: if y = cos Ø, then Ø = cos⁻¹y; if y = sin Ø, then Ø = sin⁻¹y; or if y = tan Ø, then Ø = tan⁻¹y. The inverse keys are usually labeled as cos⁻¹, sin⁻¹, and tan⁻¹, but may be shown as arccos, arcsin, or arctan. For secant, cosecant, or cotangent, if the inverse keys are not present, use the reciprocal relationships in the calculations.

3.3 Examples and Applications of Right Triangles

• There are many questions that can be answered by setting up a model involving a right triangle. In this section we will describe methods for measuring the distance to a star, the distance across a canyon, the angle of elevation of the Sun, the distance of a ship from a lighthouse, the height of a mountain, the distance of a UFO from the radar towers using bearing, and the distance across a lake. First, consider a couple of simple examples that describe angle and side measurements of right triangles.

• **Example:** Find x, y, and α.

Because the sum of the angles in a planar triangle is 180° and a right triangle has one 90° angle: 40° + α = 90°, therefore α = 50°. To find x and y, use trigonometric ratios involving the known and unknown values. To find y, use sin 40° = y/r = y / 100 ft.

Therefore, y = (100)(sin 40°) ≈ 64.3 ft.

To find x, use cos 40° = x / 100 ft

Therefore, x = (100)(cos 40°) ≈ 76.6 ft.

In summary the solution of this right triangle is: Ø = 40°, α = 50°, x = 76.6 ft, y = 64.3 ft, and r = 100 ft.

To check the side calculations, use the Pythagorean Theorem:

$$x^2 + y^2 = r^2.$$

Is it true that $76.6^2 + 64.3^2 \approx 100^2$? Yes.

- **Example:** Find x, Ø, and α.

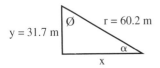

To find α, use $\sin \alpha = y/r = 31.7 \text{ m} / 60.2 \text{ m} \approx 0.527$.
Therefore, $\alpha = \sin^{-1}(0.527) \approx 31.8°$
To find Ø, use $90° - \alpha = $ Ø.
Therefore, $Ø = 90° - 31.8° \approx 58.2°$

Note that a right, 90°, angle is an exact angle and does not limit significant digit accuracy.

To find x, use $\tan \alpha = y/x$, or $\tan 31.8° = 31.7 \text{ m} / x$.
Therefore, $x = (31.7 \text{ m}) / (\tan 31.8°) \approx 51.1 \text{ m}$.

In summary, the solution of this right triangle is: $Ø = 58.2°$, $\alpha = 31.8°$, $x = 51.1$ m, $y = 31.7$ m, and $r = 60.2$ m.

To check the side calculations, use the Pythagorean Theorem:

$$x^2 + y^2 = r^2.$$

Is it true that $51.1^2 + 31.7^2 \approx 60.2^2$? Yes.

Note that we could have calculated x, given r and y, using the Pythagorean Theorem.

- The following are examples of applications that use right triangles.

- **Example:** How can we measure the *distance to a star*? The most direct measurements of distance for stars are made with *trigonometric parallax*. The trigonometric parallax is the apparent angular displacement of an object's position when viewed from two different locations. This technique works because when an object is viewed from two different locations, it appears to shift position. (Think about looking at an object through binoculars and closing one eye and then the other.) As the Earth moves in its orbit around the Sun, we observe stars from different points of view, and they appear to shift back and forth by differing amounts, depending on how far they are from us. Objects farther away exhibit a smaller shift. The parallax method measures the apparent displacement or dislocation of a star relative to a background field of much more distant objects out in space.

If we know the distance between the two locations where the observations are made (location of Earth in opposite sides of its orbit around the Sun), and if we can measure the angle through which a star appears to shift position (the angle of parallax) against background stars, then by using trigonometry the distance from the Earth to the star can be calculated. Unfortunately, for stars that are very far away, the parallax method breaks down because the angle of parallax is so small. Other methods that rely on relative distances are used to estimate distances of stars, galaxies, etc.

To use the parallax method to determine a star's distance from Earth, the star's location is observed (using a telescope) against background stars at two opposite points in the Earth's orbit around the Sun. The annual parallax is defined as half of the angular shift (or angle of parallax) in a star's position against background stars when the Earth is at its two opposite locations in its orbit around the Sun, and is depicted by θ.

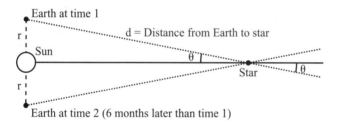

Earth at time 1

r

Sun

d = Distance from Earth to star

θ

Star

θ

r

Earth at time 2 (6 months later than time 1)

The distance from the star to the Earth, d, is calculated using the right triangle formed by Earth, the star, and the Sun, with r being the distance from Earth to Sun. To find d we can use the simple trigonometric relationship for a right triangle: $\sin \theta = $ opposite / hypotenuse $= r/d$, and rearranging,

$$d = r / \sin \theta$$

Given that r = 150,000,000 km, and if θ is measured as, for example, θ = 0.314″, then we can calculate d. (In the early 1800s Friedrich Bessel found that the star 61 Cygni in the constellation Cygnus had a parallax of 0.314 arc seconds.)

Using the conversion 1 second = 1/60 of a minute = 1/3600 of a degree, 0.314″ / 3600 ≈ 0.00008722°.

(Note that there are 1/3600 = 0.00027778 degrees per second, so θ = 0.314″ times 0.00027778 degrees/second equals 0.00008722°.)

Therefore, the distance from the Earth to the star is:

$$d = r/\sin\theta = 150{,}000{,}000 \text{ km} / \sin(0.00008722°) \approx 9.85 \times 10^{13} \text{ km}$$

To convert to light years, divide km by speed of light times number of seconds in a year:

$$9.85 \times 10^{13} \text{ km} /[(3 \times 10^5 \text{ km/s})(3{,}600 \text{ s/hr} \times 24 \text{ hr/day} \times 365 \text{ day/yr})]$$

resulting in the distance to the star being approximately 10 light years. More recent measurements of the parallax of 61 Cygni resulted in 0.294″ and a distance of 11.1 light years.

• **Example:** A geologist uses a surveyor's transit, or theodolite, which is an instrument having telescopic sight that is used to measure horizontal and vertical angles. To measure the distance across a canyon, the geologist first measures a known length of 50 feet along one side of a canyon. She uses the transit at point B, sets a 90° angle to a point C, and then uses the transit at point A and measures a 40° angle from B to C with its vertex at A.

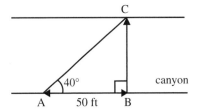

Knowing two angles and a side, she determines the width of the canyon (length *BC*) using tan A = opposite / adjacent.

$\tan 40° = BC / 50$ ft, rearranging, gives the width BC:

$\quad BC = (50$ ft$)(\tan 40°) \approx 42$ feet

• Problems involving **angle of elevation** and **angle of depression** can be solved using a right triangle. The angle of elevation is an angle measured from the *horizontal* upward, and an angle of depression is an angle measured from the *horizontal* downward.

• Example of *angle of elevation*: What is the approximate angle of elevation of the Sun, if a 100 foot telescope dome casts a shadow 70 feet long?

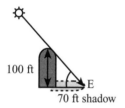

100 ft

E

70 ft shadow

To find the angle of elevation, E, use the known sides of the formed right triangle and the tangent trigonometric ratio:

$\quad \tan E = $ opposite/adjacent $= 100$ ft $/ 70$ ft

Therefore the angle of elevation, $E = \tan^{-1}(100/70) \approx 55°$

• Example of *angle of depression*: Suppose you are in a lighthouse communicating with a ship offshore and you want to determine how far the ship is from the cliff that the lighthouse is on. The top of the lighthouse is 200 feet above sea level.

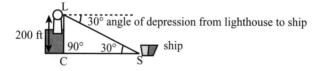

The angle of depression from the lighthouse to the ship is 30°; therefore, angle S is 30° because they are alternate interior angles of parallel lines. To find the distance from the cliff, C, to the ship, S, use the trigonometric ratio for a right triangle:

tan S = opposite / adjacent = 200 ft / CS

Therefore, CS = 200 ft / tan 30° ≈ 346 feet

• Example of *angle of elevation*: Suppose you need to find the height of a mountain using a right triangle model in which you can make two angle of elevation measurements at two distances, point A and point B.

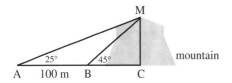

You measure the distance between point A and B as 100 meters, but the terrain between B and C is too treacherous to measure. We know that angle C is 90° for both triangles, AMC and BMC.

To find the height of the mountain, MC, we need to know the distance AC and therefore BC. Because $AB + BC = AC$, then $AC = 100 + BC$.

Begin by developing an equation for MC and then using right triangle ratios to find BC and AC:

tan 25° = MC/AC

Therefore, MC = AC tan 25° = (100 + BC) tan 25°

= 100 tan 25° + BC tan 25°

tan 45° = MC/BC

Therefore, MC = BC tan 45° = 100 tan 25° + BC tan 25°

Rearranging,

BC tan 45° – BC tan 25° = 100 tan 25°

BC (tan 45° – tan 25°) = 100 tan 25°

BC = 100 tan 25° / (tan 45° – tan 25°) ≈ 87.4 meters

Therefore, AC = 100 m + BC ≈ 187.4 meters.

To determine the height MC use: $\tan 25° = MC / AC$

Therefore, the height of the mountain,

$$MC = AC \tan 25° = (187.4 \text{ m}) \tan 25° \approx 87.4 \text{ meters.}$$

Alternatively, we can find the height of the mountain using:

$$MC = BC \tan 45° = (87.4 \text{ m}) \tan 45° \approx 87.4 \text{ m.}$$

Note that triangle BMC is a 45°:45°:90° triangle, which has both legs of equal length. Therefore, sides BC and MC are equal lengths.

• **Bearing** is generally used in navigation and aeronautics, and calculations can be made using right triangle relationships. Bearing specifies a direction and is expressed in two ways: (1) with a direction being designated beginning with a north-south line toward an east or west direction with N, S, E, W specified, or (2) with a direction being designated beginning with due north and measured in a clockwise direction. See figures.

1. Bearing beginning with the north-south line toward an east or west as specified:

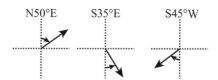

N50°E   S35°E   S45°W

2. Bearing beginning with due north and measured clockwise:

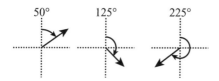

50°   125°   225°

• **Bearing** example: Two radar towers 5 miles apart are tracking a UFO. Tower A tracks the UFO at a bearing of 50° and tower B tracks it at 320°. The triangle formed between the two towers and the UFO is a right triangle. How far is the UFO from the radar towers?

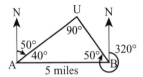

A right triangle can be drawn for triangle ABU with U as the 90° angle. Angle A of the right triangle at tower A can be obtained from the bearing of 50° as: 90° – 50° = 40°.

Angle B at tower B can be found using the fact that the sum of the angles in a planar triangle is 180° and the two acute angles of a right triangle sum to 90°. Therefore, angle B is 90° – 40° = 50°.

The distance from tower A to the UFO, or side *AU*, can be found using:

cos 40° = adjacent/hypotenuse = *AU* / 5 mi

Therefore, *AU* = (5 mi)(cos 40°) ≈ 3.8 mi.

(We can also use sin 50° = *AU* / 5 mi, or *AU* = (5 mi)sin 50° ≈ 3.8 mi.)

The distance from tower B to the UFO, or side *BU*, can be found using:

cos 50° = adjacent/hypotenuse = *BU* / 5 mi

Therefore, *BU* = (5 mi)(cos 50°) ≈ 3.2 mi.

(We can also use sin 40° = *BU*/5 mi, or *BU* = (5 mi)sin 40° ≈ 3.2 mi.)

The Pythagorean Theorem could also have been used to compute the third side, or as a test of the result. Does $3.8^2 + 3.2^2 ≈ 5^2$? Yes.

In summary, the distance of the UFO to the radar towers is 3.8 miles from tower A and 3.2 miles from tower B.

3.4 Oblique Triangles and the Law of Sines and Law of Cosines

• *Oblique triangles* are triangles in planes that are not right triangles and therefore do not contain a right angle. They are described and solved using primarily the *Law of Sines* and *Law of Cosines*, and occasionally the *Law of Tangents*. Solving oblique triangles is helpful in situations where measuring distances and angles is required. Oblique triangles may have all acute angles (measuring less than 90°) or may contain an obtuse angle (measuring more than 90°).

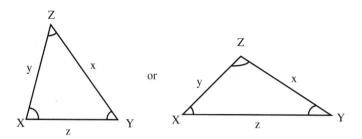

- The **Law of Sines** (along with the Law of Cosines) is used to determine unknown side and angle measurements of an oblique triangle. The **Law of Sines** is given by:

$$\frac{x}{\sin X} = \frac{y}{\sin Y} = \frac{z}{\sin Z}$$

Where x, y, and z represent sides of a triangle and X, Y, and Z represent the angles opposite to the sides x, y, and z, respectively. By rearranging using algebra, the following forms of the Law of Sines are obtained:

$$\frac{x}{y} = \frac{\sin X}{\sin Y}, \quad \frac{y}{z} = \frac{\sin Y}{\sin Z}, \quad \frac{z}{x} = \frac{\sin Z}{\sin X}$$

$$\frac{x}{z} = \frac{\sin X}{\sin Z}, \quad \frac{y}{x} = \frac{\sin Y}{\sin X}, \quad \frac{z}{y} = \frac{\sin Z}{\sin Y}$$

$$x \sin Y = y \sin X, \quad y \sin Z = z \sin Y, \quad z \sin X = x \sin Z$$

$$x = y \sin X / \sin Y, \quad x = z \sin X / \sin Z$$
$$y = z \sin Y / \sin Z, \quad y = x \sin Y / \sin X$$
$$z = x \sin Z / \sin X, \quad z = y \sin Z / \sin Y$$

$$\sin X = x \sin Y / y, \quad \sin X = x \sin Z / z$$
$$\sin Y = y \sin Z / z, \quad \sin Y = y \sin X / x$$
$$\sin Z = z \sin X / x, \quad \sin Z = z \sin Y / y$$

- The **Law of Cosines** (along with the Law of Sines) is used to determine unknown side and angle measurements of an oblique triangle. The **Law of Cosines** is given by:

$$x^2 = y^2 + z^2 - 2yz \cos X$$
$$y^2 = z^2 + x^2 - 2zx \cos Y$$
$$z^2 = x^2 + y^2 - 2xy \cos Z$$

where x, y, and z represent sides of a triangle and X, Y, and Z represent the angles opposite to the sides x, y, and z, respectively. (For a right triangle where $\cos 90° = 0$, this reduces to the Pythagorean Theorem.) By rearranging, the following forms of the Law of Cosines are obtained:

$$\cos X = (y^2 + z^2 - x^2) / 2yz$$
$$\cos Y = (z^2 + x^2 - y^2) / 2zx$$
$$\cos Z = (x^2 + y^2 - z^2) / 2xy$$

To find the angles X, Y, and Z, use the inverse cosine key $\cos^{-1}$ on your calculator.

• Note that the Law of Sines and the Law of Cosines can be used to solve right triangles as well as oblique triangles, although the methods previously used to solve right triangles are easier to use when confronted with a right triangle. The Law of Sines and the Law of Cosines are derived from the principles of right triangle relations.

• ***Derivation of Law of Sines.***

 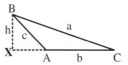

To derive the Law of Sines, we can use either an acute triangle or an obtuse triangle. The first step in either case is to create two right triangles out of the obtuse or acute triangle by extending a perpendicular line or altitude line, h, from angle B perpendicular to side b to vertex X. This results in:

Right triangle ABX, where sin A = opposite/hypotenuse = h/c

Therefore, h = c sin A

Right triangle BCX, where sin C = opposite/hypotenuse = h/a

Therefore, h = a sin C

Because h = h, c sin A = a sin C

Rearranging results in the Law of Sines: $\dfrac{c}{\sin C} = \dfrac{a}{\sin A}$

By drawing a perpendicular line, h, from angle A to side a or from angle C to side c, derivations can also be shown for the other Law of Sines relationships:

$$\frac{a}{\sin A} = \frac{b}{\sin B} \text{ and } \frac{b}{\sin B} = \frac{c}{\sin C}$$

Written together we obtain the standard form of the *Law of Sines*:

$$\frac{a}{\sin A} = \frac{b}{\sin B} = \frac{c}{\sin C}$$

• *Derivation of Law of Cosines*. There is more than one approach to derive the Law of Cosines. See textbooks for various derivations. Following is one method for deriving the Law of Cosines.

Consider an acute triangle with an altitude line, h, drawn from angle C to side c:

Two right triangles are formed, ACX and BCX.

For right triangle ACX, cos A = adjacent/hypotenuse = x/b

Therefore, x = b cos A

For each right triangle, ACX and BCX, the Pythagorean Theorem can be written:

ACX: $x^2 + h^2 = b^2$
BCX: $(c-x)^2 + h^2 = a^2$

Rearranging the BCX equation gives: $c^2 - 2cx + x^2 + h^2 = a^2$ or

BCX: $x^2 + h^2 = a^2 - c^2 + 2cx$

Combine right triangle equations by substituting $x^2 + h^2 = b^2$ from ACX into BCX equation:

$b^2 = a^2 - c^2 + 2cx$, or
$a^2 = b^2 + c^2 - 2cx$

For the large triangle ABC substitute x = b cos A to obtain the Law of Cosines: $a^2 = b^2 + c^2 - 2bc \cos A$

By drawing different perpendicular h lines from the other two angles perpendicular to their opposite sides, the other two formulas for the Law of Cosines can be derived using this same procedure. The three equations for the *Law of Cosines* are:

$$a^2 = b^2 + c^2 - 2bc \cos A$$
$$b^2 = a^2 + c^2 - 2ac \cos B$$
$$c^2 = a^2 + b^2 - 2ab \cos C$$

• The *Law of Cosines* can also be derived using an obtuse triangle.

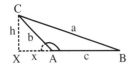

Note that angle A is indicated by the dark arc and angle (180° – A) is indicated by the dashed arc.

The following is true for right triangle AXC:

$$\cos (180° - A) = x/b$$

Therefore, x = b cos (180° – A)

Because cos (180° – A) = – cos A, then x = –b cos A

Proceeding with Pythagorean Theorem for both right triangles:

Triangle ACX: $x^2 + h^2 = b^2$
Triangle BCX: $(x + c)^2 + h^2 = a^2$
Rearranging gives: $x^2 + 2xc + c^2 + h^2 = a^2$ or $a^2 = x^2 + h^2 + c^2 + 2xc$
Substitute $x^2 + h^2 = b^2$ form triangle ACX: $a^2 = b^2 + c^2 + 2xc$

Substituting x = –b cos A gives the Law of Cosines with respect to angle A:

$$a^2 = b^2 + c^2 - 2bc \cos A$$

By drawing different perpendicular h lines, the other two formulas for the Law of Cosines can be derived.

Law of Tangents

• The *Law of Tangents*, which was stated in modern form by Viète about 1580, is included here for completeness, although it is rarely used today. It is given by:

$$\frac{x+y}{x-y} = \frac{\tan \frac{1}{2}(X+Y)}{\tan \frac{1}{2}(X-Y)}$$

The Law of Tangents can be used to solve triangles when two sides and the angle in between are given (SAS case). Today the Law of Cosines is used to find the missing side, and then one of the unknown angles is found using the Law of Sines.

3.5 Solving Oblique Triangles

• Right triangles and oblique triangles are used to determine lengths, distances, and angles for various applications and problems. Solving a triangle generally means finding the three angle measurements and the three side lengths. In many problems, however, not all side or angle measurements will be required. The *Law of Cosines* and the *Law of Sines* are used to find side and angle values and to solve oblique triangles. To solve a triangle and find all six measures, at least three measures must be known and at least one of these known measures must be a side length. In other words, to solve an oblique triangle we need to know at least one side length and any other two measurements.

• There are five possible scenarios of what we may know to solve an oblique triangle. These scenarios are:

3 sides are known, which is a SSS triangle;

2 sides and 1 angle are known, which can be a SAS or SSA triangle; or

1 side and 2 angles are known, which can be an ASA or AAS triangle.

Following are the five possible cases with X, Y, and Z representing the angles and x, y, and z representing the sides opposite to the angles.

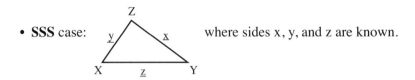

• **SSS** case: where sides x, y, and z are known.

To solve, we need to find the three angles X, Y, and Z.

First, find the largest angle Z (opposite to the largest side z) using the Law of Cosines: $\cos Z = (x^2 + y^2 - z^2) / 2xy$.

Note that if one of the other angles was largest, we would use $\cos X = (y^2 + z^2 - x^2) / 2yz$ or $\cos Y = (z^2 + x^2 - y^2) / 2zx$.

Once you know cos X, cos Y, or cos Z, use the inverse cosine key $\cos^{-1}$ on your calculator find the angles X, Y, or Z.

Next, find either remaining angle using the Law of Sines or the Law of Cosines. Remember: The Law of Sines is: x /sin X = y /sin Y = z/sin Z.

Finally, find the third angle using the sum of angles in a triangle rule:

$X° + Y° + Z° = 180°$.

• **SAS** case: given sides x and y and angle Z in between.

To solve, we need to find the third side z and angles X and Y.

First, find third side z opposite given angle Z using the Law of Cosines:

$z^2 = x^2 + y^2 - 2xy \cos Z$

Next, find the smaller unknown angle using the Law of Sines or the Law of Cosines.

Finally, find the third angle using the sum of angles in a triangle rule:

$X° + Y° + Z° = 180°$

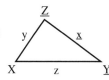

- **ASA** case: given two angles Y and Z and the side x between.

To solve, we need to find the third angle X and sides y and z.

First, find the third angle X using the sum of angles in a triangle rule:

$$X° + Y° + Z° = 180°$$

Next, solve the remaining two sides y and z using the Law of Sines:

$$y = x \sin Y / \sin X$$
$$z = x \sin Z / \sin X$$

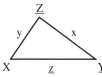

- **AAS** case: given two angles Y and Z and side z *not* between.

To solve, we need to find the third angle X and sides x and y. Note that this is similar to ASA case.

First, find the third angle X using the sum of angles in a triangle rule:

$$X° + Y° + Z° = 180°$$

Next, find the remaining two sides x and y using Law of Sines:

$$x = z \sin X / \sin Z$$
$$y = z \sin Y / \sin Z$$

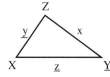

- **SSA** case: given two sides y and z and an angle Y *not* between.

This is known as the "ambiguous case" because SSA triangles may not always have a unique solution and therefore represent one triangle. In the ambiguous case there may be:

1. No solution for the angle so that no triangle exists;
2. Two solutions—the angle and its supplement—so two triangles exist; or
3. One solution for the angle so that one triangle exists.

To solve, we need to find the third side x and angles X and Z.

First, solve the angle opposite the other known side using the Law of Sines. In this case find angle Z:

$$\sin Z = (z \sin Y) / y$$

(Use $\sin^{-1}$ on your calculator to find angle Z.)

Next, find the third angle X using the sum of angles in a triangle rule:

$$X° + Y° + Z° = 180°$$

Finally, use the Law of Sines to find the third side x:

$$x = z \sin X / \sin Z$$

If two triangles exist, then steps need to be repeated to solve both triangles. Following is a discussion of the ambiguous case.

• Ambiguities of the SSA case become obvious when we attempt to draw the triangle after we begin to solve the triangle in question. As a reference, consider the following possible triangle of the SSA case where we are given sides x and y and angle X. Note that h is the perpendicular drawn to create two right triangles.

Consider if X is an acute angle:

When side x < h, then *no* triangle exists.
Note h = y sin X.

When side x = h, then *one* triangle can exist, and it is a right triangle.

When side x > y, then *one* triangle can exist.

When side y > x > h, then *two* triangles may exist, one with side x_1 and one with side x_2.

Consider if angle X is obtuse:

When side x ≤ y, then *no* triangle exists.

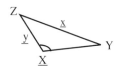

When side x > y, then *one* triangle can exist.

- In general, when solving triangles remember the following:

1. The sum of the lengths of any two sides must be greater than the length of the third side.

$$a + b > c$$
$$b + c > a$$
$$a + c > b$$

2. The smallest angle is opposite to the smallest side, the middle size angle is opposite the middle length side, and the largest angle is opposite to the largest side.

3. Supplementary angles have the same sine value, and the sine of an angle cannot be greater than $|1|$.

$$\sin \theta = \sin(180° - \theta)$$ $(180° - \theta)$ θ

4. The sum of the angles in a planar triangle is always 180°.

3.6 Examples and Applications of Oblique Triangles

• Example 1: Find all measures of a triangle given that the lengths of the three sides are a = 10 m, b = 15 m, and c = 20 m.

This is an SSS triangle. We need to find angles A, B, and C. First, use the Law of Cosines to solve for the largest angle, C, (which is opposite the largest side). This will tell us if C is acute or obtuse. Note that when cos C < 0, C is obtuse. The Law of Cosines is: $c^2 = a^2 + b^2 - 2ab \cos C$.

Rearranging:

$$C = \cos^{-1}[(a^2 + b^2 - c^2)/2ab]$$
$$= \cos^{-1}[(10^2 + 15^2 - 20^2)/2(10)(15)] = C \approx 104.5°$$

Next, find angle B using the Law of Sines: c / sin C = b / sin B.

Rearranging:

$$\sin B = b \sin C/c$$
$$B = \sin^{-1}[(15 \sin 104.5°)/20]$$
$$B \approx 46.6°$$

Finally, find angle A using A = 180° – (B° + C°).

$$A = 180° - (46.6° + 104.5°)$$
$$A = 28.9°$$

Therefore, the solution of the triangle is:

a = 10 m, b = 15 m, c = 20 m, A = 28.9°, B = 46.6°, and C = 104.5°.

Note: To check results, repeat calculations using different formulas.

• Example 2: Find all measures of a triangle given the side lengths

   a = 24 feet and b = 9 feet, and angle B = 54°.

This is an SSA triangle (the ambiguous case). We need to find angles A and B and side c.

If we try to draw the triangle, it looks questionable:

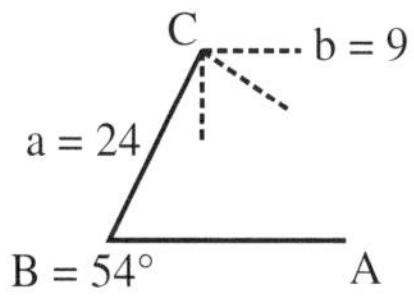

Use the Law of Sines to see if sin A < 1:

   $\sin A = (a \sin B) / b$

   $\sin A = (24 \sin 54°) / 9 \approx 2.157$, which is greater than 1

Sine cannot be greater than 1 (if you press $\sin^{-1}$ on your calculator an error message should appear). Therefore, no such angle A can exist and no triangle can be formed.

• Example 3: Find all measures of a triangle given the side lengths a = 10 meters and b = 8 meters, and angle A = 42.0°.

This is an SSA triangle (the ambiguous case). We need to find angles B and C and side c.

Find sin B using Law of Sines, $b / \sin B = a / \sin A$, to see if sin B < 1:

   $\sin B = (b \sin A) / a = (8 \sin 42.0°) / 10 \approx 0.5353$

which is less than 1, so angle B exists. Taking $\sin^{-1}$:

   $B \approx 32.4°$

Is it possible that an obtuse B (or its supplement) also exists?

The supplement would be: $180° - 32.4° = 147.6°$

To see if this is possible, add angle A + 147.6°, or

   $42.0° + 147.6° = 189.6°$

Because $189.6° > 180°$, a triangle cannot exist as all angles in a triangle must sum to $180°$. Therefore, two triangles cannot exist.

Continue with angle B $= 32.4°$, and find angle C using triangle sum rule:

$$C = 180° - (A + B) = 180° - (42.0° + 32.4°)$$
$$C = 105.6°$$

Finally, find side c using Law of Sines, c / sin C $=$ a / sin A:

$$c = (a \sin C) / \sin A = (10 \sin 105.6°) / \sin 42.0°$$
$$c \approx 14 \text{ meters}$$

Therefore, the measurements of the triangle are:

a $= 10$ meters, b $= 8$ meters, c $= 14$ meters, angle A $= 42.0°$, B $= 32.4°$, and C $= 105.6°$.

The triangle looks something like this:

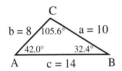

• Example 4: Find all measures of a triangle given the side lengths a $= 10$ meters and b $= 17$ meters, and angle A $= 28°$.

This is an SSA triangle (the ambiguous case). We need to find angles B and C and side c.

Find sin B using the Law of Sines, b / sin B $=$ a / sin A:

$$\sin B = (b \sin A) / a = (17 \sin 28°) / 10 \approx 0.7981$$

which is less than 1, so angle B exists. Taking $\sin^{-1}$:

$$B \approx 53°$$

Is it possible that an obtuse B (or its supplement) also exists?

The supplement would be: $180° - 53° = 127°$

To see if this is possible, add angle A + 127°, or

   28° + 127° = 155°

Because 155° < 180°, a triangle can exist.

Therefore, two triangles with B = 53° and B = 127° can exist.

Solve Triangle 1: B = 53°, A = 28°, a = 10 meters, and b = 17 meters.

Find C using the triangle sum rule:

   C = 180° – (28° + 53°) = 99°

Find c using the Law of Sines, c / sin C = a / sin A:

   c = a sin C / sin A = 10 sin 99° / sin 28° ≈ 21 meters

Therefore, Triangle 1 has measures B = 53°, A = 28°, C = 99°, a = 10 meters, b = 17 meters, and c = 21 meters. Draw this triangle to see if it looks reasonable:

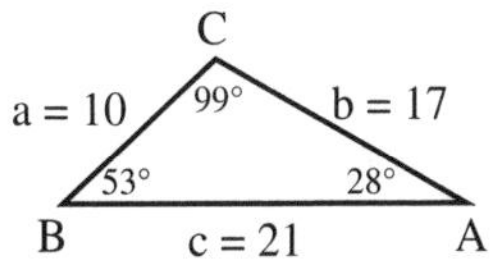

Solve Triangle 2: B = 127°, A = 28°, a = 10 meters, and b = 17 meters.

Find C using the triangle sum rule:

   C = 180° – (28° + 127°) = 25°

Find c using Law of Sines, c / sin C = a / sin A:

   c = a sin C / sin A = 10 sin 25° / sin 28° ≈ 9 meters

Therefore, Triangle 2 has measures B = 127°, A = 28°, C = 25°, a = 10 meters, b = 17 meters, and c = 9 meters. Draw this triangle to see if it looks reasonable:

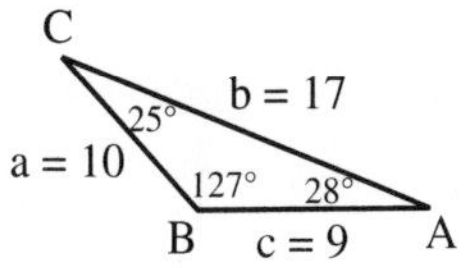

• Example 5: Find the distance to a point C across a canyon from two selected points, A and B, 100 feet from each other. (Note that in section 3.3 we used a right-triangle model to solve a canyon problem requiring that we find the distance directly across the canyon.)

First, measure out the 100 feet along the side of the canyon between two points A and B. Next measure angles A and B using a surveyor's transit device, or theodolite, pointing toward point C on the other side of the canyon.

Angle A measures 100° and angle B measures 60°.

This is an ASA triangle with A = 100°, B = 60°, and side c = 100 feet. We need to find angle C and sides a and b.

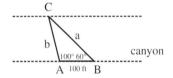

Calculate angle C using the triangle angle sum rule:

$$C = 180° - (100° + 60°) = 20°$$

Use the Law of Sines to find b, b / sin B = c / sin C:

$$b = c \sin B / \sin C = 100 \sin 60° / \sin 20° \approx 253 \text{ feet}$$

Use Law of Sines to find a, a / sin A = c / sin C:

$$a = c \sin A / \sin C = 100 \sin 100° / \sin 20° \; a \approx 288 \text{ feet}$$

Therefore, the distance from A to C is 253 feet and the distance from B to C is 288 feet, which seems correct.

• Example 6: Find the distance across a lake.

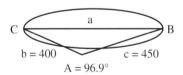

Create an oblique triangle model and measure accessible distances and angles.

   Measure distance b = 400 feet

   Measure distance c = 450 feet

   Measure angle A using a surveyor's transit as A = 96.9°

This is an SAS triangle with A = 96.9°, side b = 400 feet, and side c = 450 feet.

We need to find the distance a across the lake between points C and B.

Use the Law of Cosines to find a: $a^2 = b^2 + c^2 - 2bc \cos A$.

We need to calculate the square root of a^2 to find a:

$a = (b^2 + c^2 - 2bc \cos A)^{1/2}$

$a = [(400^2 + 450^2 - 2(400)(450) \cos 96.9°)]^{1/2}$

$a \approx 637$ feet

Judging from the drawing, the distance of 637 feet across the lake seems reasonable.

3.7 Finding the Area of a Triangle

• Several methods can be used to determine the area of a triangle, and the method used will depend on what information is known. Methods used to find area described in this section are based on knowing the following:

1. Base and height;
2. Two sides and the angle in between (SAS);
3. Two angles and one side (ASA or AAS);
4. Two sides and an angle not in between (SSA); and
5. Three sides (SSS).

Following are descriptions of each of these cases.

1. **Find the area of a triangle given *base and height.***

height = 3 in

base = 6 in

Area of a triangle = (1/2)(base)(height)

The *height* is an altitude line drawn from an angle perpendicular to the side opposite that angle. The height or altitude is the length of that line, and the base is the side to which the height is perpendicular. To obtain the height, draw a perpendicular line from the base to the opposite angle.

In this triangle area is:

(1/2)(6 inches)(3 inches) = 9 inches^2 or 9 square inches

2. **Find the area of a triangle given** *two sides and the angle in between (SAS).*

• If we know two sides and the angle in between them, we can draw an altitude or height line, h, perpendicular to the opposite base and determine the length of h using trigonometric functions for right triangles. Drawing an altitude line from an angle perpendicular to the opposite side forms two right triangles. When calculating area, the height line can be drawn from any angle. The angle chosen will depend on which triangle dimensions are known. Consider the following triangles:

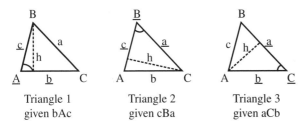

Triangle 1   Triangle 2   Triangle 3
given bAc   given cBa   given aCb

Right triangles formed by the h line are used to determine the value of h using right triangle trigonometric functions:

Triangle 1: sin A = opposite/hypotenuse = h/c; therefore h = c sin A

Triangle 2: sin B = opposite/hypotenuse = h/a; therefore h = a sin B

Triangle 3: sin C = opposite/hypotenuse = h/b; therefore h = b sin C

Because the area of a triangle is: area = (1/2)(base)(height), we can substitute for h from the information obtained from the right triangle functions. Remember: The base is the side of the triangle that h is drawn perpendicular to. Substitute into the area equation for the above three triangles: area = (1/2)(base)(height).

Triangle 1: Area = (1/2)(b)(c sin A)

Triangle 2: Area = (1/2)(c)(a sin B)

Triangle 3: Area = (1/2)(a)(b sin C)

Therefore, area equals one-half the product of two sides and the sine of the angle in between them. This is true for acute, obtuse, oblique, and right triangles.

Note that the area of an equilateral triangle is $(1/4)(\text{side}^2)\sqrt{3}$.

• **Example:** Find the area of the following triangle with h drawn perpendicular to side *AC*.

For right triangle ABX, sin A = opposite/hypotenuse, or sin 40° = h/10

Therefore, h = 10 sin 40° ≈ 6.4 meters

Now use the area formula:

Area = (1/2)(base)(height) = (1/2)(14m)(6.4m) = 45 meters^2

• **Example:** For an obtuse triangle, the height h can be drawn perpendicular to the opposite side but outside of the triangle. Find the area of the following triangle with h drawn perpendicular to side *AC*.

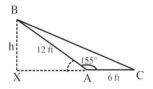

For right triangle ABX, sin(180° – 155°) = opposite/hypotenuse = h/12, sin(180° – 155°) = h/12 = sin 25° = h/12

Therefore, h = 12 sin 25° ≈ 5.1 feet

Now use the area formula:

Area = (1/2)(base)(height) = (1/2)(6ft)(5.1ft) = 15 feet^2

3. **Find the area of a triangle given** *two angles and one side (ASA or AAS).*

• If two angles and one side of a triangle are known, the area can be found using the triangle sum rule, the area formulas we derived in the previous case (2), and the Law of Sines. Consider the following triangles:

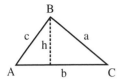 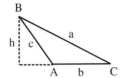

If two angles are known we can calculate the third angle using the triangle sum angle rule: $A° + B° + C° = 180°$.

Next, use the area formulas derived in the previous case (2) for two sides and the angle in between:

Area = (1/2)(a)(b sin C)

Area = (1/2)(b)(c sin A)

Area = (1/2)(c)(a sin B)

Combine with the Law of Sines: $\dfrac{a}{\sin A} = \dfrac{b}{\sin B} = \dfrac{c}{\sin C}$

Rearrange to obtain three equations:

$a \sin B = b \sin A; \quad b \sin C = c \sin B; \quad a \sin C = c \sin A$

$$a = \frac{b \sin A}{\sin B}; \quad b = \frac{c \sin B}{\sin C}; \quad c = \frac{a \sin C}{\sin A}$$

Substitute for a, b, and c into the three *area formulas*:

$$\textbf{\textit{Area}} = (1/2)(a)(b \sin C) = (1/2)(\frac{b \sin A}{\sin B})(b \sin C) = \frac{b^2 \sin A \sin C}{2 \sin B}$$

$$\textbf{\textit{Area}} = (1/2)(b)(c \sin A) = (1/2)(\frac{c \sin B}{\sin C})(c \sin A) = \frac{c^2 \sin A \sin B}{2 \sin C}$$

$$\textbf{\textit{Area}} = (1/2)(c)(a \sin B) = (1/2)(\frac{a \sin C}{\sin A})(a \sin B) = \frac{a^2 \sin B \sin C}{2 \sin A}$$

Select the area formula to use depending on which side, a, b, or c, is known.

• **Example:** Find the area in this AAS triangle, given angles A = 55° and C = 50° and side c = 25 meters.

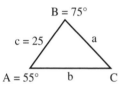

First, find the third angle B: B $= 180° - (55° + 50°) = 75°$

Find area using the area formula where side c is known:

$$\text{Area} = \frac{c^2 \sin A \sin B}{2 \sin C} = \frac{25^2 \sin 55° \sin 75°}{2 \sin 50°} \approx 323 \text{ meters}^2$$

• **Example:** Find the area in this ASA triangle, given angles A $= 55°$ and B $= 75°$ and side c $= 25$ meters.

First, find the third angle C: C $= 180° - (55° + 75°) = 50°$

Find area using the area formula where side c is known:

$$\text{Area} = \frac{c^2 \sin A \sin B}{2 \sin C} = \frac{25^2 \sin 55° \sin 75°}{2 \sin 50°} \approx 323 \text{ meters}^2$$

4. **Find the area of a triangle given *two sides and an angle not in between (SSA).***

• If we know two sides and the angle opposite one of the known sides (not the angle in between), the area can be found using the Law of Sines, the triangle sum rule, and the area formulas derived in the previous case (3). To determine area:

First, find the second angle (and check its supplement) using the Law of Sines:

$$\frac{a}{\sin A} = \frac{b}{\sin B} = \frac{c}{\sin C}$$

Next, use the triangle sum rule to find the third angle.

Then, calculate area using the applicable formula from (3):

$$\text{Area} = \frac{a^2 \sin B \sin C}{2 \sin A}, \text{Area} = \frac{b^2 \sin A \sin C}{2 \sin B}, \text{Area} = \frac{c^2 \sin A \sin B}{2 \sin C}$$

Note that there may exist two triangles, because there may be two solutions for the second angle (the angle and its supplement).

• **Example:** Find the area in a SSA triangle given angle A = 25° and sides a = 100 inches and b = 200 inches.

Find angle B using the Law of Sines: a / sin A = b / sin B. Rearranging:

$$B = \sin^{-1}[b \sin A / a] = \sin^{-1}[200 \sin 25° / 100] \approx 58°$$

Does angle B have a supplement that could represent a second triangle? The supplement of B is 180° − 58° = 122°

To see if a second triangle exists, add angle A = 25° plus supplement angle B = 122°: 25° + 122° = 147°

Because 147° < 180°, a triangle can exist. Therefore, we must consider two triangles with B = 58° and B = 122°.

Triangle 1: A = 25°, B = 58°, a = 100 inches, and b = 200 inches.

Find the third angle C using triangle sum rule:

$$C = 180° − (25° + 58°) = 97°$$

Use the area formula with the known values to obtain the area of Triangle 1:

$$\text{Area} = \frac{a^2 \sin B \sin C}{2 \sin A} = \frac{100^2 \sin 58° \sin 97°}{2 \sin 25°} \approx 9{,}958 \text{ inches}^2$$

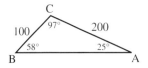

Triangle 2: A = 25°, B = 122°, a = 100 inches, and b = 200 inches.

Find the third angle C using triangle sum rule:

$$C = 180° - (25° + 122°) = 33°$$

Use the area formula with the known values to obtain the area of Triangle 2:

$$\text{Area} = \frac{a^2 \sin B \sin C}{2 \sin A} = \frac{100^2 \sin 122° \sin 33°}{2 \sin 25°} \approx 5,465 \text{ inches}^2$$

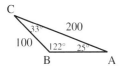

5. **Find the area of a triangle given *three sides (SSS)*.**

• If three sides of a triangle are known, **Heron's formula** can be used to determine the area of the triangle. Heron's formula is:

$$\text{Area} = \sqrt{s(s-a)(s-b)(s-c)}$$

where s is one-half of the perimeter, called the *"semiperimeter,"* and is given by: s = (1/2)(a + b + c).

• Heron's formula can be derived as follows.

Begin with the area formula for angle A, the Law of Cosines, and the half-angle identities for sine and cosine, which are discussed in Chapter 7.

Area = (1/2)(b)(c)sin A, as determined in SAS triangle,

Law of Cosines for angle A: $\cos A = \dfrac{b^2 + c^2 - a^2}{2bc}$

Half-angle identity for sine in terms of angle A: $\sin^2 \dfrac{A}{2} = \dfrac{1 - \cos A}{2}$

Half-angle identity for cosine in terms of angle A: $\cos^2 \dfrac{A}{2} = \dfrac{1 + \cos A}{2}$

Note that this derivation provides a good example of the usefulness of trigonometric identities, which are discussed in Chapter 7.

Substitute the Law of Cosines into the half-angle formula for sine:

$$\sin^2 \frac{A}{2} = \frac{1 - \frac{b^2 + c^2 - a^2}{2bc}}{2}$$

Multiply by 2bc/2bc:

$$\sin^2 \frac{A}{2} = \frac{(2bc)[1 - \frac{b^2 + c^2 - a^2}{2bc}]}{(2bc)2} = \frac{2bc - b^2 - c^2 + a^2}{4bc}$$

Factor the numerator:

$$\sin^2 \frac{A}{2} = \frac{(a + b - c)(a - b + c)}{4bc}$$

Next, we want to get the two factors in the form of the semiperimeter, s = (1/2)(a + b + c).

First rearrange factor (a + b – c):

$$a + b - c = a + b - 2c + 1c = a + b + c - 2c =$$

$$(2/2)(a + b + c - 2c) = 2[\frac{(a + b + c)}{2} - c]$$

Next, rearrange factor (a – b + c):

$$a - b + c = a - 2b + 1b + c = a + b + c - 2b =$$

$$(2/2)(a + b + c - 2b) = 2[\frac{(a + b + c)}{2} - b]$$

Substitute factors back into $\sin^2 \frac{A}{2} = \frac{(a + b - c)(a - b + c)}{4bc}$:

$$\sin^2 \frac{A}{2} = \frac{2[\frac{(a + b + c)}{2} - b]2[\frac{(a + b + c)}{2} - c]}{4bc}$$

$$= \frac{[\frac{(a + b + c)}{2} - b][\frac{(a + b + c)}{2} - c]}{bc}$$

Substitute s = (1/2)(a + b + c):

$$\sin^2 \frac{A}{2} = \frac{(s - b)(s - c)}{bc}$$

Take the square root to obtain sin(A/2):

$$\sin \frac{A}{2} = \sqrt{\frac{(s - b)(s - c)}{bc}}$$

Work through a similar procedure beginning with the half-angle identity for cosine with respect to angle A:

$$\cos^2 \frac{A}{2} = \frac{1 + \cos A}{2}$$

Substitute the Law of Cosines into the half-angle formula for cosine:

$$\cos^2 \frac{A}{2} = \frac{1 + \dfrac{b^2 + c^2 - a^2}{2bc}}{2}$$

Multiply by 2bc/2bc:

$$\cos^2 \frac{A}{2} = \frac{(2bc)[1 + \dfrac{b^2 + c^2 - a^2}{2bc}]}{(2bc)2} = \frac{2bc + b^2 + c^2 - a^2}{4bc}$$

Factor the numerator:

$$\cos^2 \frac{A}{2} = \frac{(a - b - c)(a + b + c)(-1)}{4bc}$$

Next, we want to get the factors in the form of the semiperimeter,

s = (1/2)(a + b + c)

First, rearrange factor (a + b + c):

$$a + b + c = 2[\frac{(a + b + c)}{2}]$$

Next, rearrange factors (a − b − c)(−1):

(a − b − c)(−1) = −a + b + c = −2a + 1a + b + c = a + b + c − 2a

$$= (2/2)(a + b + c - 2a) = 2[\frac{(a + b + c)}{2} - a]$$

Substitute factors back into $\cos^2 \dfrac{A}{2} = \dfrac{(a - b - c)(a + b + c)(-1)}{4bc}$:

$$\cos^2 \frac{A}{2} = \frac{2[\dfrac{(a + b + c)}{2}]2[\dfrac{(a + b + c)}{2} - a]}{4bc}$$

$$= \frac{[\dfrac{(a + b + c)}{2}][\dfrac{(a + b + c)}{2} - a]}{bc}$$

Substitute $s = (1/2)(a + b + c)$:

$$\cos^2 \frac{A}{2} = \frac{(s)(s - a)}{bc}$$

Take the square root to obtain $\cos(A/2)$:

$$\cos \frac{A}{2} = \sqrt{\frac{(s)(s - a)}{bc}}$$

Now we can combine the formulas for $\sin(A/2)$ and $\cos(A/2)$ using the *double-angle identity* $\sin A = 2 \sin(A/2) \cos(A/2)$.

Substitute $\sin(A/2)$ and $\cos(A/2)$ into the double angle identity:

$$\sin A = 2 \sqrt{\frac{(s - b)(s - c)}{bc}} \sqrt{\frac{(s)(s - a)}{bc}} = \frac{2}{bc} \sqrt{s(s - a)(s - b)(s - c)}$$

Finally, to obtain the area substitute this formula for $\sin A$ into the area formula, area $= (1/2)(b)(c)\sin A$:

$$\text{area} = (1/2)(b)(c) \frac{2}{bc} \sqrt{s(s - a)(s - b)(s - c)}$$

$$\text{area} = \sqrt{s(s - a)(s - b)(s - c)}$$

which is *Heron's formula for area of a triangle*.

Note: If this derivation seems laborious and non-obvious, remember that Heron would not have had it named after him if it had been obvious.

• **Example:** Find the area of a triangle given three sides a = 4 meters, b = 6 meters, and c = 8 meters.

a = 4   b = 6

c = 8

First, find the semiperimeter s = (1/2)(a + b + c):

s = (1/2)(4 + 6 + 8) = 9

Substitute into Heron's area formula, area = $\sqrt{s(s - a)(s - b)(s - c)}$:

area = $\sqrt{9(9 - 4)(9 - 6)(9 - 8)}$ = $\sqrt{9(5)(3)(1)}$ = $\sqrt{135}$

area ≈ 12 meters^2.

3.8 Chapter 3 Summary and Highlights

• *Triangles* enable us to determine distances, heights, and angles. By drawing a model of a system, drawing a triangle in the model with its sides and angles representing key features, we can use the properties of triangles to calculate side lengths and angle measurements. Triangles can be used to determine such values as distance to a star, distance across a canyon or lake, or the angle of elevation of the Sun.

• The *six trigonometric functions* can be defined according to the *ratios of the sides of a right triangle*. This makes right triangles especially useful. Right triangles are triangles with one angle equal to 90°. For the right triangle below:

the *six trigonometric functions* are:

cos α = x/r,   sin α = y/r,   tan α = y/x,
sec α = r/x,   csc α = r/y,   cot α = x/y

The Pythagorean Theorem is $r^2 = x^2 + y^2$

- Important relationships include the
reciprocal identities $\cos \alpha = 1/\sec \alpha$, $\sin \alpha = 1/\csc \alpha$, and $\tan \alpha = 1/\cot \alpha$,
the *quotient identities* $\tan \alpha = \sin \alpha / \cos \alpha$, and $\cot \alpha = \cos \alpha / \sin \alpha$,
and the *cofunction identities* for sine and cosine,
$\cos \alpha = \sin(90° - \alpha)$ and $\sin \alpha = \cos(90° - \alpha)$.

- There are special *right triangles* commonly used in calculations, which include the 30:60:90 and 45:45:90 triangles and the 3:4:5, 5:12:13, and 7:24:25 triplet triangles.

- To solve right triangles and find the unknown sides and angles, we can use the six trigonometric functions, the Pythagorean Theorem, and the fact that the two acute angles in a right triangle sum to 90° (because all three angles in a triangle sum to 180°).

- *Oblique triangles* are planar triangles that do not have a 90° angle and therefore are not right triangles. Oblique triangles may have all acute angles (<90°) or two acute angles and one obtuse angle (>90°). Like right triangles, oblique triangles can be used to model problems that require measurements of distances, lengths, and angles, such as determining the distance across a lake or canyon.

- To solve an oblique triangle and find all six measurements, the *Law of Sines*, the *Law of Cosines*, and the fact that the angles in a triangle sum to 180° are used. For the oblique triangle:

The sum of the angles is: $X + Y + Z = 180°$

The Law of Sines is: $(x/\sin X) = (y/\sin Y) = (z/\sin Z)$

The Law of Cosines is: $x^2 = y^2 + z^2 - 2yz \cos X$, $y^2 = z^2 + x^2 - 2zx \cos Y$, and $z^2 = x^2 + y^2 - 2xy \cos Z$

- The *area of a triangle* can be determined using several approaches depending on what information is known. For the triangle:

The simplest equation for area is: area = (1/2)(base)(height), or area = (1/2)(b)(h), where h is a line perpendicular to the base and extending to the opposite angle.

An extension of this equation calculates the height:

area = (1/2)(b)(c sin A)

A further extension of this equation is:

area = (c^2 sin A sin B) / (2 sin C)

If three sides are known, Heron's formula can be used to find area:

area = [s(s − a)(s − b)(s − c)]$^{1/2}$, where s = (1/2)(a + b + c).

• A general note: When solving any problem in mathematics, science, or engineering, including trigonometry, look at your final answer and any drawings that may result and check to see if the answer seems correct with respect to the problem. For example, if you calculate the distance across a stream to be 10,000 miles, you know you have made an error in the way you set up the problem or in one or more of the calculations.

Chapter 4

Trigonometric Functions in a Coordinate System and Circular Functions

• Trigonometry involves measurements pertaining to triangles, angles, distances, arc lengths, circles, planes, and spheres. Trigonometry is used in engineering, navigation, the study of electricity, light and sound, and in any field involving the study of periodic and wave properties. The six trigonometric functions were defined in Chapter 3 for triangles as *ratios of the sides of right triangles*. In this chapter, we define the six trigonometric functions in a *coordinate system using angles in standard position*, and then define them as *arc lengths on a unit circle*, called *circular functions*. Circular functions have as their domains sets real numbers rather than exclusively angles or triangles. Circular functions allow the use of trigonometric functions beyond angles and triangles into the study of electricity, sound, music, projectile motion, and other phenomena that exhibit a periodic nature.

4.1 Review of Functions and Their Properties

• Functions are an integral part of mathematics and reflect the fact that one or more properties can depend on another property. For example, how fast a trolley cart can carry a rock up a hill is a function of how much the rock weighs, the slope of the hill, and the horsepower of the motor. Common functions include algebraic functions represented as mathematical *operations* such as addition, subtraction, multiplication, division, powers, and roots, as well as trigonometric functions, inverse trigonometric functions, logarithmic functions, and exponential functions.

• A *function* is a relation, rule, expression, or equation that associates each element of a *domain set* with its corresponding element in the *range set*. For a relation, rule, expression, or equation to be a function, there must be *only one element or number in the range set for each element or number in the domain set*. The domain set of a function is the set of possible values of the independent variable, and the range set is the corresponding set of values of the dependent variable.

• The *domain set* is the initial set and the *range set* is the set that results after a function is applied:

   domain set → function f( ) *→ range set*

For example, x^2 is a function that is applied to the domain set resulting in the range set:

   domain set x = $\{2, 3, 4\}$
   through function $f(x) = x^2$, $f(2) = 2^2$, $f(3) = 3^2$, $f(4) = 4^2$
   to range set $f(x) = \{4, 9, 16\}$

• The domain set and range set can be expressed as $(x, f(x))$ pairs. In the previous example, the function is $f(x) = x^2$ and the pairs are $(2, 4)$, $(3, 9)$, and $(4, 16)$.

• For each member of the domain set, there must be only one corresponding member in the range set. For example:

   F = $(2, 4), (3, 9), (4, 16)$ where F is a function.
   M = $(2, 5), (2, -5), (4, 9)$ where M is not a function.

M is not a function because the number 2 in the domain set corresponds to more than one number in the range set.

• Functions can be expressed in the form of a graph, a formula, or a table. To *graph functions*, the values in the domain set correspond to the X-axis and the related values in the range set correspond to the Y-axis. For example:

domain set x = –2, –1, 0, 2

through function f(x) = x + 1

to range set f(x) = –1, 0, 1, 3

resulting in pairs (x, y) = (–2, –1), (–1, 0), (0, 1), (2, 3).

When graphed these resulting pairs are depicted as:

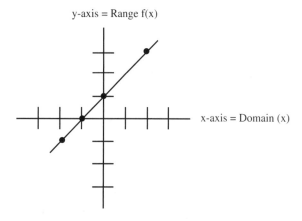

• Graphs of functions only have one value of y for each x value:

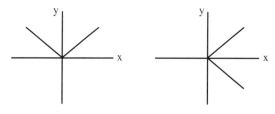

Graph is a function　　　　　Graph is not a function

Graph is not a function

If a vertical line can be drawn that passes through the graph more than one time, there is more than one y value for a given x value and the graph is not a function. This is called the **vertical line test**.

• In general, a *function is increasing* when $y = f(x)$ increases as x increases, and a *function is decreasing* when $y = f(x)$ decreases as x increases.

• The following are examples of (a) *addition*, (b) *subtraction*, (c) *multiplication*, and (d) *division* of *functions*. In these examples the functions $f(x)$ and $g(x)$ are given by $f(x) = 2x$ and $g(x) = x^2$:

(a) $f(x) + g(x) = (f + g)(x) = 2x + x^2$
(b) $f(x) - g(x) = (f - g)(x) = 2x - x^2$
(c) $f(x) \times g(x) = (f \times g)(x) = 2x \times x^2 = 2x^3$
(d) $f(x) \div g(x) = (f \div g)(x) = 2x \div x^2 = (2x)/x^2 = 2/x$

4.2 Types of Functions, Including Composite, Inverse, Linear, Nonlinear, Even, Odd, Exponential, Logarithmic, Identity, Absolute Value, Squaring, Cubing, Square Root, Cube Root, Reciprocal, and Functions with More Than One Variable

• *Composite*, or *compound, functions* are functions that are combined, and the operations specified by the functions are combined. Compound functions are written $f(g(x))$ or $g(f(x))$ where there is a function of a function. For example, if $f(x) = x + 1$ and $g(x) = 2x - 2$, then the compound functions for $f(g(x))$ and $g(f(x))$ are:

$f(g(x)) = f(2x - 2) = (2x - 2) + 1 = 2x - 1$
$g(f(x)) = g(x + 1) = 2(x + 1) - 2 = 2x + 2 - 2 = 2x$

• *Inverse functions* are functions that result in the same value of x after the operations of the two functions are performed. In inverse functions, the operations of each function are the reverse of the other function. Notation for inverse functions is $f^{-1}(x)$. If $f(x) = y$, then $f^{-1}(y) = x$. If function f is the inverse of function g then function g is the inverse of function f. A function has an inverse if its graph intersects any *horizontal line* no more than once.

• An inverse of a function has its domain and range equal to the range and domain, respectively, of the original function. If function $f(x) = y$, then $f^{-1}(y) = x$. For a function $f(x, y)$ that has only one y value for each x value, then there exists an inverse function represented by $f^{-1}(y, x)$. For example, reversing the ordered pairs in function $f(x, y) = \{(0, 3), (2, 4), (3, 5)\}$ results in the inverse function $f^{-1}(y, x) = \{(3, 0), (4, 2), (5, 3)\}$. Therefore, the domain of f equals the range of f^{-1}, and the range of f equals the domain of f^{-1}.

• If function f is represented by $f(x) = u$, then its inverse f^{-1} can be found by solving $f(x) = u$ for x in terms of u: $f^{-1}u = f^{-1}(f(x)) = x$. Therefore, if $f(x) = u$ then $f^{-1}(u) = x$, or if $f^{-1}(u) = x$ then $f(x) = u$. For more complicated or *composite functions*, if $y = f[u(x)]$, then the inverse can be written in the opposite order: $x = u^{-1}(f^{-1}(y))$.

• When two functions are inverse functions, then they will return to the first value. For example, if $y = f(x) = 2x - 1$ and $x = f^{-1}(y) = (y + 1)/2$ are inverses, and if $x = 3$, then substituting $x = 3$ into $y = f(x) = 2x - 1$ results in:

$$f(3) = 2(3) - 1 = 5$$

If we then substitute 5 into inverse function $x = f^{-1}(y) = (y + 1)/2$:

$$f^{-1}(5) = (5 + 1)/2 = 3$$

we return in the starting value of x.

• Not all functions have inverses. If a function has more than one solution, it does not have an inverse. If $u(x) = z$, only one x can result from $x = u^{-1}(z)$. If there is more than one solution for $u^{-1}(z)$, it will not be the inverse of $u(x) = z$.

• *Graphs of inverse functions* are mirror images. For example, if $z = u(x) = 2x$, then $x = (1/2)z$. Note that the slopes are the derivatives $(dz/dx) = 2$ and $(dx/dz) = 1/2$.

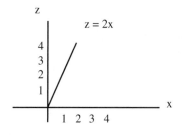
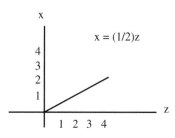

- Following are examples of functions and their inverses:

$z = x^2$ is the inverse of: $x = \sqrt{z}$ or $x = z^{1/2}$

$z = e^x$ is the inverse of: $x = \ln z$

$z = a^x$ is the inverse of: $x = \log_a z$

- **Inverses of trigonometric functions**, described in Chapter 6, Inverse Trigonometric Functions, exist in defined intervals. For example, the **inverse of sine** is $\sin^{-1} y = x$ for $1 \geq y \geq -1$, which pertains to $\sin x = y$ for $\pi/2 \geq x \geq -\pi/2$. The inverse brings y back to x.

The graph of $y = \sin x$ is a mirror image of $\sin^{-1} y = x$.

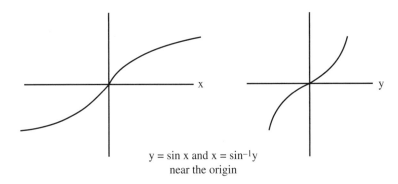

y = sin x and x = sin⁻¹y
near the origin

Only certain intervals of the sine function have inverses:

In the interval $\pi/2 \geq x \geq -\pi/2$, $\sin^{-1}(\sin x) = x$.

In the interval $1 \geq y \geq -1$, $\sin(\sin^{-1} y) = y$.

There are many points on the sine function where $\sin x = 0$.

All of the trigonometric functions have inverses in defined intervals. (See Chapter 6 for a complete discussion.)

- Functions can be **linear** or **nonlinear**. Remember: **Linear equations** are equations in which the variables do not have any exponents other than 1. These equations, if plotted, will produce a straight line. A general form of a linear equation is $Ax + By = C$, where A, B, and C are constants, and x and y are variables. Another general form of a linear equation is $y = mx + b$, where m is the slope of the line and b is where the line intercepts the Y-axis on a coordinate system. The equation for the slope of a line passing though point (x_1, y_1) can be written

$$y - y_1 = m(x - x_1)$$

• A *linear function* can have the form $y = f(x) = b + mx$, where m is the slope of the line and represents the rate of change of y with respect to x, and b is the vertical intercept where the line intercepts the Y-axis on a coordinate system that is the value of y when x equals zero. The *slope* m of a linear function can be calculated at two points $(x_1, f(x_1))$ and $(x_2, f(x_2))$ using the equation:

$$f(x_2) - f(x_1) = m(x_2 - x_1) \text{ or } m = \frac{f(x_2) - f(x_1)}{x_2 - x_1}$$

The quantity $(f(x_2) - f(x_1))/(x_2 - x_1)$ is the quotient of the two differences and is referred to as a *difference quotient*.

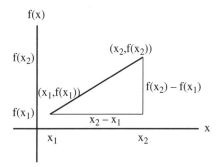

• *Nonlinear functions* have variables with exponents other than 1. Remember that *nonlinear equations* are equations containing variables that have exponents other than 1. Graphs of *nonlinear functions* form curved lines and surfaces.

• A function can be an *even function* or an *odd function*. A function is even if $f(x) = f(-x)$ for all x, and a function is odd if $f(x) = -f(-x)$ for all x.

Examples of *even functions* include:

$$f(x) = c, f(x) = x^2, f(x) = x^4, f(x) = x^{2n}$$
$$f((-x)^2) = (-x)(-x) = x^2$$

Cosine is an even function such that: $\cos(-x) = \cos x$

Examples of odd functions include:

$$f(x) = x, f(x) = x^3, f(x) = x^5, f(x) = x^{2n+1}$$
$$f((-x)^3) = (-x)(-x)(-x) = (-x)^3$$

Sine is an odd function such that: $\sin(-x) = -\sin x$

• By observing the graph of a function, it is clear whether the function is even or odd. If the area between the curve and the X-axis on the section of the function to the left of the Y-axis is equivalent to the area between the curve and the X-axis on the section to the right of the Y-axis, the function is *even*. Therefore, in an *even function*, the area for negative values along the X-axis is equal to the area for positive values along X-axis. Alternatively, if a function is *odd*, the area between the curve and the X-axis on the section of the function to the left of the Y-axis is equivalent but opposite to the area between the curve and the X-axis on the section to the right of the Y-axis. Therefore, in an *odd function*, the area for negative values along the X-axis is equal but opposite to the area for positive values along the X-axis, and the two areas subtract and cancel each other out.

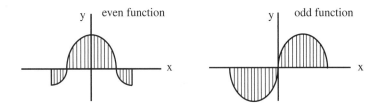

• **Exponential functions** form curved lines and contain variables in their exponents. Examples of exponential functions include e^x, a^x, and 2^x, where a is a constant. Some properties of e^x or a^x include:

$$e^x e^y = e^{x+y}, \quad e^x/e^y = e^{x-y}, \quad (e^x)^y = e^{xy}, \quad e^0 = 1,$$

where $e \approx 2.71828\ 18284\ 59045\ 23536\ 02874\ 71353$.

• The inverse of e^x is $\ln x$, or the **natural logarithm** of x. Some properties of $\ln x$ include:

$$\ln(xy) = \ln x + \ln y, \quad \ln(x/y) = \ln x - \ln y, \quad \ln x^y = y \ln x, \quad \ln(e^x) = x,$$

$$e^{\ln x} = x, \quad e^{-\ln x} = e^{\ln(1/x)} = 1/x, \quad \ln x = \log_e x = (2.3026)\log x$$

• *Logarithms* can have any base. Base 10 logarithms are the most common and are written $\log_{10} x$ or just log x. The inverse of log x is 10^x. Some properties of log x include:

$$10^{\log x} = x, \quad 10^{-\log x} = 1/x, \quad \log(xy) = \log x + \log y,$$
$$\log(x/y) = \log x - \log y, \quad \log x^y = y \log x, \quad \log(10^x) = x$$

• It is important to remember that when a number has an exponent, the logarithm is the exponent. For example:

$$\log(10^x) = x, \quad \ln(e^x) = x, \quad \log(10^3) = 3,$$
$$\log(10^{-2}) = \log(1/10^2) = -2,$$
$$\log(b^x) = x, \text{ where b represents any base.}$$

• The *exponential function* e^x and the *natural logarithm* ln x are depicted with e^x as the thicker curve and ln x with the thin curve.

Graph of e^x (thick curve) and ln x (thin curve)

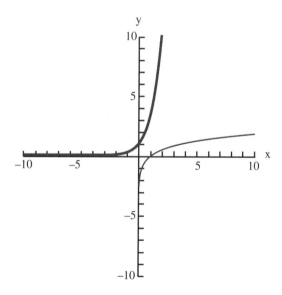

See *Master Math: Basic Math and Pre-Algebra* Chapters 7 and 8 for additional information on exponents and logarithms.

• The *identity function* is an increasing linear function given by $f(x) = x$. In the equation for a line, $f(x) = mx + b$, if $m = 1$ and $b = 0$, the result is the identity function. The identity function pairs each real number with itself. The *absolute value function*, $f(x) = |x|$, pairs each real number with its absolute value. The absolute value function decreases from − infinity to 0 and increases from 0 to + infinity and exists above the X-axis. In the following graph, the *identity function* $f(x) = x$ is depicted as the thick black line existing above and below the X-axis, and the *absolute value function* $f(x) = |x|$ is the v-shaped curve with the upper part on the positive side of the X-axis overlapping $f(x) = x$.

Graph of $f(x) = x$ (thick line) and $f(x) = |x|$ (v-line)

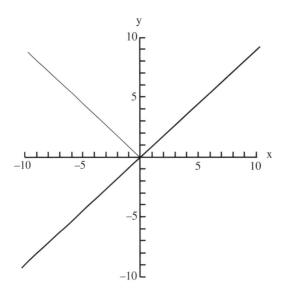

• The *squaring function*, $f(x) = x^2$, pairs each real number with its square, and its graph forms a parabola. The graph is symmetric with respect to the Y-axis, so that $f(x) = f(-x)$. The *cubing function*, $f(x) = x^3$, pairs each real number with its cube, or third power. The graph has an inflection point and is symmetric with respect to the origin, so that $f(-x) = -f(x)$. The *squaring function* $f(x) = x^2$ is depicted as the thicker curve and the *cubing function* $f(x) = x^3$ is the thin curve.

Graph of f(x) = x² (thick curve) and f(x) = x³ (thin curve)

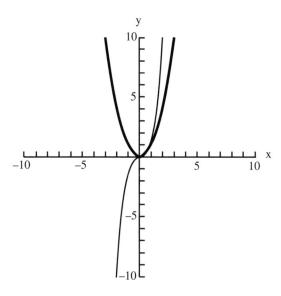

• The *square root function* is $f(x) = \sqrt{x} = x^{1/2}, x \geq 0$. The *cube root function* is $f(x) = \sqrt[3]{x} = x^{1/3}$, x is a positive, zero, or negative real number.

Graph of f(x) = ±x^{1/2} (thick curve) and f(x) = x^{1/3} (thin curve)

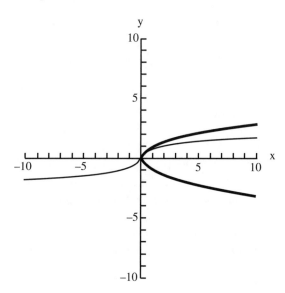

• For a function, f(x) = x, the **reciprocal function** is f(x) = 1/x. The graph of a reciprocal function forms a vertical asymptote at every point where the graph of the original function, f(x) = x, intersects the X-axis.

Graph of f(x) = x (thin line) and f(x) = 1/x (thick curve)

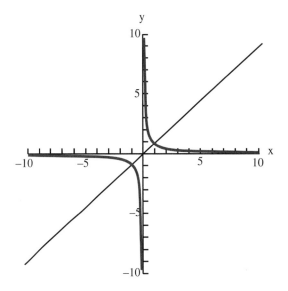

Functions with more than one variable
• Many functions depend on more than one variable. A *function that depends on two variables* can be written as z = f(x,y), where z is called the *dependent variable*, x and y are called the *independent variables*, and f represents the function. For example, the volume V of a pyramid depends on the height h and the area of its base A_b, which are independent variables. The function describing this is: V = (1/3)hA_b.

• Another example of a function that depends on more than one variable is the ideal gas law, PV = nRT, where P is pressure, V is volume, n is the number of moles in the sample, R is the universal gas constant (8.314 J/mol·k), and T is temperature. Pressure P = nRT/V can be studied by changing one variable at a time while holding the others constant. The data for functions that depend on more than one variable can be represented in tables or graphs in two or three dimensions. Temperature values for a system modeled by the ideal gas law can be listed on one axis and values for volume listed on another axis, such that resulting pressure values that correspond to a given temperature and volume will be within the table or on the third axis of the coordinate system.

• The *graph* of function y = f(x) represents all of its points with coordinates (x, y) and is generally comprised of curves or lines. The graph of a function z = f(x, y) that depends on *two variables* represents the points with coordinates (x, y, z) and generally represents a *surface* in three-dimensional space. In general, *graphs of one-variable functions* form curves or straight lines, whereas *graphs of two-variable functions* form planes or surfaces represented in three-dimensional space (which comprises a family of *level curves* in the form of f(x, y) = constant). *Graphs of three-variable functions* form solids in four-dimensional space (which comprises a *family of level surfaces* in the form of f(x, y, z) = constant).

• For example, the graph of a *linear two-variable function* forms a *plane* in which the slopes of the lines parallel to the X-axis are the same, and the slopes of lines on the plane parallel to the Y-axis are the same.

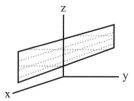

The graph of $f(x, y) = z = \sqrt{x^2 + y^2}$ forms a curved surface in three-dimensional space:

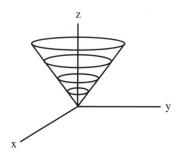

4.3 Coordinate Systems, Radians, Degrees, and Arc Length

• Trigonometric functions are used in ***coordinate systems*** to describe the position, location, and distances of points with reference to the axes of the coordinate system. A point P in a coordinate system can be identified and located by its distance from the axes, which is called its coordinates.

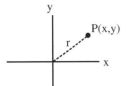

The distance from P to the origin is a radius vector $r = \sqrt{x^2 + y^2}$.

• The axes of a planar Cartesian coordinate system divide the plane into *quadrants* I, II, III, and IV, which are typically labeled counterclockwise beginning with the top right quadrant I.

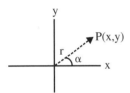

• A point in a coordinate system can also be described in terms of the angle α that its *radius vector* makes from the X-axis. Angle α can be measured in degrees or radians.

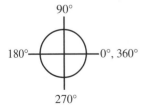

• As discussed in Chapter 2, an *angle* is formed by a rotation of a ray about its endpoint. Angles are measured in **degrees** or **radians**. Degrees can be divided into *minutes* (denoted by ') and *seconds* (denoted by "). The **degrees of a circle** are:

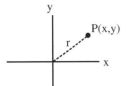

A circle always measures 360° around, equivalent to 2π radians. Half of the circle measures 180°, which is equivalent to π radians. A quarter of a circle measures 90°, which is equivalent to $\pi/2$ radians. Following are *important definitions* to remember when working with coordinate systems, angles, and circles:

1° = 1/360th of a circle

1° = 2π radians/360° = π radians/180° = 0.017453292519943 radian

1 minute = 1' = 1/60th of 1°, or 0.0167°

1 second = 1" = 1/60th of 1' = 1/3600th of 1°, or 0.0002778°

1° = 60' = 3600"

1 radian = 360°/2π = 180°/π = 57.2957795131°

2π radians = 360 degrees

The relationship, 180 degrees = π radians, can be used to convert between degrees and radians.

Other degree-radian equivalents include:

$\pi/6 = 30°, \pi/4 = 45°, \pi/3 = 60°$, and $\pi/2 = 90°$

• A *radian* is the measure of the central angle subtending an arc of a circle that is equal to the radius r of the circle. In other words, a central angle subtending (opposite to) an *arc* equal in length to the radius of a circle is defined as a *radian*.

In a *circle having a radius of one*, or a *unit circle*, a radian is equal to the angle at the center that cuts across an arc of length 1.

• *Arc length*: By the definition of radian measure, we see that the *length of an arc*, s, of a circle is given by $s = r\theta$, where θ is measured in radians and r is the radius of the circle. Therefore, the angle θ is the arc length divided by the radius, or $\theta = (s/r)$ radians.

if θ is 180°, then it is equal to π radians because it is one-half of the circumference, which is 2π radians. Therefore, for θ equal to 180°, θ = s/r = πr/r = π radians. In general, the central angle θ subtended by an arc of length *s* is determined by the number of radius r lengths that are contained in the arc length *s*.

• A section of a circle's perimeter defined by two or more points is called an *arc*. The central angle can be measured in degrees or radians. (There are 2π radians in 360° and π radians in 180°.) *Arc length* is defined as follows:

$$\text{Arc length} = (\text{radius})(\text{central angle measure in radians})$$
$$= r\theta \text{ with } \theta \text{ measured in radians}$$
$$= (\pi/180°)r\theta°, \text{ with } \theta \text{ the central angle measured in degrees}$$

• Example using arc length: Estimate the *diameter of the Sun* given that the distance between the Earth and the Sun is approximately 150,000,000 kilometers and that the angle subtended by the Sun on the Earth is measured as 0.0093 radians.

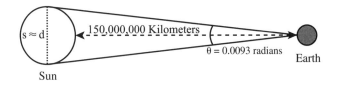

We can estimate that the arc length *s* subtended by the Sun on the Earth is approximately equal to the diameter of the Sun. Therefore, the Sun's *diameter* can be approximated to be arc length *s*:

$$s = r\theta = (150,000,000 \text{ km})(0.0093 \text{ radians}) \approx 1,400,000 \text{ kilometers}$$

4.4 Angles in Standard Position and Coterminal Angles

• When an angle is drawn in a rectangular coordinate system with its vertex at the origin (x = 0, y = 0) and its initial side coincident with the positive X-axis, it is called a ***standard position angle***. In other words, an angle is said to be in *standard position* if its vertex is at (0, 0) of the X-Y coordinate system and if its *initial* side lies on the positive part of the X-axis. If the standard position angle is measured in a counterclockwise direction, it is *positive*. If the standard position angle is measured in a clockwise direction, it is *negative*.

• The *initial side* of an angle is where its measurement begins and the *terminal side* of an angle is where its measurement ends. For a standard position angle, the initial side is always coincident with the positive X-axis. The following figures are examples of standard position angles with their terminal sides in different quadrants:

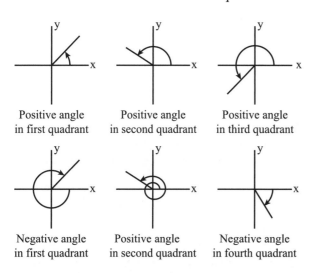

| Positive angle in first quadrant | Positive angle in second quadrant | Positive angle in third quadrant |

| Negative angle in first quadrant | Positive angle in second quadrant | Negative angle in fourth quadrant |

• *Coterminal angles*: Two angles are *coterminal* if their terminal sides coincide when both angles are placed in their standard positions in the same coordinate system. These angles have the same initial side because they are in standard position. There are an unlimited number of coterminal angles with any given angle, because coterminal angles represent integer multiples of 360°, or 2π. Following are examples of coterminal angles:

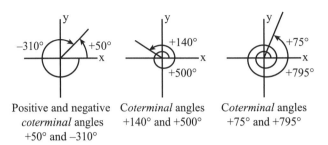

Positive and negative   *Coterminal* angles   *Coterminal* angles
   coterminal angles   +140° and +500°   +75° and +795°
   +50° and −310°

For any angle α, coterminal angles exist in radians with angles $(2\pi + \alpha)$, $(4\pi + \alpha)$, $(6\pi + \alpha)$, and so on, or in degrees, $((1)360° + \alpha)$, $((2)360° + \alpha)$, and so on. Formulas representing coterminal angles are $(2n\pi + \alpha)$ and $(n360° + \alpha)$. The value of the trigonometric functions sine, cosine, tangent, cotangent, cosecant, and secant are the same for all angles that are coterminal:

$$\sin(n360° + \alpha) = \sin\alpha, \quad \csc(n360° + \alpha) = \csc\alpha$$
$$\cos(n360° + \alpha) = \cos\alpha, \quad \sec(n360° + \alpha) = \sec\alpha$$
$$\tan(n360° + \alpha) = \tan\alpha, \quad \cot(n360° + \alpha) = \cot\alpha$$
$$\sin(2n\pi + \alpha) = \sin\alpha, \quad \csc(2n\pi + \alpha) = \csc\alpha$$
$$\cos(2n\pi + \alpha) = \cos\alpha, \quad \sec(2n\pi + \alpha) = \sec\alpha$$
$$\tan(2n\pi + \alpha) = \tan\alpha, \quad \cot(2n\pi + \alpha) = \cot\alpha$$

4.5 The Trigonometric Functions Defined in a Coordinate System in Standard Position, Quadrant Signs, and Quadrantal Angles

• In Chapter 3, the trigonometric functions were defined with respect to right triangles. Trigonometric functions can also be defined in a coordinate system for an angle of any size in standard position. The six trigonometric functions can be defined using a **standard position angle** having its vertex at the origin and its initial side coincident with the positive X-axis. Following are positive standard position angles depicted in the four quadrants:

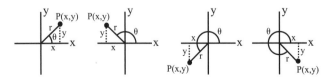

The six trigonometric functions can be defined as ratios of the quantities x, y, and r, with point P(x, y) as the point on the terminal side of the standard position angle and r as the distance from the point to the origin. Because r is the square root of the sum of the squares of x and y, $r = \sqrt{x^2 + y^2}$, if the coordinates of point (x, y) are known, r and the six trigonometric function values can be determined.

Therefore, the *trigonometric functions defined in standard position* in terms of the angle Ø, x, y, and radius r are:

$$\sin Ø = y/r, \quad \csc Ø = r/y$$
$$\cos Ø = x/r, \quad \sec Ø = r/x$$
$$\tan Ø = y/x, \quad \cot Ø = x/y$$

• **Example:** If the terminal side of a standard position angle is through point P(–2, –2), calculate the six trigonometric functions.

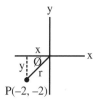

We are given x = –2, y = –2, so we need only r to make the calculations.

$$r = \sqrt{x^2 + y^2} = \sqrt{(-2)^2 + (-2)^2} = \sqrt{8} = \sqrt{(2)(2)(2)} = 2\sqrt{2}$$
$$\sin Ø = y/r = -2 / 2\sqrt{2} = -1/\sqrt{2}, \quad \csc Ø = r/y = 2\sqrt{2} / -2 = -\sqrt{2}$$
$$\cos Ø = x/r = -2 / 2\sqrt{2} = -1/\sqrt{2}, \quad \sec Ø = r/x = 2\sqrt{2} / -2 = -\sqrt{2}$$
$$\tan Ø = y/x = -2 / -2 = 1, \quad \cot Ø = x/y = -2 / -2 = 1$$

• *Quadrant signs*: In a coordinate system the signs of the trigonometric functions depend on the *signs of x and y in the quadrant* in which the terminal side of the standard position angle lies. Because the signs of the six trigonometric functions correspond with the quadrants of the terminal side of the standard position angle, the sign can be determined if it is unknown. The signs of x, y, and r for each quadrant are depicted:

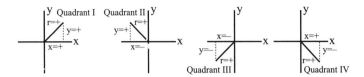

Note that r is always positive.

• The *signs of the six trigonometric* functions are determined by the quadrant in which the terminal side of the angle lies. These values are depicted in the following figure:

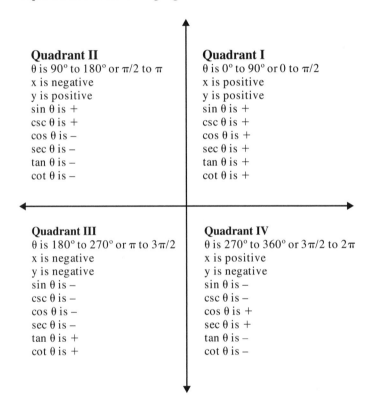

Quadrant II
θ is 90° to 180° or π/2 to π
x is negative
y is positive
sin θ is +
csc θ is +
cos θ is −
sec θ is −
tan θ is −
cot θ is −

Quadrant I
θ is 0° to 90° or 0 to π/2
x is positive
y is positive
sin θ is +
csc θ is +
cos θ is +
sec θ is +
tan θ is +
cot θ is +

Quadrant III
θ is 180° to 270° or π to 3π/2
x is negative
y is negative
sin θ is −
csc θ is −
cos θ is −
sec θ is −
tan θ is +
cot θ is +

Quadrant IV
θ is 270° to 360° or 3π/2 to 2π
x is positive
y is negative
sin θ is −
csc θ is −
cos θ is +
sec θ is +
tan θ is −
cot θ is −

If you remember the signs of x and y (r is always positive), the signs of the trigonometric functions in each quadrant can be obtained using the definitions sin Ø = y/r, csc Ø = r/y, cos Ø = x/r, sec Ø = r/x, tan Ø = y/x, and cot Ø = x/y.

In addition, a mnemonic can be used to remember the signs of the trigonometric functions for each quadrant:

All	*All* functions positive in *Quadrant I*
*S*tudents	*Sine* (and its reciprocal csc) positive in *Quadrant II*
*T*ake	*Tangent* (and its reciprocal cot) positive in *Quadrant III*
*C*alculus	*Cosine* (and its reciprocal sec) positive in *Quadrant IV*

• **Quadrantal angles**: A *quadrantal angle* is an angle in which the *terminal side coincides with one of the axes of the coordinate system*. These angles have special trigonometric function values, some of which may be undefined.

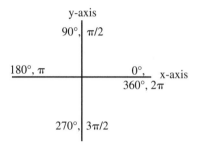

• If the terminal side of a standard position angle coincides with the X-axis and it is therefore a quadrantal angle, then it has a y-coordinate of zero for all points on the terminal side. Similarly, if the terminal side of a standard position angle coincides with the Y-axis and it is therefore a quadrantal angle, then it has an x-coordinate of zero for all points on the terminal side. Because x and y are in the denominators of some of the trigonometric functions, and fractions with zero denominators are "undefined," then some of the trigonometric functions of quadrantal angles are undefined. (Remember that as a denominator approaches zero, the value of a fraction approaches infinity until the denominator equals zero and the fraction becomes undefined.)

Table of the six trigonometric functions of quadrantal angles

θ	angle terminates	$\sin\theta$ y/r	$\csc\theta$ r/y	$\cos\theta$ x/r	$\sec\theta$ r/x	$\tan\theta$ y/x	$\cot\theta$ x/y
$0°, 0$	x-axis	0	und	1	1	0	und
$90°, \pi/2$	y-axis	1	1	0	und	und	0
$180°, \pi$	x-axis	0	und	−1	−1	0	und
$270°, 3\pi/2$	y-axis	−1	−1	0	und	und	0
$360°, 2\pi$	x-axis	0	und	1	1	0	und

• **Example:** Find values of the quadrantal angles (a.) sin π/2 and
(b.) tan (–π), with r = 1.

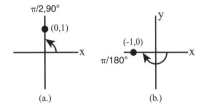

(a.) For a quadrantal angle with its terminal side at point (0, 1) and r = 1,
sin π/2 = y/r = 1/1 = 1

(b.) For a quadrantal angle with its terminal side at point (–1, 0) and r = 1,
tan(–π) = y/x = 0/–1 = 0

4.6 Reference Angles and Reference Triangles

• Reference angles and reference triangles are drawn to assist in calculations involving trigonometric functions. A *reference angle R* for a standard
position angle *is the positive acute angle between the X-axis (never the
Y-axis) and the terminal side of the standard position angle.* The values
of the six trigonometric functions of the reference angle *R* are the same as
the trigonometric functions of the standard position angle, except in some
cases where the sign is different. For the following standard position angles
θ in the four quadrants, the reference angles *R* are depicted:

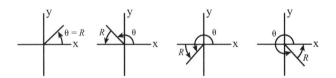

• **Example:** What is the reference angle for standard position angle 225°?

θ = 225°

$R = 225° – 180° = 45°$

• **Example:** What is the reference angle for standard position angle 315°?

$$R = 360° - 315° = 45°$$

‘ **Example:** What is the reference angle for standard position angle –300°?

$$R = 360° - 300° = 60°$$

• If a standard position angle is greater than 360°, then reference angle R is associated with the coterminal angle of the standard position that is between 0° and 360°. If the standard position angle is a quadrantal angle a reference angle is not useful.

• It is possible to find the value of a trigonometric function using *reference angles* for nonquadrantal standard position angles.

This can be done as follows:

1. If the angle θ is greater than $\pm 360°$ (or 2π), find the coterminal angle by subtracting 360° (or 2π) until θ is between 0 and 360°.
2. Make a drawing and find the reference angle R by measuring the angle between the X-axis (horizontal axis) and the terminal side of θ.
3. Determine the value of the trigonometric function in question for the reference angle R using x, y, and r.
4. Determine the correct sign for the trigonometric functions for θ using the sign of the trigonometric function dictated by the quadrant in which the terminal side of θ lies. (See the quadrant chart that follows.)

Quadrant II	Quadrant I
x is −, y is +	x is +, y is +
sin θ is +	sin θ is +
csc θ is +	csc θ is +
cos θ is −	cos θ is +
sec θ is −	sec θ is +
tan θ is −	tan θ is +
cot θ is −	cot θ is +
Quadrant III	**Quadrant IV**
x is −, y is −	x is +, y is −
sin θ is −	sin θ is −
csc θ is −	csc θ is −
cos θ is −	cos θ is +
sec θ is −	sec θ is +
tan θ is +	tan θ is −
cot θ is +	cot θ is −

• Reference triangles are formed by extending a line from a point P(x, y) on the terminal side of the standard position angle to the X-axis. A *reference triangle* may be helpful for determining trigonometric functions of a standard position angle particularly for special triangles, such as 30:60:90 and 45:45:90 triangles.

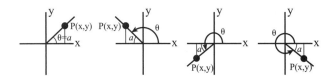

• *Special angles and triangles used as reference angles and triangles*: The 30:60:90 and 45:45:90 triangles are examples of special triangles that can be used as reference triangles. These special triangles can be defined in a coordinate system with a standard position angle.

Recall that the 30:60:90 triangles have side ratios:

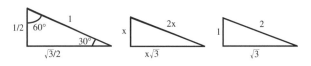

where $30° = \pi/6, 60° = \pi/3, 90° = \pi/2$

The 45:45:90 triangles have side ratios:

where 45° = π/4 and 90° = π/2

• The *trigonometric functions* can be written or calculated for *triangles of a standard position angle* and can be expressed in degrees or radians. For a 30:60:90 triangle in standard position, the trigonometric functions are:

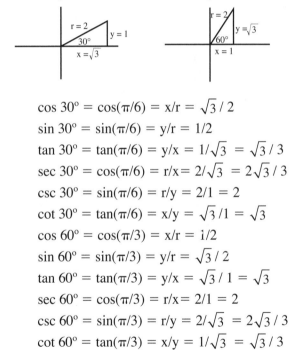

$\cos 30° = \cos(\pi/6) = x/r = \sqrt{3}/2$

$\sin 30° = \sin(\pi/6) = y/r = 1/2$

$\tan 30° = \tan(\pi/6) = y/x = 1/\sqrt{3} = \sqrt{3}/3$

$\sec 30° = \cos(\pi/6) = r/x = 2/\sqrt{3} = 2\sqrt{3}/3$

$\csc 30° = \sin(\pi/6) = r/y = 2/1 = 2$

$\cot 30° = \tan(\pi/6) = x/y = \sqrt{3}/1 = \sqrt{3}$

$\cos 60° = \cos(\pi/3) = x/r = 1/2$

$\sin 60° = \sin(\pi/3) = y/r = \sqrt{3}/2$

$\tan 60° = \tan(\pi/3) = y/x = \sqrt{3}/1 = \sqrt{3}$

$\sec 60° = \cos(\pi/3) = r/x = 2/1 = 2$

$\csc 60° = \sin(\pi/3) = r/y = 2/\sqrt{3} = 2\sqrt{3}/3$

$\cot 60° = \tan(\pi/3) = x/y = 1/\sqrt{3} = \sqrt{3}/3$

For a 45:45:90 triangle in standard position, the trigonometric functions are:

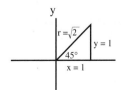

$$\cos 45° = \cos(\pi/4) = x/r = 1/\sqrt{2} = \sqrt{2}/2$$
$$\sin 45° = \sin(\pi/4) = y/r = 1/\sqrt{2} = \sqrt{2}/2$$
$$\tan 45° = \tan(\pi/4) = y/x = 1/1 = 1$$
$$\sec 45° = \cos(\pi/4) = r/x = \sqrt{2}/1 = \sqrt{2}$$
$$\csc 45° = \sin(\pi/4) = r/y = \sqrt{2}/1 = \sqrt{2}$$
$$\cot 45° = \tan(\pi/4) = x/y = 1/1 = 1$$

• The following are examples of *special reference triangles*. The correct sign for the trigonometric functions can be determined according to the quadrant in which the terminal side of angle θ is located.

Quadrant I (both sine and cosine are positive):

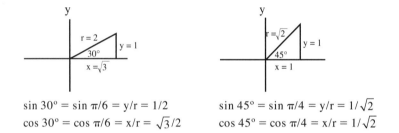

$\sin 30° = \sin \pi/6 = y/r = 1/2$ $\sin 45° = \sin \pi/4 = y/r = 1/\sqrt{2}$
$\cos 30° = \cos \pi/6 = x/r = \sqrt{3}/2$ $\cos 45° = \cos \pi/4 = x/r = 1/\sqrt{2}$

Quadrant II (sine is positive and cosine is negative):

$\sin 2\pi/3 = y/r = \sqrt{3}/2$
$\cos 2\pi/3 = x/r = -1/2$

• The following chart summarizes the **trigonometric functions for special angles and quadrantal angles** (which are standard position angles in which the terminal side coincides with an axis). Because x and y are in the denominators of some of the trigonometric functions and fractions with zero denominators are "undefined," then some of the trigonometric functions are undefined.

θ	quadrantal angle terminates	sin θ y/r	csc θ r/y	cos θ x/r	sec θ r/x	tan θ y/x	cot θ x/y
0°, 0	x-axis	0	und	1	1	0	und
30°, π/6		1/2	2	$\sqrt{3}/2$	$2/\sqrt{3}$	$1/\sqrt{3}$	$\sqrt{3}$
45°, π/4		$1/\sqrt{2}$	$\sqrt{2}$	$1/\sqrt{2}$	$\sqrt{2}$	1	1
60°, π/3		$\sqrt{3}/2$	$2/\sqrt{3}$	1/2	2	$\sqrt{3}$	$1/\sqrt{3}$
90°, π/2	y-axis	1	1	0	und	und	0
180°, π	x-axis	0	und	−1	−1	0	und
270°, 3π/2	y-axis	−1	−1	0	und	und	0
360°, 2π	x-axis	0	und	1	1	0	und

- **Example:** Find sin(−240°) and cos(−240°). First draw the reference angle and triangle:

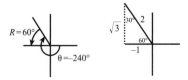

The reference angle is: $R = 240° − 180° = 60°$

For the reference triangle, sine and cosine are:

$$\sin 60° = y/r = \sqrt{3}/2$$
$$\cos 60° = x/r = −1/2$$

Determine the correct sign for these trigonometric functions according to the quadrant in which the terminal side of θ is located. Because the terminal side of θ lies in Quadrant II, sine is positive and cosine is negative. Therefore, $\sin(−240°) = \sqrt{3}/2$ and $\cos(−240°) = −1/2$.

Note that we can verify these results using a calculator estimate. Using a calculator:

$$\sin 60° \approx 0.8660254, \sin(−240°) \approx 0.8660254, \text{ and } \sqrt{3}/2 \approx 0.8660254$$
$$\cos 60° = 0.5000, \cos(−240°) = −0.5000, \text{ and } −1/2 = −0.5000$$

Note that when we used the reference angle 60° for cosine, it was important to adjust the sign to match the quadrant.

• **Example:** Find sin(2π/3).

The reference triangle is a 30:60:90 triangle in Quadrant II.

From the reference triangle sine is:

$$\sin(2\pi/3) = y/r = \sqrt{3}/2$$

Now determine the correct sign for these trigonometric functions by the quadrant in which the terminal side of θ lies. Because the terminal side of θ lies in Quadrant II, sine is positive. Therefore,

$$\sin(2\pi/3) = \sqrt{3}/2$$

4.7 Negative Angles

• When solving problems using trigonometric functions, negative angles may be involved. The trigonometric functions can be written for the *negative of an angle* and are defined as follows:

$$\cos(-\theta) = \cos\theta, \quad \sec(-\theta) = \sec\theta$$
$$\sin(-\theta) = -\sin\theta, \quad \csc(-\theta) = -\csc\theta$$
$$\tan(-\theta) = -\tan\theta, \quad \cot(-\theta) = -\cot\theta$$

These are called *negative number identities*, or *negative angle identities*. These identities can be obtained using the difference identities discussed in Chapter 7.

$$\cos(A - B) = \cos A \cos B + \sin A \sin B$$
$$\sin(A - B) = \sin A \cos B - \cos A \sin B$$
$$\tan(A - B) = [\tan A - \tan B] / [1 + \tan A \tan B]$$

When angle A = 0 and angle B = θ:

$$\cos(0 - \theta) = \cos 0 \cos\theta + \sin 0 \sin\theta = \cos\theta$$
$$\sin(0 - \theta) = \sin 0 \cos\theta - \cos 0 \sin\theta = -\sin\theta$$
$$\tan(0 - \theta) = [\tan 0 - \tan\theta] / [1 + \tan 0 \tan\theta] = -\tan\theta$$

Secant, cosecant, and cotangent negative angle identities can be obtained using the *reciprocal identities* described in the next section:

sec θ = 1 / cos θ

csc θ = 1 / sin θ

cot θ = 1 / tan θ

4.8 Reciprocal Functions and Cofunction Relationships

• The *reciprocal functions* are often used when solving problems involving trigonometric functions and side ratios of triangles. The six trigonometric functions provide for three pairs of reciprocal functions. These reciprocal relationships are called the *reciprocal identities*.

The reciprocal identities are found by considering that because

sin $\emptyset$ = y/r and csc $\emptyset$ = r/y, then

sin $\emptyset$ = 1 / csc $\emptyset$ and csc $\emptyset$ = 1 / sin $\emptyset$

Similarly, for all of the six trigonometric functions, the *reciprocal identities* are:

sin $\emptyset$ = 1 / csc $\emptyset$,   csc $\emptyset$ = 1 / sin $\emptyset$

cos $\emptyset$ = 1 / sec $\emptyset$,   sec $\emptyset$ = 1 / cos $\emptyset$

tan $\emptyset$ = 1 / cot $\emptyset$,   cot $\emptyset$ = 1 / tan $\emptyset$

By rearranging, the *reciprocal identities* can also be written:

sin $\emptyset$ csc $\emptyset$ = 1, cos $\emptyset$ sec $\emptyset$ = 1, and tan $\emptyset$ cot $\emptyset$ = 1

• **Example:** Given cot $\emptyset$ = 1/2, find tangent and $\emptyset$, and compare with *cofunction identities*.

Because tan $\emptyset$ = 1 / cot $\emptyset$, then tan $\emptyset$ = 2. Angle $\emptyset$ = arctan 2 ≈ 63.43.

Because tan α = cot(90° – α), then cot(26.57) = tan 63.43 ≈ 2.

• *Cofunction relationships* (or cofunction identities) are often useful when solving problems involving trigonometric functions. The sine and cosine functions are *complementary functions*, meaning that the sine of some angle α is equal to the cosine of the complement of α and vice versa. The complement of angle α is the angle (90° – α) in degrees or (π/2 – α) in radians. The *cofunction identities* are:

$\cos \alpha = \sin(90° - \alpha),\quad \sec \alpha = \csc(90° - \alpha)$

$\sin \alpha = \cos(90° - \alpha),\quad \csc \alpha = \sec(90° - \alpha)$

$\tan \alpha = \cot(90° - \alpha),\quad \cot \alpha = \tan(90° - \alpha)$

$\cos \alpha = \sin(\pi/2 - \alpha),\quad \sec \alpha = \csc(\pi/2 - \alpha)$

$\sin \alpha = \cos(\pi/2 - \alpha),\quad \csc \alpha = \sec(\pi/2 - \alpha)$

$\tan \alpha = \cot(\pi/2 - \alpha),\quad \cot \alpha = \tan(\pi/2 - \alpha)$

- The cofunction identities for 30:60:90 and 45:45:90 triangles are:

$\cos 30° = \sin 60°,\quad \sec 30° = \csc 60°$

$\sin 30° = \cos 60°,\quad \csc 30° = \sec 60°$

$\tan 30° = \cot 60°,\quad \cot 30° = \tan 60°$

$\cos 45° = \sin 45°,\quad \sec 45° = \csc 45°,\quad \tan 45° = \cot 45°$

4.9 Circular Functions and the Unit Circle

- *Trigonometric functions* can be defined using ratios of sides of a right triangle (as discussed in Chapter 3) and can also be defined in a coordinate system using angles in standard position as described in this chapter. In addition, trigonometric functions can be defined as **arc lengths** on a circle *of radius one*, or a **unit circle**, and are called **circular functions**. Circular functions have as their domains sets of real numbers. Trigonometric functions are referred to as **circular functions** because their domains are lengths of arcs on a circle and are defined as real numbers, which may include angles. Circular functions allow the use of trigonometric functions beyond angles and triangles into the study of electricity, sound, music, projectile motion, and other phenomena that exhibit a periodic nature.

- Remember: The *domain set* of a function is the set of all possible values of the independent variable, and the *range set* is the corresponding set of values of the dependent variable. For each member of the domain set, there must be only one corresponding member in the range set. The domain set is the initial set and the range set is the set that results after a function is applied: domain set $\rightarrow$ function f( ) $\rightarrow$ range set. For example, if the domain set is $x = \{2, 3, 4\}$, and it goes through function $f(x) = x^2$, the result is the range set $f(x) = \{4, 9, 16\}$.

• The six trigonometric functions, *sine, cosine, tangent, cotangent, secant,* and *cosecant,* can be defined such that their domains are real numbers rather than only angles. These defined functions are called *circular functions* and are defined according to the *unit circle.* A ***unit circle*** is a circle having a radius of one that is centered at the origin of a coordinate system. A unit circle is given by the equation $x^2 + y^2 = 1$ and depicted as:

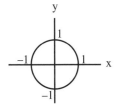

• The *unit circle* can be described in terms of any *real number* x where the size of the angle has x units, which is equal to the arc length. On a circle having a radius of one, a point P on the circle has coordinates defined by the angle of the arc formed from the X-axis:

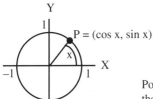

Point P has coordinates (cos x, sin x), and the arc distance of the angle has x units of length and is measured in radians.

The distance around the circle is $2\pi r$, and the distance to any point on the circle is defined by the length of the *arc* beginning from the X-axis and extending to the point on the circle that is given by the angle x multiplied by the radius r, or xr. Positive angles are measured from the positive X-axis in a counterclockwise fashion. Angles measured clockwise are negative.

• The *relationship between circular functions and trigonometric functions* can be observed in the comparison between the *length of the arc on the unit circle being associated with circular functions* and the *standard position angle that subtends an arc in a coordinate system being associated with trigonometric functions.* The terminal side of an arc at point P on a unit circle corresponds to the terminal side of a standard position angle in radians in a coordinate system.

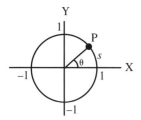

The angle θ subtending an arc of *s* units has a radian measure of *s*. Therefore, each real number on a unit circle corresponds with an arc of *s* units and a standard position angle of *s* radians.

• Consider an arc *s* on a unit circle that begins on the positive X-axis at point (x=1, y=0) or (1, 0) and ends at point P(x, y). The coordinates of point P(x, y) can be written in terms of the real number *s*.

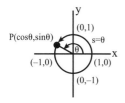

We know that for any angle θ in a coordinate system, cos θ = x/r and sin θ = y/r. Because r = 1 on a unit circle, then *on a unit circle*: cos θ = x and sin θ = y. We also know that *arc length* equals radius multiplied by the central angle measure in radians, or $s = r\theta$, and for a unit circle $s = \theta$. Therefore, on a unit circle: cos θ = cos *s* = x and sin θ = sin *s* = y. From this we can describe the position or *coordinates of point P* on a unit circle as: P(cos *s*, sin *s*).

Therefore, on a unit circle, each ordered pair for point P(cos *s*, sin *s*) is determined by the arc length *s* for that point. The arc length *s*, as well as sin *s* and cos *s*, are real numbers that define the *circular sine function* and *circular cosine function*.

• The *domain* of a circular function is the set of all real numbers for which the function is defined. Any point P on the unit circle is defined by P(cos *s*, sin *s*). Therefore, the domain for sin *s* and cos *s* is the set of all real numbers. The *range* of cos *s* and sin *s* is determined by the points on the unit circle. The X-axis coordinate x = cos *s* and the Y-axis coordinate y = sin *s* vary between −1 and +1. Therefore, the range of sin *s* and cos *s* is the set of real numbers including and between −1 and +1.

- The six *circular functions* are written:

  $\cos s = x$

  $\sin s = y$

  $\tan s = \sin s / \cos s = y/x$

  $\cot s = \cos s / \sin s = x/y$

  $\sec s = 1 / \cos s = 1/x$

  $\csc s = 1 / \sin s = 1/y$

Note that r = 1 on a unit circle.

Also note that because the *equation for a unit circle* is $x^2 + y^2 = 1$, and for a unit circle x = cos *s* and y = sin *s*, then the equation for a unit circle can be written: $(\cos s)^2 + (\sin s)^2 = 1$, or $\cos^2 s + \sin^2 s = 1$.

- On a unit circle each real number corresponds with $|s|$ units, where *s* is the length of the arc to point P, which is between the X-axis at point $(1, 0)$ and point P on the circle where $P(x, y) = P(\cos s, \sin s)$. When *s* is measured in the counterclockwise direction it is *positive*, and when *s* is measured in the clockwise direction it is *negative*. For negative values of *s* where $s = |-s|$:

  $\sin(-s) = -\sin s = -y$

  $\cos(-s) = -\cos s = x$

  $\tan(-s) = -\tan s = -y/x$

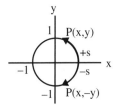

- The values of circular functions of real numbers are determined the same way the values of trigonometric functions for angles measured in radians are determined. Values can be found using points on a unit circle that correspond to ratios of a reference triangle, and they can also be found using a calculator. When using a calculator to determine values of circular functions, it must be in *radian mode* as they are real numbers.

• *Special values of angles and arc lengths* are often mapped on a unit circle to represent the location of a point on a unit circle and used as a reference when solving problems. In special cases the arc lengths correspond to integer multiples of π/4 and π/6. The corresponding coordinates of the points can be determined using special reference angles and triangles (such as 30:60:90 and 45:45:90 triangles). The following figure depicts coordinates of points of integer multiples of π/4 and π/6 on a unit circle up to 2π. (This figure can be expanded to coterminal arcs.) Notice the symmetry in the figure and that, for example, π/2 located in the counter-clockwise direction corresponds to −3π/2 in the clockwise direction.

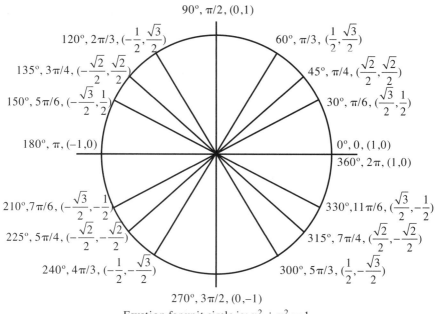

Equation for unit circle is: $x^2 + y^2 = 1$

• **Example:** Use the unit circle figure above to determine the exact value of $\cos(2\pi/3)$ and $\sin(2\pi/3)$.

The arc length has a terminal point at $(-\frac{1}{2}, \frac{\sqrt{3}}{2})$, therefore $\cos(2\pi/3) = -1/2$ and $\sin(2\pi/3) = \sqrt{3}/2$.

This result can also be illustrated using a reference triangle:

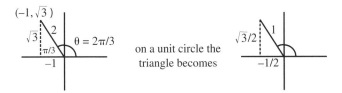

where $\cos(2\pi/3) = x/r = -1/2$ and $\sin(2\pi/3) = y/r = \sqrt{3}/2$
(Remember r on a unit circle is 1.)

• **Example:** Estimate the circular functions $\cos(500)$ and $\sin(-3)$ using a calculator. First put calculator in radian mode. Then calculate $\cos(500)$ and $\sin(-3)$ to 4 significant figures:

$\cos(500) \approx -0.8839$ and $\sin(-3) \approx -0.1411$.

• The unit circle also depicts the periodic nature of circular functions, which is described in detail in Chapter 5. Because the circumference of a circle is $c = 2\pi r$ and for a unit circle $c = 2\pi$, for any point on the unit circle, adding or subtracting 2π will result in the same point on the circle. Therefore, $\sin(s \pm 2\pi) = \sin s$ and $\cos(s \pm 2\pi) = \cos s$. In addition, adding or subtracting 2π will result in the same point on the circle. Therefore, $\sin(s \pm n2\pi) = \sin s$ and $\cos(s \pm n2\pi) = \cos s$, where n = any integer $(..., -5, -4, -3, -2, -1, 0, 1, 2, 3, 4, 5, ...)$. These repetitive functions are called *periodic functions*, and for sine and cosine the period is 2π.

4.10 Linear and Angular Velocity

• Circular motion can be described using trigonometric or *circular functions*. **Linear velocity on a circle** and **angular velocity** have numerous applications in physics, astronomy, and engineering.

• We know that, in general, distance equals rate times time, where velocity is the rate. In equation form this is written $d = vt$, where d is distance, v is rate or velocity, and t is time. If a particle or point is moving at a constant speed around a circle, then the distance traveled is an arc, s. The rate is the speed or velocity of the point around the circle, and time is the time it takes to get from the beginning of the arc to the end of the arc. Therefore, for a point moving around a circle at a constant velocity the following is true:

$s = vt$

Rearranging gives the **linear velocity** around the circle as: $v = s/t$, which is the change in distance per unit time of a point moving around a circle, or of a point on a rotating circle.

The linear forward velocity corresponds to a point moving along a circle and describes how fast the point is moving. As the point moves along the circle, the *positive angle θ in standard position* also changes. The rate at which the angle changes is the **angular velocity**. The angular velocity, denoted by ω, is a measure of the change in angle θ as point P moves around the circle. The **angular velocity** ω is given by: $\omega = \theta/t$, where angle θ is in radians and is the measure of where P is at time t.

• We know that the *arc length s* of a circle is given by: $s = r\theta$, where r is the radius and θ is the central angle in radians subtended by the arc. Therefore, we can relate the *linear and angular velocity* formulas as:

$$v = s/t = r\theta/t$$

Also, because $\omega = \theta/t$, then the following relations for *linear and angular velocity* can be written:

$$v = r\theta/t, \, v = r\omega, \, v = s/t, \, \omega = \theta/t, \, \omega = s/rt, \, \omega = v/r$$

• **Example:** If a person on a skateboard is traveling 10 miles/hour and the skateboard wheels are 2 inches in diameter, what is the angular velocity of each wheel in radians per minute? What is the angular velocity in revolutions per minute?

First, because the wheels are measured in inches and the end result is in radians/minute, change the velocity units for 10 miles/hour to velocity in inches/minute.

$$v = 10\frac{\text{mi}}{\text{hr}} \times \frac{5280 \text{ ft}}{1 \text{ mi}} \times \frac{12 \text{ in}}{1 \text{ ft}} \times \frac{1 \text{ hr}}{60 \text{ min}} = 10,560\frac{\text{in}}{\text{min}}$$

The angular velocity is given by $\omega = v/r$, where r is the radius of each wheel and is 1 inch. Therefore, the angular velocity of each wheel in radians per minute is:

$$\omega = \frac{v}{r} = \frac{10{,}560 \text{ in/min}}{1 \text{ in}} = 10{,}560 \frac{\text{radians}}{\text{min}}$$

Remember: A radian is the measure of the central angle subtending an arc of a circle that is equal to the radius of the circle, and 1 radian = $360°/2\pi$, which is equivalent to 1 revolution/2π, or 1 revolution is 2π radians. To convert radians per minute to revolutions per minute:

$$\omega = 10{,}560 \frac{\text{radians}}{\text{min}} \times \frac{1 \text{ revolution}}{2\pi \text{ radians}} = 1681 \text{ revolutions/minute}$$

Therefore, the angular velocity in revolutions per minute is 1681 revolutions/minute.

• **Example:** If we estimate that the orbit of the Earth around the Sun is approximately circular (even though it is slightly elliptical), and we know that the distance from Earth to Sun is approximately 150,000,000 kilometers or 93,000,000 miles, estimate the speed of the Earth as it travels around the Sun in miles per hour.

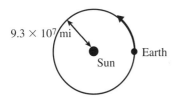

In one year the Earth makes one revolution around the Sun, which is 2π radians. Therefore, $\omega = \theta/t = 2\pi$ radians / 1 year.

In hours, 1 year is $\dfrac{24 \text{ hours}}{1 \text{ day}} \times \dfrac{365 \text{ days}}{1 \text{ year}} = 8760$ hours

Therefore, the angular velocity is $\omega = \theta/t$:

$$\omega = 2\pi \text{ radians/year} = 2\pi/8760 \text{ hours} = 7.173 \times 10^{-4} \text{ radians/hour}$$

and the linear velocity in the orbit or the speed at which Earth travels around the Sun is:

$$v = r\omega = (93{,}000{,}000 \text{ miles})(7.173 \times 10^{-4} \text{ radians/hour}) \approx 66{,}700 \text{ miles/hour}$$

4.11 Chapter 4 Summary and Highlights

• The six trigonometric functions are not only defined using the ratios of the sides of a right triangle (as described in Chapter 3), but they can also be defined in a coordinate system using angles in standard position, and also as arc lengths on a unit circle, called circular functions. The latter two are described in this chapter. This chapter describes trigonometric and circular functions and also includes a brief review of general functions.

• A *standard position angle* is an angle drawn in a rectangular coordinate system with its vertex at the origin and its initial side coincident with the positive X-axis. Coterminal angles are standard position angles that have the same terminal side. The *six trigonometric functions* can be defined in a coordinate system using a standard position angle as ratios of the quantities x, y, and r, with point P(x, y) as the point on the terminal side of the angle and r as the distance from the origin to the point.

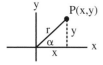

The six *trigonometric functions* are:

$$\cos \alpha = x/r, \quad \sin \alpha = y/r, \quad \tan \alpha = y/x,$$
$$\sec \alpha = r/x, \quad \csc \alpha = r/y, \quad \cot \alpha = x/y$$

• In a coordinate system, the *signs* of the trigonometric functions depend on the signs of x and y in the quadrant in which the terminal side of the standard position angle lies. The signs for sine, cosine, and tangent in each quadrant are:

Quad II	Quad I
cos α is −	cos α is +
sin α is +	sin α is +
tan α is −	tan α is +

Quad III	Quad IV
cos α is −	cos α is +
sin α is −	sin α is −
tan α is +	tan α is −

• *Reference angles* and *reference triangles* are used to assist in calculations involving trigonometric functions. A reference angle *R* for a standard position angle is the positive acute angle θ between the X-axis and the terminal side of the standard position angle. Reference triangles are formed by extending a line from a point P(x, y) on the terminal side of the standard position angle to the X-axis. For example:

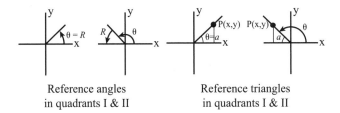

Reference angles
in quadrants I & II

Reference triangles
in quadrants I & II

When trigonometric function calculations are made using reference angles and reference triangles, the sign of the function depends on the quadrant in which the terminal side of the angle or triangle lies.

• When solving problems using trigonometric functions, certain identities are used. They include *negative angle identities* $\cos(-\alpha) = \cos \alpha$, $\sin(-\alpha) = -\sin \alpha$, and $\tan(-\alpha) = -\tan \alpha$, *reciprocal and quotient identities* $\sec \alpha = 1/\cos \alpha$, $\csc \alpha = 1/\sin \alpha$, $\cot \alpha = 1/\tan \alpha$, $\tan \alpha = \sin \alpha/\cos \alpha$, and $\cot \alpha = \cos \alpha/\sin \alpha$, and the *cofunction identities* $\cos \alpha = \sin(90° - \alpha)$ and $\sin \alpha = \cos(90° - \alpha)$.

• The *six trigonometric functions* can be defined as arc lengths on a circle of radius one, a unit circle, and are called *circular functions*. Circular functions have as their domains lengths of arcs on a unit circle and are defined as real numbers rather than being limited to angles. Circular functions allow the use of trigonometric functions beyond angles and triangles into the study of phenomena with a periodic nature, such as electricity and sound.

• The coordinates of point P on a unit circle can be described as P(cos *s*, sin *s*),

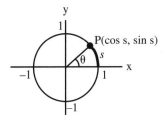

where s is the arc length, the X-axis coordinate is $x = \cos s$, and the Y-axis coordinate is $y = \sin s$. The six circular functions are:

$$\cos s = x, \qquad \sin s = y, \qquad \tan s = y/x,$$
$$\sec s = 1/x, \qquad \csc s = 1/y, \qquad \cot s = x/y$$

where $r = 1$ on a unit circle.

Chapter 5

Graphs of Trigonometric and Circular Functions and Their Periodic Nature

5.1 Circular Motion

• In circular motion, a point or particle moving in a circular path around the perimeter of a circle of radius 1 can be mapped using *cosine* and *sine*. The coordinates of the point or particle can be represented by (x = cos t, y = sin t), where t represents time. One complete revolution of a particle around the circle corresponds to 2π radians. For a particle moving at a constant speed, if it takes one second for the particle to move around the circle, then it is moving at an *angular rate* of 1 revolution per second. The particle therefore moves around the circle with an *angular velocity* of 2π radians per second. When angles are measured in radians, sin x and cos x have *period* 2π. The position of the particle is given by the angle ϕ, which is measured in radians. The rate of change of ϕ is the angular velocity of the particle. The angular velocity, ω, is the change in ϕ divided by the change in t, or $\omega = \Delta\phi/\Delta t$. If the motion is uniform, then $\phi = \omega t$.

• For each point around a circle the six functions, cos x, sin x, tan x, csc x, sec x, and cot x, can be drawn as six graphs of the corresponding *waveforms*. A particle moving around a circle can be compared to a particle moving along the sine curve. Before drawing the graphs for the trigonometric functions, it is helpful to examine the correlation between the motion of a particle around a circle and a sine wave graph.

• Because the coordinates of the particle traveling around a unit circle (at constant velocity) are given by (cos t, sin t), (or (cos s, sin s) as described in section 4.9), as a particle moves around the circle, a point reflected onto the cosine axis (X-axis) that is following the movement of the particle will oscillate from side to side between +1 and –1, and a point reflected onto the sine axis (Y-axis) that is following the movement of the particle will oscillate up and down between +1 and –1.

• Consider the movement of a particle that begins at $t = \phi = 0$, $x = \cos \phi = 1$, and $y = \sin \phi = 0$. (See figure below.) As the particle moves upward and to the left toward the Y-axis, the particle reaches the top where $t = \pi/2 = \phi$. At this point $\cos \pi/2 = 0$ and $\sin \pi/2 = 1$. Then the particle moves in the negative x direction and downward to $\phi = t = \pi$, where $x = \cos \pi = -1$ and $y = \sin \pi = 0$. The particle then moves down and to the right to $\phi = t = 3\pi/2$, where $\cos 3\pi/2 = 0$ and $\sin 3\pi/2 = -1$. Finally, the particle moves to the right and upward to $\phi = 2\pi = t = 0$, where $\cos 2\pi = 1$ and $\sin 2\pi = 0$.

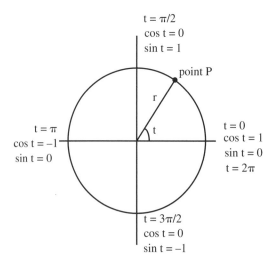

• As a point moves around a circle, at any given position the direction of motion of the particle is *tangent* to the circle. For any particle on the circle the following figure can be drawn. The *velocity tangent* to the circle at point P has a cosine component, cos t, and a sine component, sin t, and points in the direction that the particle is moving. The acceleration this particle experiences is *centripetal acceleration*, which points inward along the radius line.

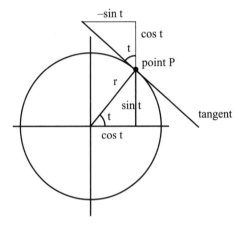

- In simple *harmonic motion*, or *oscillatory motion*, a particle or object moves back and forth between two fixed positions in a straight line. The connection between simple harmonic motion and uniform *circular motion* can be visualized by projecting the image of a particle moving in a circular path onto a screen (perpendicular to the plane of the circle). By projecting the circular path from its side, the projected image looks like a particle moving back and forth (or up and down) in a straight line.

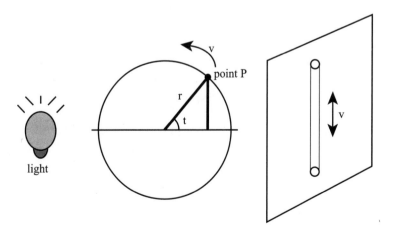

The shadow of the particle translates to simple harmonic motion.

• Another means to visualize the correlation between a particle moving around a circle in circular motion and its corresponding harmonic motion is to project the image of the particle as it moves around the circle onto the Y-axis.

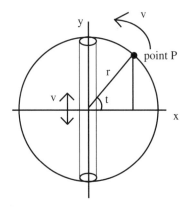

As the particle moves around the circle counterclockwise, the projection oscillates up and down. The position of the particle

begins at the center on the right at ($\cos 0 = 1, \sin 0 = 0$),

at the top is ($\cos \pi/2 = 0, \sin \pi/2 = 1$),

in the center on the way down is ($\cos \pi = -1, \sin \pi = 0$),

at the bottom is ($\cos 3\pi/2 = 0, \sin 3\pi/2 = -1$), and

in the center on the way up is ($\cos 2\pi = \cos 0 = 1, \sin 2\pi = \sin 0 = 0$).

• By comparing the motion of this particle moving around the circle with its projection, it is evident that even when the velocity of the particle is constant, the velocity of the projection of the particle slows to a stop at each end (top and bottom). The *velocity of the oscillatory motion* can be related to a **sine wave pattern**. The velocity is the rate of change of distance, and the slope of a curve at a given point represents the velocity at that point. By rotating a right triangle around a circle the relationships between sine, cosine, distance, and velocity can be visualized. When the *distance* equals sin t, then the *velocity* equals cos t, and when the distance equals cos t, then the velocity equals –sin t.

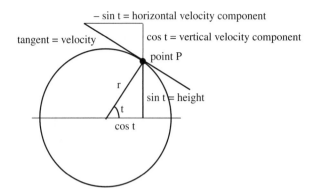

These relationships will be seen using the graphs of sine and cosine.

• A *particle moving around a circle* can be translated into a *particle moving along the sine curve.* For example, the rate of change of position (slope) is zero at the top and bottom of the circle and also on the sine curve where $t = \pi/2$ and $t = 3\pi/2$. The *slope (velocity)* of the sine curve at these points is zero because the curve is flat. At these points the particle in straight line motion on the projected image in the circle is changing directions and comes to a stop as it turns where $v = 0$.

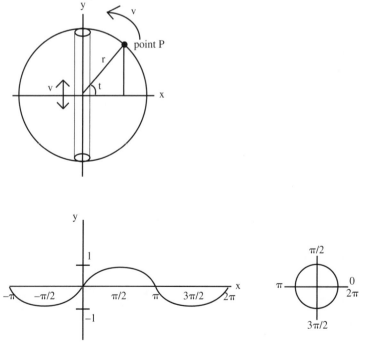

Graph of sine curve

At t = 0, the particle is in the center of its projected straight line on the *circle* and is moving upward and has a corresponding slope on the sine curve *graph* of 1; therefore, the velocity at this point is equal to 1. At t = p, the particle is in the center of its projected image on the circle and going straight down. At this position the corresponding slope on the *sine curve* is –1 and v = –1. The velocity of the particle at the center is at its greatest. Remember: Velocity is the rate of change of distance, and the slope of a curve at a given point represents the velocity at that point.

• Another interesting fact is that the slope at each point on the **sine curve graph** is given by the corresponding value of the **cosine curve graph** at that point.

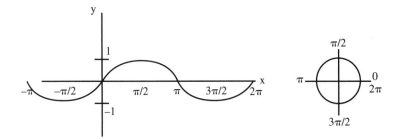

Graph of sine curve – projection of circular motion onto Y-axis

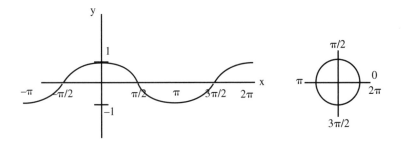

Graph of cosine curve – projection of circular motion onto X-axis

• *Cosine* and *sine* can be described by the equations:

$\cos x = \cos(x + 2n\pi)$

$\cos x = \sin(\pi/2 + x)$

$\sin x = \sin(x + 2n\pi)$

Where n is any integer and x is any real number.

5.2 Graphs of Sine and Cosine

• The graphs of certain functions possess a repeating pattern where the values of the function repeat themselves over and over. These functions are called ***periodic functions***. Sine and cosine are often considered to be the most important periodic functions. A periodic function can be written in the form f(x) = f(x + np), where x is a real number in the domain of f, n is an integer, and p is the period.

• The *graphs of sine and cosine* shown above are often referred to a ***sinusoids*** and describe numerous physical phenomena such as sound, music, electricity, motion of a vibrating object, harmonic waves, water waves, temperature variation, a mass on a spring, and electromagnetic radiation including light, radio waves, and X-rays.

Graph of y = sin x

• The graph of y = sin x is the set of all ordered pairs of real numbers that satisfy the equation. The *domain*, or values for x, include all real numbers and the *range*, or values for y, include real numbers including and between −1 and +1.

• The graph can be created by choosing values of x and calculating y = sin x to create (x, y) ordered pairs to plot. This is laborious, and the graphs of sine and cosine are often created by taking into consideration how sine and cosine vary on the unit circle. Graphs are also created using a graphing utility. Consider a unit circle with 30:60:90 triangles. The length of the *vertical sides* of the triangle represents the value of sin θ, and the length of the horizontal sides represent cos θ.

For the upper part of the unit circle between 0° and 180°:

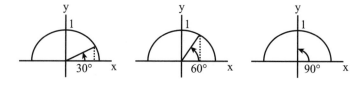

When $\theta = 0$, $\sin \theta = 0$.

When $\theta = \pi/6$ or $30°$, the vertical side of the triangle is $1/2$, or $\sin \theta = 1/2$.

When $\theta = \pi/3$ or $60°$, the vertical side of the triangle is $\sqrt{3}/2$, or 0.866, and $\sin \theta = \sqrt{3}/2$, or 0.866.

When $\theta = \pi/2$ or $90°$, $\sin \theta = 1$.

Continuing around:

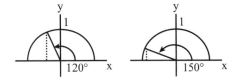

When $\theta = 2\pi/3$ or $120°$, the vertical side of the triangle is $\sqrt{3}/2$, or 0.866, and $\sin \theta = \sqrt{3}/2$, or 0.866.

When $\theta = 5\pi/6$ or $150°$, the vertical side of the triangle is $1/2$, or $\sin \theta = 1/2$.

When $\theta = \pi$ or $180°$, the vertical side is $\sin \theta = 0$.

Plotting these values on a graph of angles, or domain values, versus $\sin \theta$ we have:

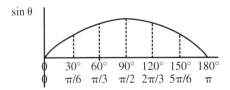

For the lower part of the unit circle between $180°$ and $360°$, we can use the standard position angle and draw triangles that have their vertical sides in the negative direction, below the X-axis.

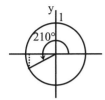

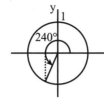

$y = \sin 210° = -1/2$

$y = \sin 7\pi/6 = -1/2$

$y = \sin 240° = -0.866$

$y = \sin 4\pi/3 = -\sqrt{3}/2$

$y = \sin 270° = -1$

$y = \sin 3\pi/2 = -1$

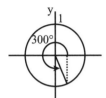

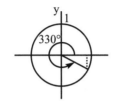

$y = \sin 300° = -0.866$ $y = \sin 330° = -1/2$

$y = \sin 5\pi/3 = -\sqrt{3}/2$ $y = \sin 11\pi/6 = -1/2$

Plotting these values on a graph of angles, or domain values, versus sin θ continuing the graph from 180° to 360°, or π to 2π:

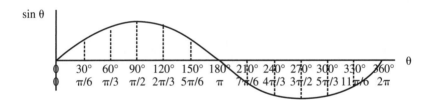

This graph represents one ***period***, which is 0° to 360°, or 0 to 2π radians. The period of y = sin x is 2π or 360°. In addition, the maximum distance of y = sin θ above and below the horizontal X-axis is 1, which represents the ***amplitude***, so that for this graph the amplitude is 1. The amplitude is half of the difference between the maximum and minimum values of the function, or half of the span of the graph along the vertical axis. Therefore, for sine the amplitude is $(1/2)(+1 - (-1)) = (1/2)(2) = 1$.

• To the left of the Y-axis, the section to the right of the axis is repeated with the same pattern, period, and amplitude. Therefore, from −360° or −2π to 0 the graph has the same characteristics as from 0 to +360° or 2π.

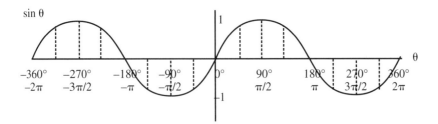

One cycle or period of the graph of y = sin x repeats itself in both directions along the X-axis. The period is 2π or $360°$. In addition, the graph of the sine function, y = sin x, is symmetric with respect to the origin (so that if it is flipped over the Y-axis and then over the X-axis, it will be the same). This corresponds to the negative identity for sine: sin(–x) = –sin x. Note that the graph of the cosine function, y = cos x, is symmetric with respect to the Y-axis, which corresponds to the negative identity for cosine: cos(–x) = cos x. Using a graphing calculator or graphing software, the graph of sine is:

Graph of y = sin x

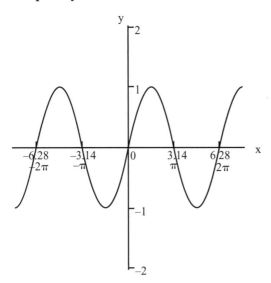

Graph of y = cos x

• Like the graph of sine, cosine is a sinusoid. The graph of y = cos x is the set of all ordered pairs of real numbers that satisfy the equation. The domain, or values for x, includes all real numbers, and the range, or values for y, includes real numbers including and between −1 and +1.

• The graph of y = cos x can be created by choosing values of x and calculating y = cos x to create (x, y) ordered pairs to plot. This is laborious, and the graphs of sine and cosine are often created by taking into consideration how sine and cosine vary on the unit circle. (Graphs are also created using a graphing utility.) Using the same unit circle diagrams with the 30:60:90 triangles that we used for y = sin x (in the previous paragraphs), we can plot y = cos x. In y = sin x, the length of the vertical sides of the triangles represent the value of sin θ, whereas the *horizontal sides* of the triangles represent values of y = cos x.

For the upper part of the unit circle between 0° and 180°:

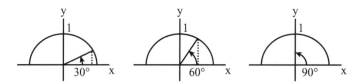

When θ = 0, cos θ = 1.

When θ = π/6 or 30°, the horizontal side of the triangle is cos θ = $\sqrt{3}$/2, or 0.866.

When θ = π/3 or 60°, the horizontal side of the triangle is cos θ = 1/2.

When θ = π/2 or 90°, cos θ = 0.

Continuing around:

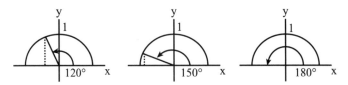

When θ = 2π/3 or 120°, the horizontal side is cos θ = –1/2.

When θ = 5π/6 or 150°, the horizontal side is cos θ = – $\sqrt{3}$ /2, or –0.866.

When θ = π or 180°, the horizontal side is cos θ = –1.

For the lower part of the unit circle between 180° and 360°, we use the standard position angle and draw triangles below the X-axis.

When θ = 7π/6 or 210°, the horizontal side is cos θ = – $\sqrt{3}$ /2, or –0.866.

When θ = 4π/3 or 240°, the horizontal side is cos θ = –1/2.

When θ = 3π/2 or 270°, the horizontal side is cos θ = 0.

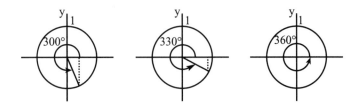

When θ = 5π/3 or 300°, the horizontal side is cos θ = 1/2.

When θ = 11π/6 or 330°, the horizontal side is cos θ = $\sqrt{3}$ /2, or 0.866.

When θ = 2π or 360°, the horizontal side is cos θ = 1.

Plotting these values on a graph of angles, or domain values, versus cos θ the graph from –360° to +360°, or –2π to 2π is:

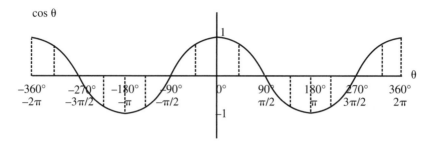

The graph of y = cos x or y = cos θ is the same as the graph of y = sin x except that it is shifted horizontally so that when x = θ = 0, y = 1.

- For y = cos x, the **amplitude**, or maximum distance above or below the horizontal X-axis, is 1 (as was y = sin x). The **period** of y = cos x is one complete cycle and is 2π or 360°. Cosine repeats along the horizontal X-axis. The graph of y = cos x is *symmetric* with respect to the Y-axis (if it is flipped over the Y-axis it is the same), so that cos(–x) = cos x, which is the *negative number identity* for cosine.

Using a graphing utility, the graph of cosine is:

Graph of y = cos x

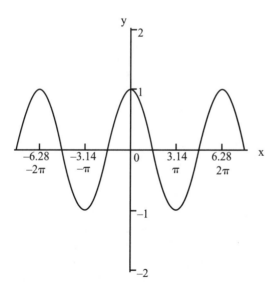

- Superimposing y = sin x and y = cos x illustrates the shift:

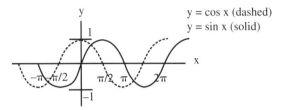

y = cos x (dashed)
y = sin x (solid)

Because sin(x + π/2) = cos x, the graph of y = cos x is the graph of
y = sin x shifted by π/2 to the left.

5.3 Transforming Graphs of Sine and Cosine Through Changes in Amplitude, Period, and Vertical and Horizontal Shifting

- Graphs of trigonometric functions can be transformed and modified by
multiplying and adding constants in the equations y = sin x and y = cos x.
If there are *coefficients* in the equations for y = cos x, y = sin x, etc., the
graph of the function will have the same general shape, but it will have a
larger or smaller *amplitude*, or it will be *elongated* or *narrower* by changing the *period*, or it will be moved to the *right* or *left* or *up* or *down*.

- For example, if there is a coefficient of 2 in front of cosine or sine,
y = 2 sin x and y = 2 cos x, the *amplitude* will change and the graph will
go to +2 and –2 (rather than +1 and –1) on the Y-axis. Similarly, if there
is a coefficient of 1/2 in front of cosine or sine, the graph will go to +1/2
and –1/2 (rather than +1 and –1) on the Y-axis. (See examples of graphs
throughout this section illustrating transformations.)

If there is a coefficient of 2 in front of x, resulting in y = cos 2x and
y = sin 2x, the *period* will change and the graph will complete each cycle
along the X-axis twice as fast. Because there is one cycle between 0 and
2π for y = cos x and y = sin x, there will be two cycles between 0 and
2π for y = cos 2x and y = sin 2x. Similarly, if there is a coefficient of 1/2
in front of x, giving y = cos x/2 and y = sin x/2, the graph will complete
each cycle along the X-axis half as fast. Because there is one cycle between
0 and 2π for y = cos x and y = sin x, there will be one-half of a cycle
between 0 and 2π for y = cos x/2 and y = sin x/2.

If a number is added or subtracted, y = 2 + cos x and y = 2 + sin x, the
graph will *shift vertically* and the function will be moved up or down on
the Y-axis (in this case up 2).

If a number is added to or subtracted from x, $y = \cos(x + 2)$ and $y = \sin(x + 2)$, the graph will have a *phase shift* and the function will be shifted to the right or left on the X-axis.

Transforming Graphs of Any Function

• Graphs of general functions can be modified by adding and/or multiplying by coefficients. Following are examples of how functions can be transformed.

• *Vertical shifting* of a function:

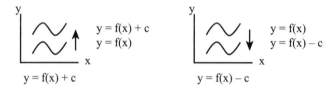

where c is a constant.

• *Horizontal shifting* of a function:

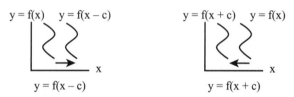

where c is a constant.

Note that the signs of horizontal shifts are tricky. A $(x + c)$ in the parentheses shifts to the left and a $(x - c)$ shifts to the right.

• *Vertical stretching* and *squeezing* of a function:

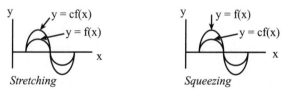

| where c is a constant | where c is a constant |
| and $c > 1$. | and $0 < c < 1$. |

Greater values of c generally result in greater vertical stretching.

• Reflecting across the axes:

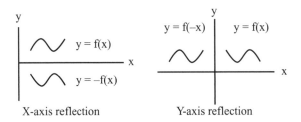

X-axis reflection   Y-axis reflection

Transformations of y = sin x and y = cos x

• The *graphs of sine and cosine* are both sinusoids and can be transformed by stretching or shrinking in the x direction changing the *period*, increasing or decreasing the *amplitude*, shifting up, down, or sideways, and reflecting across the axes. Transformations are made by multiplying, dividing, adding, or subtracting coefficients in the equations y = sin x and y = cos x. Following is a discussion of amplitude, period, vertical shifting, and horizontal, or phase shift, transformations of the graphs of sine and cosine.

Amplitude of Sine and Cosine Functions

• The **amplitude** is the maximum deviation from the centerline of the horizontal component of sine and cosine graphs. In the case of y = a sin x and y = a cos x, the centerline is the X-axis. Changing the amplitude changes the y-component of a sine or cosine graph. The amplitude of a sine or cosine function is changed by multiplying the equations by the *amplitude a*:

   y = a sin x and y = a cos x

The graphs will have the same general shape, but the maximum and minimum points on the Y-axis will be bound by ±a. The amplitude is therefore given by the absolute value of a, or $|a|$, and the range is $(-|a|, |a|)$. Negative a values will flip the graph across the X-axis as depicted in the first example below. In these equations the *period*, or length of a complete cycle along the X-axis, will remain unchanged at 2π. In addition, the graph of y = a sin x will cross the X-axis everywhere the graph of y = sin x crosses the X-axis, and similarly for y = a cos x and y = cos x.

• **Example:** Compare y = (1) cos x and y = (–1) cos x.

On the graph the negative amplitude curve (black line) is reflected across the X-axis so that y = – cos x is upside down compared to y = cos x.

Graph of y = cos x (gray curve) and y = –cos x (black curve)

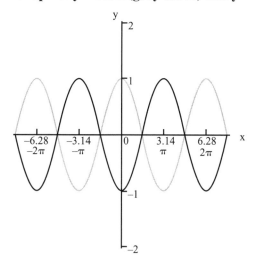

• **Example:** Compare the graphs of y = 2 sin x and y = sin x, and then compare the graphs of y = 2 cos x and y = cos x. Use a graphing calculator or graphing utility to graph y = 2 sin x and y = sin x on one graph, and y = 2 cos x and y = cos x on a second graph.

In the graph of y = 2 sin x and y = sin x, the period remains at 2π; however, the amplitude of y = 2 sin x is |2| or 2 rather than |1| for y = sin x.

Graph of y = 2 sin x (black curve) compared to y = sin x (gray curve)

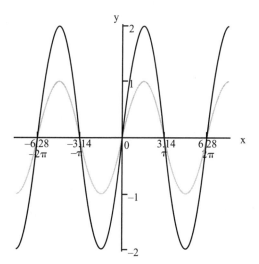

In the graph of $y = 2 \cos x$ and $y = \cos x$, the period remains at 2π. However, the amplitude of $y = 2 \cos x$ is $|2|$ or 2 rather than $|1|$ for $y = \cos x$.

Graph of $y = 2 \cos x$ (black curve) compared to $y = \cos x$ (gray curve)

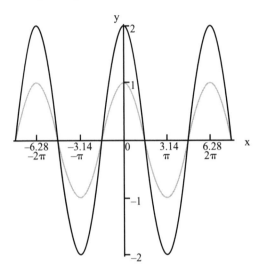

- **Example:** Compare $y = (1/2) \sin x$ and $y = -2 \sin x$ on a graph.

$y = (1/2) \sin x$ (gray curve) and $y = -2 \sin x$ (black curve)

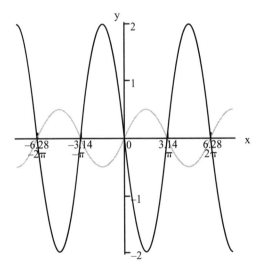

We can see from the graph that the periods of both functions are 2π; however, the amplitudes are $|1/2|$ or $1/2$ and $|-2|$ or 2. The negative sign before the amplitude in $y = -2\sin x$ reflects the graph across the X-axis, which turns the graph upside down.

Period of Sine and Cosine Functions:

• The **period** represents one complete cycle of a function along the horizontal axis. The periods of $y = \sin x$ and $y = \cos x$, and even such functions as $y = 3\sin x$ and $y = 10\cos x$, are still 2π or $360°$. The periods are changed when the functions of sine and cosine have a multiplier b of x such that:

$y = \sin bx$ and $y = \cos bx$

where the *period* is $2\pi/b$ and $b > 0$.

These graphs have the same basic shape as $y = \sin x$ and $y = \cos x$, an amplitude of 1, and a range of $[-1, 1]$.

• The period of $y = \sin bx$ and $y = \cos bx$ is obtained by considering that $\sin bx$ and $\cos bx$ each complete one cycle as bx ranges from 0 to 2π, or $bx = 0$ to $bx = 2\pi$. Therefore, x will vary from:

$x = 0/b = 0$ to $x = 2\pi/b$

or equivalently, $0 \le x \le 2\pi/b$

Therefore, the **period** *of* $y = \sin bx$ *and* $y = \cos bx$ *is* $2\pi/b$.

• When the variable x in $y = \sin x$ or $y = \cos x$ is multiplied by 2, the functions complete one cycle in half of $y = \sin x$ and $y = \cos x$. When the variable x is multiplied by 1/2, the period is doubled. If the variable x is multiplied by 1/4, the period is 4 times as long.

• **Example:** What are the periods of $y = \sin 2x$, $y = \sin 3x$, and $y = \sin x/2$?

Graph the functions using a graphing utility or calculator. Compare $y = \sin 2x$ and $y = \sin 3x$ with $y = \sin x$ on one graph, and compare $y = \sin x/2$ with $y = \sin x$ on a second graph.

In the first graph, the period of $y = \sin 2x$ is $2\pi/2 = \pi$ and therefore completes one cycle as x varies from $x = 0$ to $x = \pi$. The period of $y = \sin 3x$ is $2\pi/3$ and therefore completes one cycle as x varies from $x = 0$ to $x = 2\pi/3$.

In the second graph, the period of y = sin x/2 is 2π/(1/2) = 4π and therefore completes one cycle as x varies from x = 0 to x = 4π.

Graph of y = sin 3x (black), y = sin 2x (gray), and y = sin x (light)

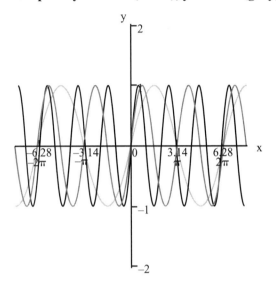

Graph of y = sin x/2 (black curve) and y = sin x (gray curve)

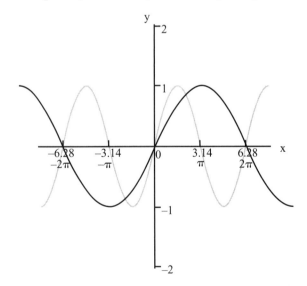

Vertical Shifting of Sine and Cosine Functions

• A vertical shift occurs when the entire graph is shifted up or down along the vertical Y-axis. The equations for a vertical shift take the form:

$$y = d + \sin x$$
$$y = d + \cos x$$

where d is the number of units the graph is shifted up or down.

When $d > 0$, the sinusoid graph shifts up.

When $d < 0$, the sinusoid graph shifts down.

• **Example:** Compare the graph $y = 3 + \cos x$ with $y = \cos x$.

The $+3$ in $y = 3 + \cos x$ shifts the graph vertically up by three so that the horizontal centerline is at $y = 3$ rather than along the X-axis.

Graph of y = 3 + cos x (black curve) and y = cos x (gray curve)

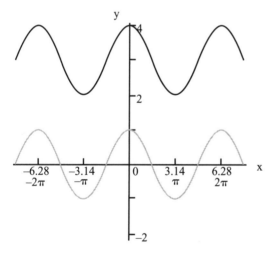

• **Example:** Compare the graph of y = –3 + 2 cos x with y = cos x.

In the graph of y = –3 + 2 cos x the horizontal centerline is shifted down from the X-axis by 3 to y = –3, and the amplitude is 2 rather than 1.

Graph of y = –3 + 2 cos x (black curve) and y = cos x (gray curve)

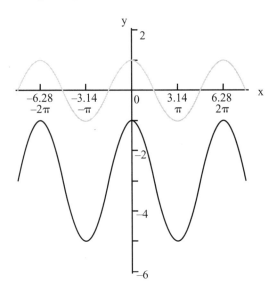

• **Example:** Compare the graph of y = 3 – 2 sin x with y = sin x.

In the graph of y = 3 – 2 sin x, the horizontal centerline is shifted up 3 to y = 3, and the amplitude is 2. In addition, the negative sign before the amplitude |2| causes the graph to be a negative sine curve and therefore upside down.

Graph of y = 3 – 2 sin x (black curve) and y = sin x (gray curve)

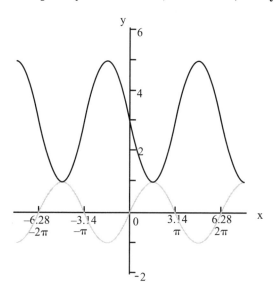

Horizontal Shift, or Phase Shift, of Sine and Cosine Functions

• The graphs of $y = \sin x$ and $y = \cos x$ can be shifted horizontally, which is called a ***phase shift***. The equations for phase shift are:

$$y = \sin(x + c) \text{ and } y = \cos(x + c)$$

The graphs of these equations are the same as $y = \sin x$ and $y = \cos x$ except that they are shifted to the right or left, depending on the sign of c.

• In the phase shift equations $y = a \sin(bx + c)$ and $y = a \cos(bx + c)$:

a is the amplitude

$2\pi/b$ is the period

The graphs of sin bx and cos bx each complete one cycle as bx varies from 0 to 2π, or bx = 0 to bx = 2π, and therefore x varies from x = 0/b = 0 to x = $2\pi/b$, resulting in $2\pi/b$ as the period.

- To determine the **phase shift** for y = a sin(bx + c), consider that the graph completes one cycle as bx + c varies from:

  bx + c = 0   to   bx + c = 2π

Therefore x varies from (solve for x in each equation):

  x = –c/b   to   x = –c/b + 2π/b

where 2π/b is the **period** and –c/b is the **phase shift**

Therefore, the graphs of y = a sin(bx + c) and y = a cos(bx + c) have a period of 2π/b and are shifted along the horizontal axis by the absolute value of the phase shift |–c/b|.

  If –c/b > 0, the graph shifts to the right.
  If –c/b < 0, the graph shifts to the left.

- **Example:** Compare the graph of y = sin x with y = cos(x – π/2).

The graph of y = sin x has an amplitude of 1, a period of 2π, and a phase shift of 0. The graph of y = cos(x – π/2) has an amplitude of 1. To find period and phase shift, remember that (bx + c) varies from

  bx + c = 0   to   bx + c = 2π

and x varies from

  x = –c/b   to   x = –c/b + 2π/b

Therefore (x – π/2) varies from

  x – π/2 = 0   to   x – π/2 = 2π
  or x = π/2   to   x = π/2 + 2π

where 2π is the period and π/2 is the phase shift.

Therefore, one period of the graph begins at x = π/2 and ends at x = π/2 + 2π = 5π/2.

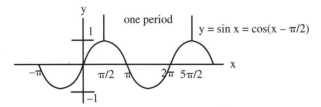

The graphs of y = sin x and y = cos(x – π/2) equal each other for all x and y values, and they perfectly overlap.

• **Example:** Graph $y = 2 - 2\cos(2x + \pi)$.

The graph has an amplitude of $|-2| = 2$.

To find phase shift and period remember that $(2x + \pi)$ varies from
$2x + \pi = 0$ to $2x + \pi = 2\pi$
and x varies from
$x = -\pi/2$ to $x = -\pi/2 + 2\pi/2 = -\pi/2 + \pi = \pi/2$
where π is the period and $-\pi/2$ is the phase shift.

The graph of $y = 2 - 2\cos(2x + \pi)$ is flipped upside down because of the negative sign in front of the amplitude. In addition, one cycle begins at $x = -\pi/2$ and ends at $x = -\pi/2 + \pi = \pi/2$.

Because of the 2, from which $2\cos(2x + \pi)$ is subtracted, the graph is shifted up vertically so that the horizontal centerline is at $y = 2$.

Graph of $y = 2 - 2\cos(2x + \pi)$

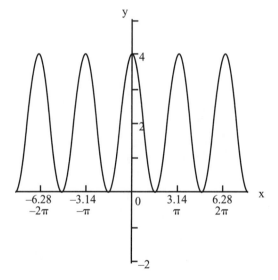

Summary of Sine and Cosine Transformations

• Transformations can be made to the generalized equations,

y = d + a sin(bx + c) and y = d + a cos(bx + c)

a is the *amplitude* given by the absolute value of a, or $|a|$.

d is the *vertical shift* where d is the number of units the graph is shifted up or down, so that when d > 0, the graph shifts up, and when d < 0, the graph shifts down.

(bx + c) is used to determine the *phase shift* and *period* as follows: Because the graph of a sinusoid completes a cycle as x varies from 0 to 2π, solve the following for the variable x.

   (bx + c) = 0   and   (bx + c) = 2π

   x = –c/b   and   x = –c/b + 2π/b

where –c/b is the *phase shift* and 2π/b is the *period.*

Therefore, the graph of a sinusoid completes one cycle as x varies from 0 to 2π, or as x varies from –c/b to –c/b + 2π/b. If the phase shift is greater than zero, or –c/b > 0, the phase shift is to the right; and if the phase shift is less than zero, –c/b < 0, the phase shift is to the left.

• Note that functions in the form

   y = a sin(bx + c) and y = a cos(bx + c)

are sometimes referred to as simple harmonics.

• In *time-dependent sine and cosine functions*:

   y = d + a sin(bt + c) and y = d + a cos(bt + c)
   $|a|$ is the *amplitude*
   d is the *vertical shift*
   –c/b is the *phase shift*
   2π/b is the *period* or the time for one complete cycle
   frequency is f = 1/period = b/2π and is the number of cycles per unit time.

- Another general formula for a sine curve is $y = a \sin(\omega t + \theta)$, where:

  $|a|$ is the amplitude

  ω is the angular frequency in radians per second

  $f = \omega/2\pi$ is the frequency in cycles per second

  $1/f = 2\pi/\omega$ is the period

  $\omega t + \theta = 0$ and $\omega t + \theta = 2\pi$

  $t = -\theta/\omega$ and $t = -\theta/\omega + 2\pi/\omega$

  $-\theta/\omega$ is the phase and $2\pi/\omega$ is period

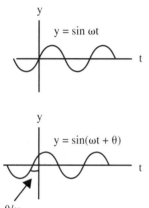

5.4 Applications of Sinusoids

- *Sinusoidal waves* are found in the study of such phenomena as electricity, electromagnetic waves, waves in water, sound waves, harmonic waves, and standing waves. For example, the electricity we use is generally *alternating current*, or AC. In addition, radio signals that travel from a transmitting station to a receiver are produced by alternating currents in the transmitting antenna.

- *Electric generators* can be used to create alternating current in a wire, which has a sinusoidal form when the wire is connected to an oscilloscope. An oscilloscope depicts a graph with current as the y-coordinate and time as the x-coordinate.

• Faraday discovered that the flow of electricity could be created by moving or turning a wire in a magnetic field. Magnetism is produced by passing an electric current through a conductor, and conversely, when a conductor is moved across a magnetic field, a current develops as a result of an induced voltage. This is the principle of *electromagnetic induction*, which is the process of inducing a voltage in a wire and is the basis for most *electric generators*.

Generators are used to convert mechanical energy to electrical energy. In a generator, a conducting wire is moved between magnets through a magnetic field and a voltage is induced in the wire, which directs current to an external circuit. In a generator when the magnetic field is held constant, rotating the armature (which is an iron core with a conducting coil, or conductor, wrapped around it) in the magnetic field varies the magnetic flux through the coil. The sign of the current produced in a generator alternates as the armature revolves, because the direction of the magnetic flux through the coil reverses twice each revolution. As the armature is rotated at a constant frequency, the current produced has a sinusoidal time dependence. This current is alternating current (AC). The standard household current used in the United States is alternating current of 60 cycles/second (hertz) and 120 volts.

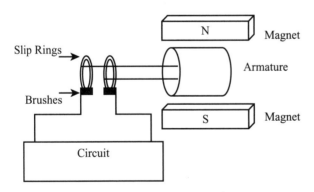

The *armature* is an iron core with a conducting coil (conductor) wrapped around it.

The current is conducted from the rotating shaft to fixed external wires by the slip rings and brushes. The slip rings mounted on the shaft provide a sliding contact so the wires do not become tangled and wound up. The slip ring makes contact with metallic wires called brushes. In practice, generators have many coils and several magnets.

• In the simplest generator, the conductor is an open coil of wire rotating between the poles of a permanent magnet. During a single rotation, one side of the coil passes through the magnetic field, first in one direction and then in the other, so that the resulting current is alternating current (AC), moving first in one direction and then in the other. Each end of the coil is attached to a separate metal slip ring that rotates with the coil. The brushes rest on the slip ring and pass the current to an external circuit.

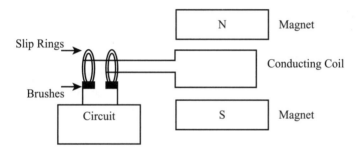

• When the conducting coil of wire is rotated, the rotation causes a continuous change in magnetic flux linking the conductor, which induces a sine wave voltage in the conductor. The change in voltage varies from zero when the conductor is horizontal to a maximum when the conductor is vertical. The induced *voltage* v is given by:

$$v = Vm \sin \omega t$$

where Vm is the maximum height the curve reaches, or amplitude; sine is the function that gives the curve its wave shape; t is time; and ω is the radian frequency, also called angular frequency or angular velocity, which describes voltage and how rapidly the curve oscillates (it is actually the rate of rotation of the generator measured in radians per second).

The function v is periodic, repeating itself at intervals of 2π radians. The rate of repetition is the frequency f, where $f = \omega/2\pi$ in cycles/second or hertz (Hz), where 1 Hz = 1 cycle/second, and $\omega = 2\pi f$.

The **current** i in amperes is given by:

$$i = I_m \sin(\omega t + \theta)$$

where θ is the phase angle between v and i.

Voltage v and current i can be depicted as:

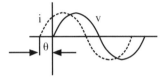

The voltage lags current by the *phase angle* θ. The sinusoids are out of phase by the amount θ.

• Generators require an *external energy source* to rotate the armature. External sources of energy used to rotate an armature include hydro-electric generators in which falling water falls through turbines, and steam-operated turbines where water is boiled and steam is heated further so that its expansion forces turbines to rotate. The steam is then allowed to condense so that the water is recycled. The fuels used to produce steam include coal and thermonuclear reactors.

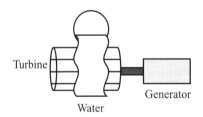

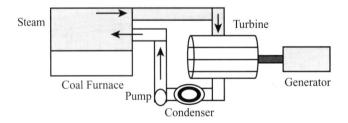

Electric Current Examples

• **Example:** A generator at a power plant at 60 cycles per second will drive alternating current into homes and businesses at 60 cycles per second, or 60 Hertz. If the voltage v is given by v = 160 sin ωt, and ω is the angular frequency in radians/second of the rotating generator, find the value of ω and graph v.

Because each cycle is 2π radians, at 60 cycles/second
ω = 2πf = 2π(60) = 120π radians/second
and therefore v = 160 sin 120πt

The graph of v = 160 sin 120πt is depicted as:

Graph of v = 160 sin 120πt
Voltage vs. Time

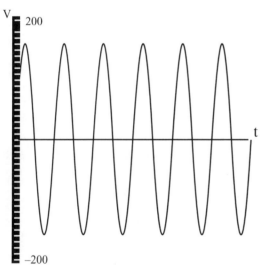

Horizontal t-axis extends from t = 0 to t = 0.10 seconds.
Vertical v-axis extends from 200 to –200.

• **Example:** An AC generator delivers current in amperes according to i = 20 sin(120πt – π), where i is the current in amperes and t is time in seconds. Determine the amplitude, frequency, and phase shift, and graph the equation for i.

Amplitude |a| = |20| = 20 amperes.

To find phase shift and period solve for t.

$120\pi t - \pi = 0$	and	$120\pi t - \pi = 2\pi$
$120\pi t = \pi$	and	$120\pi t = \pi + 2\pi$
$120t = 1$	and	$120t = 1 + 2$
$t = 1/120$	and	$t = 1/120 + 1/60$

where 1/120 is the phase shift in seconds and 1/60 is the period in seconds.

Frequency is 1/period or $\omega/2\pi$,
therefore, 1/period = 60 cycles/second, or 60 Hz,
or alternatively, using the equation given, ω is 120π, so
$f = \omega/2\pi = 120\pi/2\pi = 60$ Hz.

Graph of i = 20 sin(120πt – π)
Current vs. Time

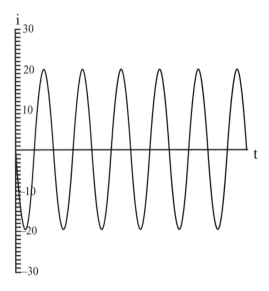

The horizontal time-axis extends from t = 0 to t = 0.10 seconds, and current i along the vertical axis is in amperes. Note there are 60 cycles/second, so there are 6 cycles/0.1 second.

• *Electromagnetic waves* include the spectrum of electromagnetic radiation from long wavelengths to short wavelengths, specifically, long wave radio, amplitude modulated wave (AM), short wave, frequency modulated wave (FM), television, radar, microwaves, infrared, visible light (red, orange, yellow, green, blue, indigo, violet), ultraviolet, X-rays, and gamma rays. The visible range of light is a small part of the electromagnetic spectrum. Electromagnetic waves travel at the speed of light (approximately 3×10^{10} centimeters per second, 3×10^8 meters per second, or 186,000 miles per second).

• *Electromagnetic radiation* travels at the speed of light and occurs in the form of quanta, or photons. Light and other electromagnetic radiation consist of oscillating electric and magnetic fields that carry both energy and momentum. Electromagnetic waves are produced by electric charges that are undergoing oscillation. When electric charges undergo acceleration, a time-varying electromagnetic field is produced and electromagnetic waves are propagated outward from the source.

• Electromagnetic waves are often studied using their characteristic wavelengths and frequencies. A relationship between wavelength and frequency is: $\lambda f = c$, where λ is wavelength, f is frequency, and c is the speed of light.

• **Example:** Radio pulsars are spinning magnetized neutron stars that produce pulses of X-rays, gamma rays, and radio waves. Suppose the gamma rays detected by an Earth-orbiting satellite have a wavelength $\lambda = 3 \times 10^{-12}$ m. The equation form for the gamma rays is $y = a \sin \omega t$. What is ω?

First determine frequency as:

$$f = c/\lambda = (3 \times 10^8 \text{ m/s})/(3 \times 10^{-12} \text{ m}) = 10^{20} \text{ Hz}$$

Because $f = \omega/2\pi$, then $\omega = 2\pi f = 2\pi \times 10^{20}$

• *Water waves* can be described by a sinusoidal moving wave equations in the form: $y = a \sin 2\pi(ft - d/\lambda)$, where d is the distance from the source, t is time, f is frequency, λ is wavelength, and a is amplitude.

• *Standing waves*, often modeled as vibrations on a string, form cosine and sine functions.

• *Sound waves* are pressure waves produced by a vibrating source or object that causes motion in the material they are transmitted through (e.g. air, water, solid materials). Tuning forks are used to produce pure tone's that can be described by: $y = a \sin 2\pi ft$. Tuning forks have two prongs of a specified size that vibrate and produce sound when tapped. Sound waves are depicted as sine or cosine functions on an oscilloscope.

Most sound is not a simple pure tone but a complex mix of tones. Complex sound is a mix of partial tones and has its smallest frequency tone as its *fundamental tone* and other tones as *overtones*. Sound waves are described by their wavelength and frequency: $\lambda = v/f$, where λ is the wavelength, f is frequency often measured in Hz, and v is the speed of sound. A period of a tone is the time for the tone to produce one complete cycle.

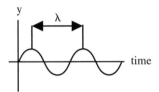

• *Simple and damped harmonic motion and resonance* also are described by sinusoids and are often exemplified by a mass and spring model, where a mass is suspended by a spring and set in motion. If the amplitude of the up and down motion remains constant (no friction), then the motion is called *simple harmonic motion*. If the amplitude decreases over time (friction loss), the motion is called *damped harmonic motion*. If the amplitude increases over time, the motion is called *resonance*.

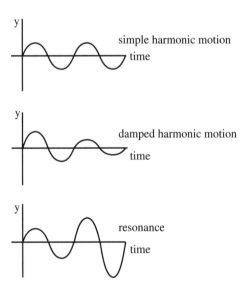

5.5 Graphs of Secant and Cosecant

• Cosecant and secant are the reciprocal functions of sine and cosine, where csc x = 1/sin x and sec x = 1/cos x. Sine and cosine are periodic functions and their graphs are continuous, with each value in the domain having a value in the range. Because cosecant and secant are reciprocal functions of sine and cosine, they are also periodic. Cosecant and secant are not continuous, and their graphs are undefined where sine and cosine have values of zero. Specifically, cosecant, $y = \csc x$, is undefined at values of x where $y = \sin x$ has a value of zero. Vertical asymptotes occur in $y = \csc x$ at x-intercepts of $y = \sin x$. Similarly, secant, $y = \sec x$, is undefined at values of x where $y = \cos x$ has a value of zero. Vertical asymptotes occur in $y = \sec x$ at x-intercepts of $y = \cos x$. Graphs of secant and cosecant have no amplitude because there are no maximum or minimum values, and there are no x-intercepts. Secant and cosecant are generally not as useful as sine and cosine.

- **Cosecant, $y = \csc x$**

$y = \csc x = 1/\sin x$

Because $y = \sin x = \cos(x + 90°) = \cos(x + \pi/2)$,

then $y = \csc x = \sec(x + 90°) = \sec(x + \pi/2)$

When $\sin x = 1$, the value of csc x is also 1.

When $0 < \sin x < 1$, then $\csc x > 1$.

When $-1 < \sin x < 0$, then $\csc x < -1$.

As $|x|$ approaches 0, $|\sin x|$ approaches 0, and $|\csc x|$ becomes large.

At $x = 0$, csc x approaches a vertical line asymptote.

Graph of $y = \csc x$ (black curve) and $y = \sin x$ (gray curve)

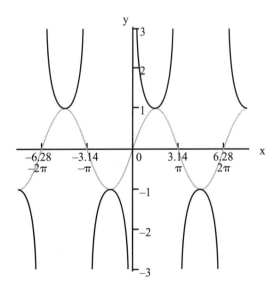

y = csc x is an odd function, and its graph is symmetric with respect to the origin (it can be rotated around the line y = –x).

The period of y = csc x is 2π.
The domain of y = csc x includes all real numbers of x, except x = nπ where n is an integer (where y = sin x crosses zero).
The range of y = csc x includes real numbers for y where y ≤ –1 and y ≥ 1.

For the interval 0 to 2π (one period), the graph of y = csc x has these characteristics:

From 0 to π/2, the graph decreases from + infinity to 1.
From π/2 to π, the graph increases from 1 to + infinity.
From π to 3π/2, the graph increases from – infinity to –1.
From 3π/2 to 2π, the graph decreases from –1 to – infinity.

Graph of y = csc x (black curve) and y = sin x (gray curve) in the interval 0 to 2π, (one period)

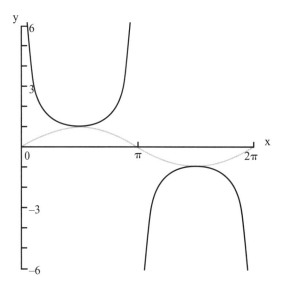

- **Secant, y = sec x**

    $y = \sec x = 1/\cos x$

Because $y = \cos x = \sin(x + 90°) = \sin(x + \pi/2)$

$y = \sec x = \csc(x + 90°) = \csc(x + \pi/2)$

The period of $y = \sec x$ is 2π.

The domain of $y = \sec x$ includes all real numbers of x, except

$x = \pi/2 + n\pi = (\pi/2)(1 + 2n)$, where n is an integer.

The range of $y = \sec x$ includes real numbers for y where

$y \le -1$ and $y \ge 1$.

$y = \sec x$ is an even function, and its graph is symmetric with respect to the Y-axis.

Graph of y = sec x (black curve) and y = cos x (gray curve)

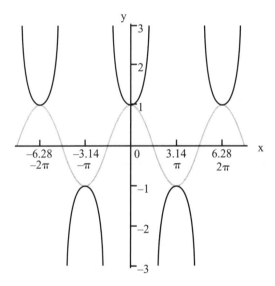

Graph of y = sec x and y = cos x for the interval 0 to 2π, (one period)

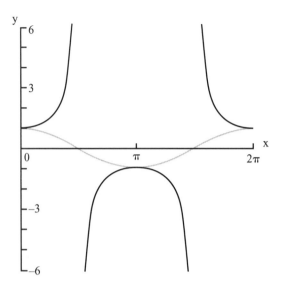

For the interval 0 to 2π (one period), the graph of y = sec x has these characteristics:

From 0 to π/2, the graph increases from 1 to + infinity.

From π/2 to π, the graph increases from – infinity to –1.

From π to 3π/2, the graph decreases from –1 to – infinity.

From 3π/2 to 2π, the graph decreases from + infinity to 1.

• **Example:** Graph $y = 5 + 0.5 \cos \theta$ and reciprocal $y = 5 + 0.5 \sec \theta$. The graph is:

Graph of $y = 5 + 0.5 \cos \theta$ (gray curve) and $y = 5 + 0.5 \sec \theta$ (black curve)

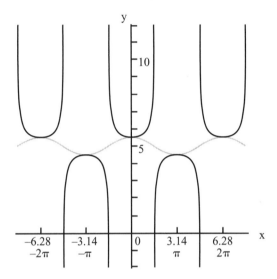

5.6 Graphs of Tangent and Cotangent

• *Tangent*, $y = \tan x$, and *cotangent*, $y = \cot x$, are reciprocal functions:

$$\tan x = 1/\cot x = \sin x/\cos x$$
$$\cot x = 1/\tan x = \cos x/\sin x$$

Because $\tan x = \sin x/\cos x$, it is undefined at $\cos x = 0$, and $\tan x$ has x-intercepts when $\sin x = 0$. Therefore, $\tan x$ is undefined when $x = \pi/2 + n\pi = (\pi/2)(1 + 2n)$, where n is an integer. The graph has vertical asymptotes at these undefined points, and $\tan x$ has x-intercepts when $x = n\pi$. *Cotangent* is the reciprocal of tangent and therefore has a value of zero where tangent is undefined, and it is undefined at values where tangent is zero. The x-intercepts for $y = \tan x$ are vertical asymptotes for $y = \cot x$, and the vertical asymptotes for $y = \tan x$ are x-intercepts for $y = \cot x$. Because $\cot x = \cos x/\sin x$, cotangent is undefined at $\sin x = 0$, or when $x = n\pi$, where n is an integer, and the graph has vertical asymptotes at these undefined points. The graph of $y = \cot x$ has x-intercepts when $\cos x = 0$, or when $x = \pi/2 + n\pi = (\pi/2)(1 + 2n)$, where n is an integer.

• **Tangent**, y = tan x, is an odd function, and its graph is symmetric with respect to the origin.

The period of y = tan x is π, or 180°.

The domain of y = tan x includes all real numbers of x, except x = π/2 + nπ = (π/2)(1 + 2n), where n is an integer. The range of y = tan x includes real numbers from – to + infinity.

The graph has vertical asymptotes at x = π/2 + nπ and x-intercepts at x = nπ. The graph has no amplitude because no maximum or minimum values exist.

Graph of y = tan x

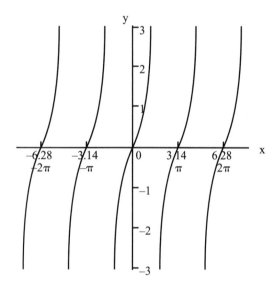

Because the *negative identity* for tangent, tan(–x) = –tan x, the graph of y = tan x can be reflected through the origin for the range 0 ≤ x ≤ π/2 to see the graph for the interval –π/2 to +π/2.

For the interval 0 to π (one period), the graph of $y = \tan x$ has these characteristics:

At 0, tan x is zero.

From 0 to $\pi/2$, the graph increases from 1 to infinity.

At $\pi/2$, the graph increases without limit.

From $\pi/2$ to π, the graph increases from – infinity to 0.

Graph of $y = \tan x$ from interval 0 to π

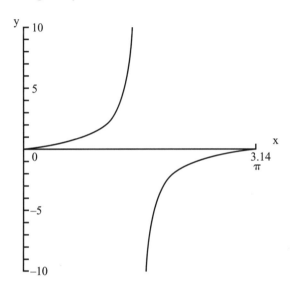

• *Cotangent*, $y = \cot x$, is an odd function, and its graph is symmetric with respect to the origin.

The period of $y = \cot x$ is π, or 180°.

The domain of $y = \cot x$ includes all real numbers of x, except $x = n\pi$, where n is an integer.

The range of $y = \cot x$ includes real numbers from – to + infinity.

The graph has vertical asymptotes at $x = n\pi$ and x-intercepts at $x = \pi/2 + n\pi = (\pi/2)(1 + 2n)$. The graph has no amplitude because no maximum or minimum values exist.

Graph of y = cot x

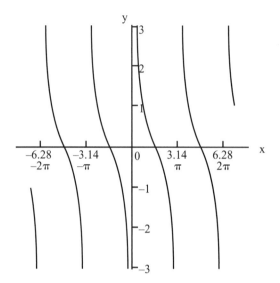

• *Transforming the graphs of y = tan x and y = cot x* can be achieved by adding or multiplying constants in the functions. These constants result in the graphs being shifted, flipped, stretched, or squeezed. For example: (1) A negative sign in front of tan x, giving y = –tan x, will flip the graph upside down; (2) Adding a constant, y = tan(x + c), will shift the graph right or left; (3) Multiplying by a constant, y = c tan x, will shrink or stretch vertically; or (4) Adding a constant to the function, y = c + tan x, will shift the graph up or down.

• **Example:** Graph y = –2 tan x.

The 2 vertically stretches the graph and the – sign flips the graph upside down. The graph is depicted as:

Graph of y = –2 tan x

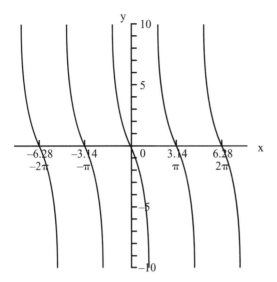

5.7 Chapter 5 Summary and Highlights

• In circular motion, a point or particle moving in a circular path around the perimeter of a circle of radius one can be mapped using cosine and sine. A particle moving around a circle can be translated into a particle moving along the *sine curve*.

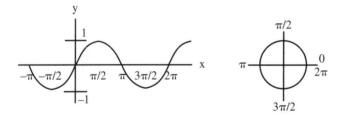

Graph of sine curve – projection of circular motion onto Y-axis

Note that the graph of the cosine curve is a projection of circular motion onto X-axis.

• Graphs of sine and the other trigonometric functions can be drawn using a graphing calculator or graphing software, in which the equation is entered and the calculator or software creates the graph.

• The graphs of periodic functions possess a repeating pattern in which the values of a function repeat over and over. Sine and cosine are often considered to be the most important periodic functions. The graphs of sine and cosine are referred to a *sinusoids* and describe numerous physical phenomena such as sound, electricity, motion of a vibrating object, harmonic waves, water waves, temperature variation, a mass on a spring, and electromagnetic waves and radiation, including light, radio waves, and X-rays.

• Superimposing $y = \sin x$ and $y = \cos x$ illustrates the shift of $\pi/2$ or 90° for the two functions:

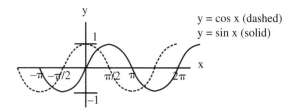

y = cos x (dashed)
y = sin x (solid)

• Graphs of trigonometric functions can be transformed and modified by multiplying and adding constants in their equations. The graphs of sine and cosine can be transformed by stretching or shrinking in the x-direction changing the period, increasing or decreasing the amplitude, shifting up, down, or sideways, and reflecting across the axes. Transformations can be made to the generalized equations
$y = d + a \sin(bx + c)$ and $y = d + a \cos(bx + c)$

    a is the *amplitude* given by the absolute value of a, or $|a|$.

    d is the *vertical shift*.

    (bx + c) is used to determine the *phase shift* and *period*, where $-c/b$ is the *phase shift* and $2\pi/b$ is the *period*.

• *Cosecant* and *secant* are the reciprocal functions of sine and cosine, where csc x = 1/sin x and sec x = 1/cos x. Cosecant and secant are also periodic; however, they are not continuous and their graphs are undefined where sine and cosine have values of zero. Vertical asymptotes occur in y = csc x at x-intercepts of y = sin x, and similarly, vertical asymptotes occur in y = sec x at x-intercepts of y = cos x.

• *Tangent* and *cotangent* are reciprocal functions such that tan x = 1/cot x = sin x/cos x and cot x = 1/tan x = cos x/sin x. The graph of tan x is undefined with vertical asymptotes at cos x = 0, or x = $\pi/2 + n\pi$, and has x-intercepts when sin x = 0. Cot x is the reciprocal of tan x and has a value of zero where tan x is undefined and is undefined at values where tan x is zero.

Chapter 6

Inverse Trigonometric Functions

• Inverses of the six trigonometric functions $y = \sin x$, $y = \cos x$, $y = \tan x$, $y = \sec x$, $y = \csc x$, and $y = \cot x$ exist within defined intervals. It is necessary to find an inverse of a trigonometric function when the value y of the function, such as $y = \sin x$, is known but the value of x, which may be an angle or a real number, is not known. To determine x, the inverse of the function must be calculated. In addition, to find an inverse of a trigonometric function, the function must be limited to a defined interval where there is a one-to-one correspondence between the domain (x) and range (y) values. As we will see in the following sections, the six trigonometric functions in their natural form are not true one-to-one functions and therefore must be limited to an interval in order to define an inverse function.

6.1 Review of General Inverse Functions

• Remember from Section 4.1: A *function* is a relation, rule, expression, or equation that associates each element of a *domain set* with its corresponding element in the *range set*. For a relation to be a function, there must be *only one element or number in the range set for each element or number in the domain set.* The domain set and range set can be expressed as (x, f(x)), or (x, y), pairs. Consider the following:

$\quad$ F = (2, 4), (3, 9), (4, 16) $\quad$ where F is *a function.*

$\quad$ M = (2, 5), (2, –5), (4, 9) $\quad$ where M is *not a function.*

M is not a function because the number 2 in the domain set corresponds to more than one number in the range set.

• The **domain set** is the initial set, and the **range set** is the set that results after a function is applied. Domain set → function f( ) → range set.

For example:

$\quad$ domain set x = {2, 3, 4}

$\quad$ through function $f(x) = x^2$, $f(2) = 2^2$, $f(3) = 3^2$, $f(4) = 4^2$

$\quad$ to range set f(x) = {4, 9, 16}.

• Graphs of functions only have *one value of y for each x value:*

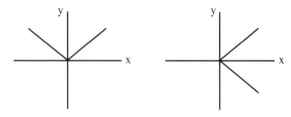

Graph is a function $\qquad\qquad$ Graph is not a function

The graph on the left is not, however, a one-to-one function. In general, a *function* is one-to-one if each domain value corresponds to only one range value, and each range value corresponds to only one domain value. For all functions, each domain value corresponds to only one range value; however, the condition that each range value corresponds to only one domain value is not true for some functions. If a vertical line can be drawn that passes through a function more than one time, there is more than one y value for a given x value and the graph is not a function. This is called the *vertical line test.*

• **Inverse functions** are functions that result in the same value of x after the operations of the two functions are performed. In inverse functions, the operations of each function are the reverse of the other function.

Notation for inverse functions is $f^{-1}(x)$, (where $^{-1}$ is *not* an exponent). If $f(x) = y$, then $f^{-1}(y) = x$. If f is the inverse of g, then g is the inverse of f. A function has an inverse if its graph intersects any *horizontal line* no more than once. (Graph of function above intersects twice in the range shown.)

• Two functions f and g are defined to be *inverse functions* if: $f(g(x)) = x$ for all x in the domain of g and $g(f(x)) = x$ for all x in the domain of f. This definition requires that the range of g is contained in the domain of f and that the range of f is contained in the domain of g. Therefore, $f(g(x)) = g(f(x)) = x$.

• An *inverse of a function has its domain and range equal to the range and domain, respectively, of the original function*. If $f(x) = y$, then $f^{-1}(y) = x$. For a function $f(x, y)$ that has only one y value for each x value, then there exists an inverse function represented by $f^{-1}(y, x)$. For example, *reversing the ordered pairs in function* $f(x, y) = \{(0, 3), (2, 4), (3, 5)\}$ results in the inverse function $f^{-1}(y, x) = \{(3, 0), (4, 2), (5, 3)\}$. Therefore, the domain of f equals the range of f^{-1}, and the range of f equals the domain of f^{-1}.

• The equation $y = x - 2$ defines a function because there is exactly one value of y for each value of x. The values of x that can be used are those real numbers that will produce a single real number for y. In this equation we can replace x with any real number and get a real-number value for y, so the domain of this function is the set of all real numbers. The range of this function is also the set of all real numbers. Because there is only one x value for any value of y, this function is a *one-to-one function*. Every one-to-one function has an inverse function that will operate the other way. The inverse function can be found by solving the original equation for x. For example, if $y = x - 2$ is the original equation, then solving for x gives the inverse: $x = y + 2$.

• In general, to find the inverse function of a one-to-one function defined by an algebraic equation in which y is a function of x:

First, solve the original equation for x.

Then, it is sometimes customary to replace every x with y and every y with x (that is, exchange x and y).

For example, for the original function $y = x - 2$,

    solving for x gives $x = y + 2$

    then replacing x with y and y with x gives the inverse function:

    $y = x + 2$

To check whether we have found the inverse function we can see if it results in the same value of x after the operations of the two functions are performed. If we choose x values of 2 and 4 in original function $y = x - 2$, the resulting y values are 0 and 2. If we substitute these values of 0 and 2 for x in the inverse function $y = x + 2$, the resulting y values are 2 and 4, which are the original input values.

• When functions f and f^{-1} are inverse functions, then they will return to the first value. Consider another example: If $y = f(x) = 2x - 1$ and $x = f^{-1}(y) = (y + 1)/2$ are inverses, and if $x = 3$, then by substituting for x: $f(3) = 2(3) - 1 = 5$
By substituting 5 into inverse function: $f^{-1}(5) = (5 + 1)/2 = 3$
which results in the starting point.

• Not all functions have inverses. If a function has more than one solution, it does not have an inverse. For a function $f(x) = y$, only one x can result, $x = f^{-1}(y)$. If there is more than one solution for $f^{-1}(y)$, it will not be the inverse of $f(x) = y$.

• *Graphs of inverse functions* are mirror images or reflections across the $y = x$ line, or in the graph below, the $x = y$ line. For example, if $y = f(x) = 2x$, then $x = (1/2)y$. The slopes are $(y_2 - y_1)/(x_2 - x_1) = (dy/dx) = 2$ and $(x_2 - x_1)/(y_2 - y_1) = (dx/dy) = 1/2$, where d represents the derivative.

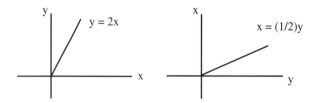

We can rewrite these equations using x and y as the independent and dependent variables as $y = 2x$, $y = x/2$, and $y = x$, and graph them:

Graph of $y = 2x$ (top line), $y = x/2$ (lower line), and $y = x$ (middle)

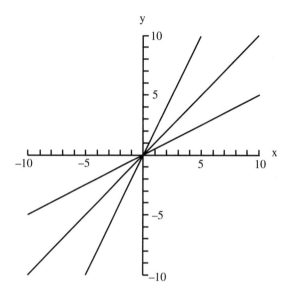

The top more vertical line is $y = 2x$, the middle line is $y = x$, and the bottom more horizontal line is $y = x/2$. It is clear that $y = 2x$ and $y = x/2$ are reflections across the $y = x$ line.

• To illustrate again that the graph of a function and its inverse are reflections of each other in the line $y = x$, consider the function $y = 2x - 1$ and its inverse $y = x/2 + 1/2$. Note that when the original equation was solved for x to obtain its inverse equation, then x was substituted for y and y substituted for x. The graph of $y = 2x - 1$ and its inverse $y = x/2 + 1/2$ is drawn using a graphing calculator type of software:

Graph of y = 2x–1 (more vertical line), y = x/2 + 1/2 (more horizontal line), and y = x (middle line)

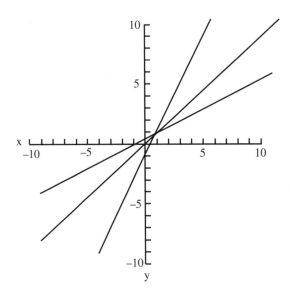

The top more vertical line is $y = 2x - 1$, the middle line is $y = x$, and the bottom more horizontal line is $y = x/2 + 1/2$. It is clear that the graph of function $y = 2x - 1$ and the graph of its inverse function $y = x/2 + 1/2$ are reflections of each other in the line $y = x$.

• In mathematical calculations it is often useful to develop an inverse function, because an inverse function does the exact opposite of the original function. Examples of inverse functions include the common logarithm function $y = \log x$ which is the inverse function for the exponential function $x = 10^y$. Examples of functions and their inverses include:

$$2 = \log 100, \qquad 100 = 10^2$$
$$5 = \log 100{,}000, \qquad 100{,}000 = 10^5$$
$$1 = \log 10, \qquad 10 = 10^1$$
$$0.3010 = \log 2, \qquad 2 = 10^{0.3010}$$
$$z = x^2 \text{ is the inverse of: } x = \sqrt{z} \text{ or } x = z^{1/2}$$
$$z = e^x \text{ is the inverse of: } x = \ln z$$
$$z = a^x \text{ is the inverse of: } x = \log_a z$$

For example, if we have the function $y = x^3$, then the inverse function is $x = \sqrt[3]{y}$, or equivalently $x = (y)^{1/3}$. Examples of these functions are:

$$8 = 2^3, \qquad 2 = \sqrt[3]{8} = (8)^{1/3}$$
$$27 = 3^3, \qquad 3 = \sqrt[3]{27} = (27)^{1/3}$$

• If two functions are inverses of each other, either function can be considered to be the original function and the other function is its inverse.

6.2 Inverse Trigonometric Functions

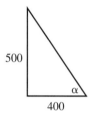

500

400

• If two sides of the above right triangle are 500 and 400, we know that the relationship between the angle α and the sides is tan α = opposite/adjacent, or tan α = 500/400. In order to find the angle α, we need the inverse of tangent, or arctan 500/400 = α, or equivalently arctan 5/4 = α. (This is discussed in detail in the following sections.) The value of α can be found using a calculator with inverse trigonometric function keys, graphs, and reference triangles, or by working backward and interpolating values in a trigonometric table.

Defined Intervals

• Like trigonometric functions, inverse trigonometric functions are *periodic* and their [unrestricted] inverses are actually **relations** that are multivalued, meaning there is more than one value in the range for certain values in the domain. This can be observed by viewing the graphs of the six trigonometric functions. For example, consider the graph of $y = \sin x$:

Graph of y = sin x

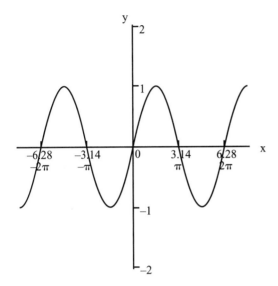

It is clear that different x values have the same y values. If a horizontal line is drawn at or between $y = 1$ and $y = -1$, the curve of $y = \sin x$ will cross the line multiple times. (Remember $\pi \approx 3.14159$.)

• For an ***inverse function*** to exist, a function must be one-to-one so that no two different domain values yield the same range value. When an inverse trigonometric function is considered, the original trigonometric function is defined over a specific ***interval*** in the domain where it is a single-valued, one-to-one function. For $y = \sin x$ the interval or region where it is a one-to-one function is $-\pi/2 \le x \le \pi/2$ and $-1 \le y \le 1$.

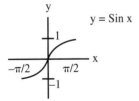

y = Sin x

In this interval each x value corresponds to only one y value.

• Because the trigonometric functions $y = \sin x$, $y = \cos x$, and $y = \tan x$ are not one-to-one functions, before we can define their inverses, their domains must be restricted so that each y-value corresponds to one and only one x-value. In order for the inverse to be a function, the range of the inverse must be limited to an *interval* where there is a one-to-one correspondence between domain and range values. The intervals of each of the six inverse trigonometric functions are generally agreed upon by mathematicians.

• The inverse trigonometric functions are each described in detail in the following sections. When trigonometric functions are defined in an interval where they are one-to-one functions, the inverse identities for sine, cosine, and tangent are written using capital letters.

• *Notation* for inverse sine, written y = arcsin x or y = $\sin^{-1}x$, and inverse cosine, written y = arccos x or y = $\cos^{-1}$ x, represents *inverse sine of x* and *inverse cosine of x*. The $^{-1}$ for an inverse function represents the inverse function and is *NOT an exponent*. When $^{-1}$ is an exponent, it designates the *reciprocal*, $(\sin x)^{-1}$ = 1/sin x, which is a completely different concept than the inverse function. It is important to distinguish between $\sin^{-1}x$, which is the arcsin x or inverse of sin x, and $(\sin x)^{-1}$, which is 1/sin x the reciprocal. *Capital letters* in the trigonometric functions and inverse trigonometric functions are often used to signify that the interval over which the function is considered is limited to a one-to-one correspondence between domain and range values. The word *arc* in arcsin, arccos, arctan, arcsec, arccsc, and arccot, the equation, such as y = arcsin x, represents *y is an angle whose sine is x*.

Determining Inverse Functions

• To *calculate* values for any of the inverse trigonometric functions, a calculator with trigonometric keys or on-line resources are often used. If these are not available, trigonometric tables can be used backwards. In some cases, you may be familiar with the value of a trigonometric function. For example, if you need to calculate arctan 1, then what you are looking for is *an angle whose tangent is equal to 1*. If you know that tan(45°) = tan(π/4) = 1, then arctan(1) = 45°, or π/4 radians. When there are values of inverse functions that are unfamiliar, such as arctan(4/3) = arctan(1.3333), and you don't have a calculator, then it is possible to look in a trigonometric table for an angle α such that tan α = 1.3333. In this case you may find values for tan 53° = 1.3270 and tan 54° = 1.3764 but no value 1.3333. Because arctan(1.3270) = 53° and arctan(1.3764) = 54°, then arctan(1.3333) must be somewhere between 53° and 54°. To be more exact, an interpolation method can be used. The results for inverse trigonometric functions may be expressed in either degrees or radians. In some cases, inverses can be found using graphs and reference triangles.

Graphs

• In the graph of an *inverse trigonometric relation*, which is the trigonometric "function" not limited to an interval of one-to-one domain and
range values, the roles of x and y are interchanged. For example, the
graph of y = arcsin x is the graph of x = sin y with the roles of x and
y interchanged. Because x and y are interchanged, the graph of y = arcsin x
is a sine curve that can be *drawn on the y axis instead of the x axis*.
Similarly, the graphs of the other inverse trigonometric relations are those
of the corresponding trigonometric functions except that the roles of x and
y are interchanged. The graphs of the trigonometric functions and their
inverse functions are limited to the intervals listed in the table below.
Consider the graph of the function y = Sin x and its inverse y = Arcsin x:

**Graph of y = Arcsin x (backward S), y = Sin x (forward S), and y = x
(straight line)**

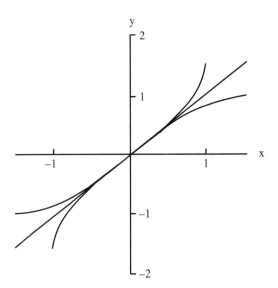

The interval of *principal values* depicted is:

Sin x $-\pi/2 \leq x \leq \pi/2$ $-1 \leq y \leq 1$
Sin^{-1}x $-1 \leq x \leq 1$ $-\pi/2 \leq y \leq \pi/2$
(note that $\pi/2 \approx 1.5708$)

If the arcsine curve was extended beyond its interval, it would continue
along the y-axis as y = sin x continues along the x-axis. The graph of an
inverse trigonometric function can also be obtained by reflecting the

graph of the trigonometric function, over its restricted domain, across the line $y = x$, which is depicted in the graph. The graphs of each of the six trigonometric functions and their inverses are in the following sections. Graphs can be easily drawn using a graphing calculator or software.

Table of Domain and Range Values

• The defined intervals of *domain* and *range* values of the inverse trigonometric functions are part of their definitions. These domain and range values of trigonometric functions and their inverses are:

Function	**Domain**	**Range**
Sin x	$-\pi/2 \leq x \leq \pi/2$	$-1 \leq y \leq 1$
Sin^{-1}x	$-1 \leq x \leq 1$	$-\pi/2 \leq y \leq \pi/2$
Cos x	$0 \leq x \leq \pi$	$-1 \leq y \leq 1$
Cos^{-1}x	$-1 \leq x \leq 1$	$0 \leq y \leq \pi$
Tan x	$-\pi/2 < x < \pi/2$	$-\infty < y < \infty$
Tan^{-1}x	$-\infty < x < \infty$	$-\pi/2 < y < \pi/2$
Cot x	$0 < x < \pi$	$-\infty < y < \infty$
Cot^{-1}x	$-\infty < x < \infty$	$0 < y < \pi$
Sec x	$0 \leq x \leq \pi$, x not $\pi/2$	$-\infty < y \leq -1, 1 \leq y < \infty$
Sec^{-1}x	$-\infty < x \leq -1, 1 \leq x < \infty$	$0 \leq y < \pi$, y not $\pi/2$

Secant and Arcsecant are also defined as:

Sec x	$-\pi \leq x \leq 0$, x not $-\pi/2$	$-\infty < y \leq -1, 1 \leq y < \infty$
Sec^{-1}x	$-\infty < x \leq -1, 1 \leq x < \infty$	$-\pi \leq y \leq 0$, y not $-\pi/2$
Csc x	$-\pi/2 \leq x \leq \pi/2$, x not 0	$-\infty < y \leq -1, 1 \leq y < \infty$
Csc^{-1}x	$-\infty < x \leq -1, 1 \leq x < \infty$	$-\pi/2 \leq y \leq \pi/2$, y not 0

• Trigonometric functions are defined using degree and radian measurements (with angle domains) and in real number values. Circular functions are defined with real number domains and real number ranges. Both definitions are closely related, and every real number in the domain of a circular function can be associated with an angle in degree or radian measure, and angles can be associated with real numbers. The confusion occurs because circular functions are also referred to as trigonometric functions, which results in two sets of trigonometric functions, one with angle domains in radian or degree measure and the other with real number domains. This situation also exists with inverse trigonometric functions. The *inverse circular functions* have real number domains and ranges, and *inverse trigonometric functions* have angle ranges in degree or radian measure.

6.3 Inverse Sine and Inverse Cosine

Unrestricted Inverse Sine and Cosine Relations in a Coordinate System

• The domain for the unrestricted *sine relation* is the set of all angles, and the range of the sine relation is the set of all real numbers including and between −1 and +1. Because sine is *multivalued*, there is not a one-to-one correspondence between domain and range values. For example, the sine of 30° is 1/2 and the sine of 150° is also 1/2, where 30° and 150° are domain values. This can be illustrated in a coordinate system with standard position angles:

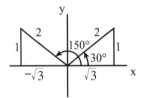

sin 150° = 1/2 and sin 30° = 1/2

Note: a triangle with hypotenuse 1, vertical leg is 1/2 = sin 30

• Because numerous coterminal angles can exist in a coordinate system, 30° plus once around counterclockwise is 390°, 30° plus twice around counterclockwise is 750°, and so on. Also, 150° plus once around counter-clockwise is 510°, 150° plus twice around counterclockwise is 870°, and so on. Therefore, numerous positive angles whose sine is 1/2 may exist, such as 30°, 390°, 750°, and 150°, 510°, 870°, and so on. This may be seen also by looking at the graph of the unrestricted function in the preceding section.

• Suppose we know the range value of 1/2 and want to find the angle α, or 1/2 = sin α. In this case we are looking for *the angle whose sine is 1/2*, or equivalently the *inverse sine of 1/2*, which is written:

arcsin 1/2 = ? or sin^{-1} 1/2 = ?

The standard position angles in a coordinate system are the same, and coterminal angles can exist.

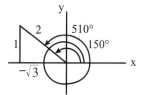 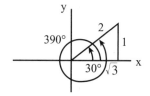

arcsin $1/2 = 150° = 510°$ and   arcsin $1/2 = 30° = 390°$

arcsin 1/2 includes numerous angles:

   arcsin $1/2 = 30°, 390°, 750°, ..., 150°, 510°, 870°, ...$

It is clear that there are multiple angles for every domain value of the inverse sine when it is unrestricted, so the unrestricted inverse sine is a relation rather than a function. The *domain* for the *arcsin relation* is the set of all real numbers including and between −1 and +1, and the *range* is the set of all angles.

• Similarly, the *domain* for the unrestricted *cosine relation* is the set of all angles, and the *range* of cosine is the set of all real numbers including and between −1 and +1. Like sine, cosine is not a one-to-one function. For example, the cosine of 60° is 1/2 and the cosine of 300° is also 1/2. This can be illustrated in a coordinate system with standard position angles:

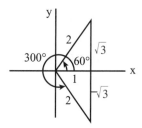

cos 60° = 1/2 and cos 300° = 1/2

Note: a triangle with hypotenuse 1, horizontal leg is 1/2 = cos 60

• Because numerous coterminal angles can exist in a coordinate system, 60° plus once around counterclockwise is 420°, 60° plus twice around counterclockwise is 780°, and so on. Therefore, there are numerous positive angles whose cosine is 1/2, such as 60°, 420°, 780°, ..., and 300°, 660°, 1020°, and so on. Suppose we want to find the angle whose cosine is 1/2, which is written arccos 1/2 = ? or cos^{-1} 1/2 = ?, the standard position angles in a coordinate system are the same. It is also clear that coterminal angles can exist.

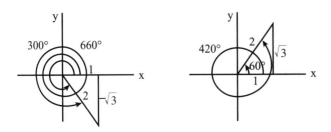

arccos $1/2 = 300° = 660°$ and arccos $1/2 = 60° = 420°$

arccos $1/2$ includes numerous angles:

   arccos $1/2 = 60°, 420°, 780°, ..., 300°, 660°, 1020°, ...$

It is clear that there are multiple angles for every domain value of the *inverse cosine*.

Limited Inverse Sine and Cosine Functions in a Coordinate System

• Generally, when the inverse sine or inverse cosine of a number is needed, only one value is preferred, and the desired result will be a *function*. To obtain the *inverse sine function or inverse cosine function*, only certain angles can be considered so there is a one-to-one correspondence between domain and range values.

• In order for the **inverse sine** to be a function, there must be only one angle for every domain value of the inverse sine, and this occurs when the range of the inverse sine is restricted to the *first and fourth quadrants* of a coordinate system. This corresponds to angles including and between $-90°$ and $+90°$ (or $-\pi/2$ to $\pi/2$) where *positive values* of the sine are associated with a *first-quadrant angle*, and *negative values* of the sine are associated with a *fourth-quadrant angle*.

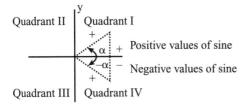

The *domain* for the **Arcsine function** is the set of all real numbers including and between -1 and $+1$, and the *range* is the set of all angles including and between $-90°$ and $+90°$ (or $-\pi/2$ to $\pi/2$). This corresponds to the *sine function* in the interval:

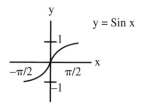

y = Sin x

- In order for **inverse cosine** to be a function, we consider only angles including and between 0° and +180° in the *first and second quadrants*, where positive values of the cosine correspond with a *first-quadrant angle*, and negative values of the cosine correspond with a *second-quadrant angle*.

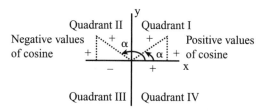

The *domain* for the **Arccosine function** is the set of all real numbers including and between −1 and +1, and the *range* is the set of all angles including and between 0° and +180° (or 0 to π). On a graph, this corresponds to the *cosine function* in the interval:

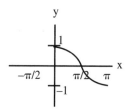

- In the example of a domain value of 1/2 for the *inverse sine function*, only a single angle including and between −90° and +90° can exist. Similarly, for a domain value of −1/2 for the inverse sine function, only a single angle including and between −90° and +90° can exist.

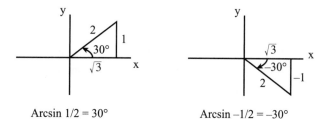

Arcsin 1/2 = 30°   Arcsin −1/2 = −30°

• Similarly, in the example of a domain value of 1/2 for the *inverse cosine function*, only a single angle including and between 0° and +180° can exist. For a domain value of −1/2 for the inverse cosine function, only the single angle including and between 0° and +180° can exist.

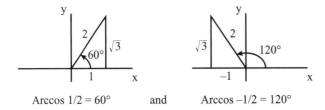

Arccos 1/2 = 60°   and   Arccos −1/2 = 120°

Inverse Sine and Cosine in Graphical Form

• In graphical form it is clear that the sine and cosine functions are not one-to-one. This can be observed in the graph of sine and cosine:

Graph of y = sin x (black curve) and y = cos x (gray curve)

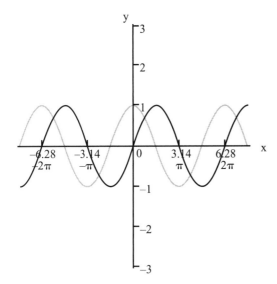

By restricting the domain of the sine and cosine functions so that only one x value corresponds to only one y value, one-to-one functions can be defined.

- In the *graph of an inverse trigonometric function*, the roles of x and y are interchanged. Because x and y are interchanged, the graph of y = arcsin x is a sine curve that can be drawn on the vertical axis instead of the horizontal axis. The graph of a function and its inverse are reflections of each other in the line y = x. In fact, the graph of an inverse trigonometric function can be found by reflecting the graph of the corresponding trigonometric function over its restricted domain across the line y = x. For example, the graph of y = Sin x is a mirror image of Sin⁻¹y = x reflected across the y = x line. For y = Sin x, the interval or region where it has a one-to-one function is $-\pi/2 \le x \le \pi/2$ and $-1 \le y \le 1$.

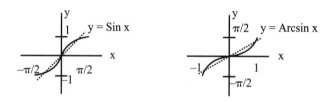

Depicted on one graph:

Graph of y = Arcsin x (backward S), y = sin x (forward S), and y = x (straight line)

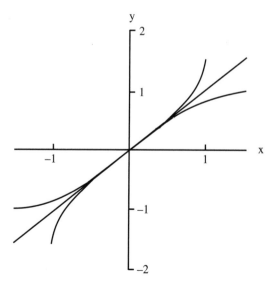

The interval of *principal values* depicted is:

Sin x $-\pi/2 \leq x \leq \pi/2$ $-1 \leq y \leq 1$

Sin^{-1} x $-1 \leq x \leq 1$ $-\pi/2 \leq y \leq \pi/2$

(note that $\pi/2 \approx 1.5708$)

If the Arcsine curve was extended beyond its interval, it would continue along the Y-axis as y = Sin x continues along the X-axis. Over the interval (–1, 1), the inverse sine function is increasing. Its x-intercept is 0 and its y-intercept is 0. Its graph is symmetric with respect to the origin. In general, $\text{sin}^{-1}(-x) = -\text{sin}^{-1} x$, and it is an odd function.

• For y = Cos x the interval or region where it is a one-to-one function is $0 \leq x \leq \pi$ and $-1 \leq y \leq 1$.

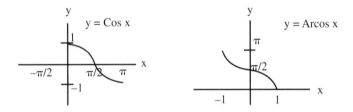

Depicted on one graph:

Graph of y = Arccos x (upper curve), y = Cos x (lower curve), and y = x (straight line)

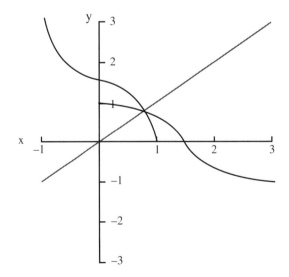

The interval of *principal values* depicted is:

Cos x $0 \le x \le \pi$ $-1 \le y \le 1$

Cos⁻¹x $-1 \le x \le 1$ $0 \le y \le \pi$

(note that $\pi \approx 3.14159$; $\pi/2 \approx 1.5708$)

Over the interval $(-1, 1)$, the inverse cosine function is decreasing. Its x-intercept is 1 and its y-intercept is $\pi/2$. The graph is neither symmetric with respect to the y-axis nor symmetric with respect to the origin.

• Graphs of inverse trigonometric functions can be obtained by using a calculator in radian mode and selecting domain values from -1 to 1, then plotting the resulting domain/range pairs. A graph can most easily be obtained using a graphing utility in a graphing calculator or computer software.

• It is also possible to graph y = Arcsin x by choosing the coordinates of a few points on the graph of the restricted sine function, reversing their order, then using the points to sketch the graph of the inverse sine function. For example, because $(-\pi/2, -1)$, $(0, 0)$ and $(\pi/2, 1)$ are on the graph of the restricted sine function, then $(-1, -\pi/2)$, $(0, 0)$, and $(1, \pi/2)$ are on the graph of the inverse sine function, and these three points can be used to draw a graph of the inverse function.

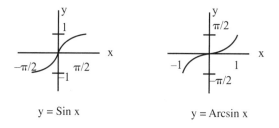

y = Sin x   y = Arcsin x

• Similarly, it is also possible to graph y = Arccos x by choosing the coordinates of a few points on the graph of the restricted cosine function, reversing their order, then using the points to sketch the graph of the inverse cosine function. For example, because $(0, 1)$, $(\pi/2, 0)$, and $(\pi, -1)$ are on the restricted cosine graph, then $(1, 0)$, $(0, \pi/2)$, and $(-1, \pi)$ are on the graph of the inverse sine function, and these three points can be used to draw a graph of the inverse function.

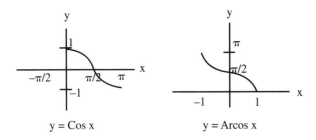

y = Cos x   y = Arcos x

• The graphs of the inverse trigonometric functions can also be revealed by drawing the function on a transparent sheet of paper, interchanging the X and Y axes, and turning it over and rotating it 90°.

• The value for an inverse sine or cosine function can be determined using a graph, a reference triangle, a calculator, on-line resources, or trigonometric tables backwards and interpolating.

• **Example:** Find y = Arccos(−1/2) and y = Arccos(1/2) using the graph of Arccosine.

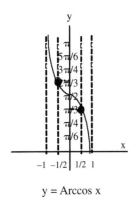

y = Arccos x

At x = −1/2 it is clear from the graph that y = 2π/3.
Therefore, Arccos(−1/2) = 2π/3.

At x = 1/2 it is clear from the graph that y = π/3.
Therefore, Arccos(1/2) = π/3.

• **Example:** Verify Arccos(−1/2) = 2π/3 using a calculator.

Set your calculator to radian mode. Calculate Arccos(–1/2) by entering –0.5 and pressing the cos⁻¹ key. The result is approximately 2.094. Next calculate $2\pi/3$. The result is also approximately 2.094. (It is important to make sure your calculator is in the proper mode.)

• **Example:** Verify Arccos(1/2) = $\pi/3$ using a reference triangle in a coordinate system.

By definition, y = Arccos(1/2) is equivalent to 1/2 = cos y, $0 \le y \le \pi$. A positive value of x and y in the range of 0 to π will fall in the first quadrant. Remember: cos y = 1/2 = adjacent / hypotenuse:

Therefore, this is a 30:60:90 triangle with the third side $\sqrt{3}$. In radians 60° is $\pi/3$ and so y = $\pi/3$. Therefore, Arccos(1/2) = $\pi/3$ radians.

• **Example:** Calculate Arcsin(0.2342).

Set your calculator to radian mode. Calculate Arcsin(0.2342) by entering 0.2342 and pressing the sin⁻¹ key. The result is 0.2364.

• Remember, **notation** for inverse sine is *arcsine* abbreviated *arcsin* and sin⁻¹ denotes *the angle whose sine is*, and the ⁻¹ in sin⁻¹ is NOT an exponent. The unrestricted relation arcsine has a lowercase a in arcsin and a lowercase s in sin⁻¹, and the *inverse function* uses capital Arcsin and Sin⁻¹. Similarly, for inverse cosine is *arccosine*, abbreviated *arccos* and cos⁻¹, where the *inverse function* uses capital Arccos and Cos⁻¹. The restricted values for inverse trigonometric functions are often referred to as *principal values*.

6.4 Inverse Tangent

• *Tangent* is equal to the ratio of sine to cosine, or tan x = sin x / cos x. If x is an angle α in a triangle or coordinate system, then tan α = sin α / cos α. When the angle α is equal to any odd-integer multiple of 90° (or $\pi/2$ where cosine is zero), such as α = ±90°, ±270°, ±450°, etc., cos α will be equal to zero, which will result in tan α being *undefined* (because division by zero is undefined). Therefore, the domain for tangent is the set of all angles other than ±90°, ±270°, ±450°, etc., and the range of tangent is the set of all real numbers.

• Like sine and cosine, tangent is not a one-to-one function. A function must always specify one unique value of the dependent variable for every value of the independent variable. Consider, for example, arctan 1; which could be equal to either $\pi/4$ or a multiple, $n\pi + \pi/4$ in radians, where n is an integer. The tangent of $\pi/4$, or 45°, is 1 and the tangent of 225° is also 1, where 45° and 225° are domain values. This can be illustrated in a coordinate system with standard position angles:

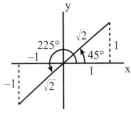

tan 225° = 1 and tan 45° = 1

• Because numerous coterminal angles can exist in a coordinate system, angles 45° plus once around counterclockwise is 405°, 45° plus twice around counterclockwise is 765°, and so on. Also, 225° plus once around counterclockwise is 585°, 225° plus twice around counterclockwise is 945°, and so on. Therefore, angles whose tangent is 1 may exist, such as 45°, 405°, 765°, ..., 225°, 585°, 945°, ...

• Suppose we know the range value of 1 and want to find the angle α, or $1 = \tan \alpha$. In this case, we are looking for *the angle whose tangent is 1*, or equivalently the ***inverse tangent of 1***, which is written: arctan 1 = ? or $\tan^{-1} 1 = ?$ The standard position angles in a coordinate system are the same for inverse tangent as for tangent, and coterminal angles can exist.

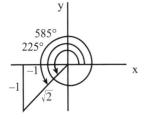   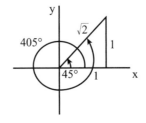

arctan 1 = 225° = 585°   and   arctan 1 = 45° = 405°

arctan 1 includes numerous angles:

arctan $1 = 45°, 405°, 765°, ..., 225°, 585°, 945°,...$

It is clear that there are multiple angles for every domain value of the inverse tangent when it is unrestricted, so the *unrestricted inverse tangent* is a *relation* rather than a function. The *domain* for the *arctan relation* is the set of all real numbers, and the *range* is the set of all angles with the exception of angles that have a cosine of zero, such as $\pm 90°, \pm 270°, \pm 450°$ (because tan α = sin α / cos α).

Limited **inverse tangent function in a coordinate system**

• Generally, when the ***inverse tangent*** of a number is needed, only one value is preferred and the desired result will be a function. In order for the *inverse tangent* to be a function there must be only one value or angle for every domain value of the inverse tangent, and this occurs when the range of the inverse tangent is restricted to the *first and fourth quadrants* of a coordinate system. This corresponds to angles including and between $-90°$ and $+90°$, where *positive values* of the tangent are associated with a *first-quadrant angle*, and *negative values* of the tangent are associated with a *fourth-quadrant angle*.

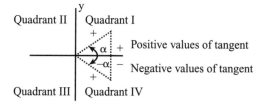

• The *domain* for the *Arctan function* is the set of all real numbers, and the range is the set of all angles between $-90°$ and $+90°$ (or $-\pi$ and $+\pi$). In the example of a domain value of 1 for the *inverse tangent function*, only a single angle between $-90°$ and $+90°$ can exist. Similarly, for a domain value of -1 for the inverse tangent function, only a single angle between $-90°$ and $+90°$ can exist.

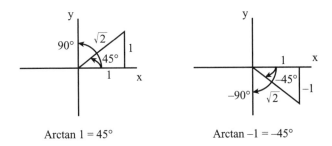

Arctan 1 = 45°   Arctan −1 = −45°

- Restricting the domain of the function y = tan x to the interval (−π/2, π/2) (or between −90° and +90°) yields a one-to-one function. In this restricted interval, the value of π/4 (or 45°) would be in the interval where arctan is a function and would be the principal value of the function Arctan 1.

- Inverse tangent or *arctangent* is abbreviated *arctan* and $\tan^{-1}$, where the *inverse function* uses capital Arctan and Tan^{-1}.

Inverse Tangent in Graphical Form

- The graph of y = tan x illustrates that the tangent function is not a one-to-one function.

Graph of y = tan x

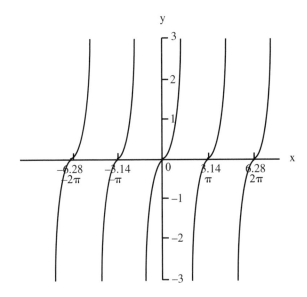

By restricting the domain so that only one x value corresponds to only one y value, a one-to-one function can be defined.

• A function must always specify one unique value of the dependent variable for every value of the independent variable. Considering arctan 1, it could be equal to either $\pi/4$ or a multiple, $n\pi + \pi/4$ in radians, where n is an integer. Restricting the domain of the function y = tan x to the interval $(-\pi/2, \pi/2)$ yields a one-to-one function. In this case, the value of $\pi/4$ would be in the interval where arctan is a function and would be the principal value of the function Arctan 1. The lines $y = \pi/2$ and $y = -\pi/2$ are *horizontal asymptotes* of the graph. (See graphs of tangent and arctangent below.) The restricted tangent function is used to define the inverse tangent function.

• In the *graph of an inverse trigonometric function*, the roles of x and y are interchanged. By interchanging the roles of x and y, we obtain the *inverse tangent function* $y = \text{Tan}^{-1} x$ or y = Arctan x. The graph of a function and its inverse are reflections of each other in the line y = x. For y = Tan x, the interval or region where it has a one-to-one function is $-\pi/2 < x < \pi/2$ and $-\infty < y < \infty$.

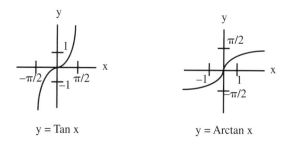

y = Tan x   y = Arctan x

Depicting y = Tan x and its inverse y = Arctan x on one graph:

Graph of y = Arctan x (forward S), y = Tan x (backward S), and y = x (straight line)

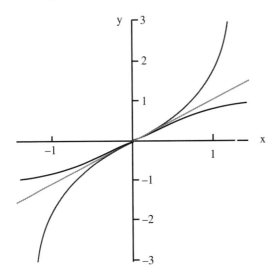

Depicting tangent, y = Tan x, and its inverse, y = Arctan x, on one graph we can see the interval of *principal values* as:

Tan x $-\pi/2 < x < \pi/2$ $-\infty < y < \infty$

Tan^{-1}x $-\infty < x < \infty$ $-\pi/2 < y < \pi/2$

(note that $\pi \approx 3.14159$; $\pi/2 \approx 1.5708$)

Tan x extends to infinity along the vertical axis in both directions, and Arctan x extends to infinity along the horizontal axis in both directions.

• Properties of the *inverse tangent function* include that it has a *domain* of negative infinity to positive infinity and a *range* of $-\pi/2$ to $\pi/2$. Over the domain $(-\infty, \infty)$, the inverse tangent function is increasing, with its x-intercept at 0 and its y-intercept at 0. The graph is symmetric with respect to the origin. As x approaches infinity, y approaches $\pi/2$ from below and the line y = $\pi/2$ forms a horizontal asymptote. As x approaches $-$ infinity, y approaches $-\pi/2$ from above, and the line y = $-\pi/2$ forms a horizontal asymptote. Arctangent is an odd function, and Arctan(–x) = – Arctan x.

• Graphs of inverse tangent and other trigonometric functions can be obtained using a calculator in radian mode, selecting domain values, and then plotting the resulting domain/range pairs. A graph can most easily be obtained using a graphing utility in a graphing calculator or computer

software. It is also possible to graph y = Arctan x by choosing the coordinates of a few points on the graph of the restricted tangent function, reversing their order, and then using the points to
sketch the graph of the inverse tangent function. For example, because
$(-\pi/4, -1), (0, 0)$, and $(\pi/4, 1)$ are on the graph of the y = Tan x, then
$(-1, -\pi/4), (0, 0)$, and $(1, \pi/4)$ are on the graph of the y = Arctan x, and
these three points can be used to draw a graph of the inverse function.
Note that the vertical asymptotes become horizontal asymptotes.

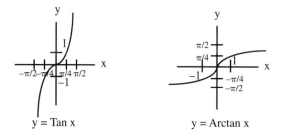

y = Tan x   y = Arctan x

• The value of an inverse tangent function can be determined using a
graph, a reference triangle, a calculator, on-line resources, or trigonometric
tables backwards.

• **Example:** Find Arctan(1) using a reference triangle in a coordinate
system.

By definition, y = Arctan(1) is equivalent to 1 = Tan y, $-\pi/2 < y < \pi/2$.
A positive value of x and y in the range of $-\pi/2$ to $\pi/2$ will fall in the first
quadrant. Remember: tangent = opposite/adjacent, and in this example
tan y = 1

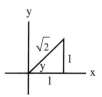

Therefore, this is a 45:45:90 triangle with the hypotenuse $\sqrt{2}$, and y is
45° and in radians y is $\pi/4$. Therefore, Arctan(1) = $\pi/4$ radians.

• **Example:** Calculate Arctan(1) in degree measure using a calculator. Set the calculator in degree mode. Calculate Arctan(1) by entering 1 and pressing the $\tan^{-1}$ key. The result is 45°.

• **Example:** Calculate Arctan(1) in radian measure using a calculator. Set the calculator in radian mode. Calculate Arctan(1) by entering 1 and pressing the $\tan^{-1}$ key. The result is 0.7854. Note that $\pi/4 \approx 0.7854$.

• **Example:** Calculate Arctan(–1) in degree measure using a calculator. Set the calculator in degree mode. Calculate Arctan(–1) by entering –1 and pressing the tan–1 key. The result is –45°. Depicted in a coordinate system this looks like:

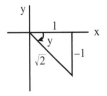

6.5 Inverse Cotangent, Inverse Secant, and Inverse Cosecant

• Remember that secant, cosescant, and cotangent are reciprocal identities of cosine, sine, and tangent:

secant = 1 / cosine, cosecant = 1/sine, cotangent = 1 / tangent

• In their natural unrestricted forms, cotangent, secant, and cosecant are not one-to-one functions. In order for cotangent, secant, and cosecant to produce inverse functions, they are restricted to defined intervals where they are one-to-one for x and y values. These intervals correspond to the arccotangent, arcsecant, and arccosecant functions in their restricted intervals:

Cot x	$0 < x < \pi$	$-\infty < y < \infty$
$\text{Cot}^{-1}x$	$-\infty < x < \infty$	$0 < y < \pi$
Sec x	$0 \leq x \leq \pi$, x not $\pi/2$	$-\infty < y \leq -1, 1 \leq y < \infty$
$\text{Sec}^{-1}x$	$-\infty < x \leq -1, 1 \leq x < \infty$	$0 \leq y < \pi$, y not $\pi/2$

Secant and Arcsecant are also defined as:

Sec x	$-\pi \leq x \leq 0$, x not $-\pi/2$	$-\infty < y \leq -1, 1 \leq y < \infty$
$Sec^{-1}x$	$-\infty < x \leq -1, 1 \leq x < \infty$	$-\pi \leq y \leq 0$, y not $-\pi/2$
Csc x	$-\pi/2 \leq x \leq \pi/2$, x not 0	$-\infty < y \leq -1, 1 \leq y < \infty$
$Csc^{-1}x$	$-\infty < x \leq -1, 1 \leq x < \infty$	$-\pi/2 \leq y \leq \pi/2$, y not 0

- The graph of **y = cot x** is not one-to-one:

Graph of y = cot x

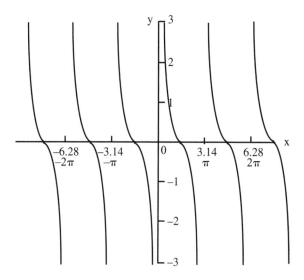

The intervals where cotangent is one-to-one and its inverse function arccotangent exists are:

Cot x	$0 < x < \pi$	$-\infty < y < \infty$
$Cot^{-1}x$	$-\infty < x < \infty$	$0 < y < \pi$

For y = Cot x, the interval or region where it is a one-to-one function is $0 < x < \pi$ and $-\infty < y < \infty$.

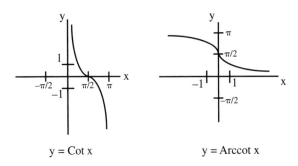

y = Cot x   y = Arccot x

• The graph of **y = sec x** is not one-to-one:

Graph of y = sec x

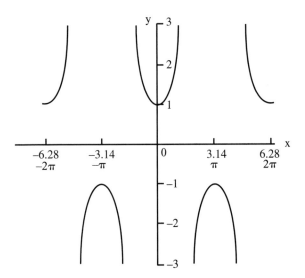

The intervals where secant is one-to-one and its inverse function arcsecant exists are:

Sec x $\qquad$ $0 \leq x \leq \pi$, x not $\pi/2$ $\qquad$ $-\infty < y \leq -1, 1 \leq y < \infty$

Sec^{-1}x $\qquad$ $-\infty < x \leq -1, 1 \leq x < \infty$ $\qquad$ $0 \leq y < \pi$, y not $\pi/2$

For y = Sec x, the interval or region where it is a one-to-one function is $0 < x < \pi$, except $\pi/2$.

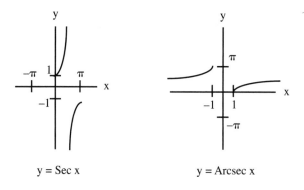

y = Sec x   y = Arcsec x

- The graph of **y = csc x** is not one-to-one:

Graph of y = csc x

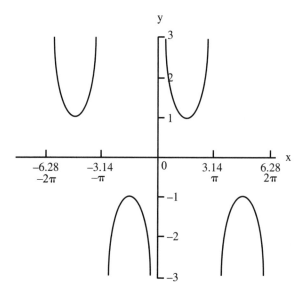

The intervals where cosecant is one-to-one and its inverse function arc-cosecant exists are:

Csc x	$-\pi/2 \leq x \leq \pi/2$, x not 0	$-\infty < y \leq -1, 1 \leq y < \infty$
Csc^{-1}x	$-\infty < x \leq -1, 1 \leq x < \infty$	$-\pi/2 \leq y \leq \pi/2$, y not 0

For y = Csc x, the interval or region where it is a one-to-one function is $-\pi/2 < x < \pi/2$, except zero.

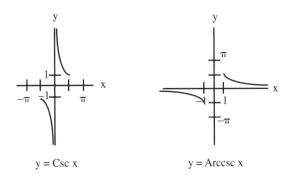

y = Csc x   y = Arccsc x

• To evaluate inverse trigonometric functions, a graph, on-line resources, or a calculator can be used. When a calculator is available that has keys for all six inverse trigonometric functions, using it is the easiest method for finding inverse values. Calculators usually have keys for sin, cos, tan, $\sin^{-1}$, $\cos^{-1}$, and $\tan^{-1}$, but many may not have keys for csc, sec, cot, $\csc^{-1}$, $\sec^{-1}$, and $\cot^{-1}$. To find sec x, csc x, and cot x using a calculator, the reciprocal identities can be used. These identities are:

$$\sec x = 1/\cos x, \qquad \csc x = 1/\sin x, \qquad \cot x = 1/\tan x$$

To evaluate the inverse functions $\csc^{-1} x$, $\sec^{-1} x$, and $\cot^{-1} x$, a calculator with $\sin^{-1}$, $\cos^{-1}$, and $\tan^{-1}$ keys can be used with the following *inverse identities*:

$\text{Cot}^{-1} x = \text{Tan}^{-1}(1/x)$ for x > 0

$\text{Cot}^{-1} x = \pi + \text{Tan}^{-1}(1/x)$ for x < 0

$\text{Sec}^{-1} x = \text{Cos}^{-1}(1/x)$ for x ≥ 1 or x ≤ −1

$\text{Csc}^{-1} x = \text{Sin}^{-1}(1/x)$ for x ≥ 1 or x ≤ −1

Derivations of the Inverse Identities

• *Derivation of inverse cotangent identities*:

$\text{Cot}^{-1} x = \text{Tan}^{-1}(1/x)$ for x > 0

$\text{Cot}^{-1} x = \pi + \text{Tan}^{-1}(1/x)$ for x < 0

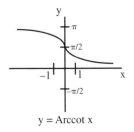

y = Arccot x

For the region of the Arccosine graph, $y = \text{Cot}^{-1} x$ where $x > 0$, the range of y is $0 < y < \pi/2$.

Rearranging $y = \text{Cot}^{-1} x$ gives: $\text{Cot } y = x$

The reciprocal identity Cotangent $= 1/\text{Tangent}$ in this region is:

$\text{Cot } y = 1/\text{Tan } y = x, 0 < y < \pi/2$

Rearranging $1/\text{Tan } y = x$: $\text{Tan } y = 1/x, 0 < y < \pi/2$

Taking the inverse to solve for y: $y = \text{Tan}^{-1}(1/x), 0 < y < \pi/2$

Therefore, $y = \text{Cot}^{-1} x = \text{Tan}^{-1}(1/x)$, for $x > 0$

For negative values of x:

$\text{Cot}^{-1} x = \pi + \text{Tan}^{-1}(1/x)$ for $x < 0$

which can be visualized on the following graph of $\text{Cot}^{-1} x$ and $\text{Tan}^{-1}(1/x)$:

Graph of $y = \text{Cot}^{-1} x$ (all of upper curve) and $y = \text{Tan}^{-1}(1/x)$ (lower + Rt. upper curve)

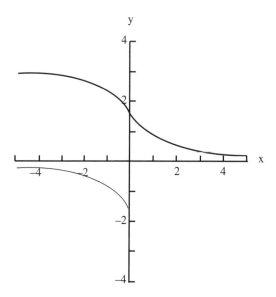

where Arccot x is all of upper curve above X-axis. Arctan(1/x) is lower curve plus right side of upper curve, which overlaps Arccot x for positive values of x and is shifted down by a value of π for negative values of x.

• *Derivation of inverse secant identity*:

$Sec^{-1} x = Cos^{-1}(1/x)$ for $x \geq 1$ or $x \leq -1$

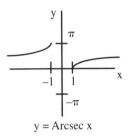

y = Arcsec x

The graph of Arcsecant $y = Sec^{-1} x$ is in the range $x \geq 1$ or $x \leq -1$.

Rearranging $y = Sec^{-1} x$ gives: $Sec\ y = x, 0 \leq y \leq \pi$, except $\pi/2$

The reciprocal identity Secant $= 1/Cosine$ is:

   $Sec\ y = 1/Cos\ y = x, 0 \leq y \leq \pi$, except $\pi/2$

Rearranging $1/Cos\ y = x$: $Cos\ y = 1/x, 0 \leq y \leq \pi$, except $\pi/2$

Taking the inverse to solve for y:

   $y = Cos^{-1}(1/x), 0 \leq y \leq \pi$, except $\pi/2$

Therefore, $y = Sec^{-1} x = Cos^{-1}(1/x), x \geq 1$ or $x \leq -1$

which can be visualized on the graph of $y = Sec^{-1} x$ and $y = Cos^{-1}(1/x)$. Note that in this graph Arcsec x is represented by the split curve and Arccos(1/x) is represented by the same curves, and they overlap each other for all values of x.

Graph of $y = Sec^{-1} x$ and $y = Cos^{-1}(1/x)$, which overlap

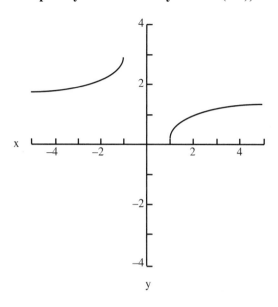

- *Derivation of inverse cosecant identity*:

$$Csc^{-1} x = Sin^{-1}(1/x) \quad \text{for } x \geq 1 \text{ or } x \leq -1$$

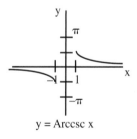

$y = \text{Arccsc } x$

The graph of Arccosecant $y = Csc^{-1} x$ is in the range $x \geq 1$ or $x \leq -1$.

Rearranging $y = Csc^{-1} x$ gives: $Csc\ y = x, -\pi/2 \leq y \leq \pi/2$, except 0

The reciprocal identity Cosecant $= 1/$Sine is:

$\quad Csc\ y = 1/Sin\ y = x, -\pi/2 \leq y \leq \pi/2$, except 0

Rearranging $1/Sin\ y = x$: $Sin\ y = 1/x, -\pi/2 \leq y \leq \pi/2$, except 0

Taking the inverse to solve for y:

$y = \text{Sin}^{-1}(1/x), -\pi/2 \leq y \leq \pi/2$, except 0

Therefore, $y = \text{Csc}^{-1} x = \text{Sin}^{-1}(1/x)$, $x \geq 1$ or $x \leq -1$

which can be visualized on the graph of $y = \text{Csc}^{-1} x$ and $y = \text{Sin}^{-1}(1/x)$:

Graph of $y = \text{Csc}^{-1} x$ and $y = \text{Sin}^{-1}(1/x)$, which overlap

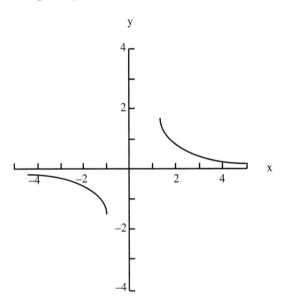

Arccsc x is represented by this split curve and Arcsin(1/x) is represented by the same curves, and they overlap each other for all values of x.

• The values of an inverse trigonometric function can be determined using a graph, a reference triangle, a calculator, on-line resources, or trigonometric tables backward and interpolating.

• **Example:** Find Arccot(−1) using a reference triangle in a coordinate system.

By definition, $y = \text{Arccot}(-1)$ is equivalent to $-1 = \text{Cot } y, 0 < y < \pi$. A negative value in the range of 0 to π will fall in the second quadrant. Remember that cotangent = adjacent/opposite, and in this example

$\text{Cot } y = -1/1 = -1$.

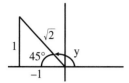

This is a 45:45:90 reference triangle with the hypotenuse $\sqrt{2}$. Angle y is measured from the positive X-axis and is $180° - 45° = 135°$ in degrees or $\pi - \pi/4 = 3\pi/4$ in radians. Therefore, Arccot$(-1) = 3\pi/4$ radians.

• **Example:** Calculate Arccot(-1) in radian measure using a calculator that does not have an cot^{-1} key.

Use the Arccot identity:

$Cot^{-1} x = Tan^{-1}(1/x)$ for x > 0
$Cot^{-1} x = \pi + Tan^{-1}(1/x)$ for x < 0

Set the calculator in radian mode. Calculate Arccot(-1) using negative x identity:

$$Cot^{-1} x = \pi + Tan^{-1}(1/x) = \pi + Tan^{-1}(1/(-1)) = \pi + Tan^{-1}(-1)$$

Enter -1 and press the tan^{-1} key, then add π. The result is 2.356. Note that $3\pi/4 \approx 2.356$.

• Example: Calculate Arccsc(1.150) in radian measure using a calculator that does not have an csc^{-1} key.

Use the Arccsc identity:

$Csc^{-1} x = Sin^{-1}(1/x)$ for x ≥ 1 or x ≤ −1

Set the calculator in radian mode. Calculate Arccsc(1.150) using identity:

$$Csc^{-1} x = Sin^{-1}(1/x) = Sin^{-1}(1/1.150)$$

Enter $(1/1.150)$ and press the sin^{-1} key. The result is 1.054.

• **Example:** Calculate Arcsec(-2.005) in radian measure using a calculator that does not have an cot^{-1} key.

Use the Arcsec identity:

$\text{Sec}^{-1}x = \text{Cos}^{-1}(1/x)$ for $x \geq 1$ or $x \leq -1$

Set the calculator in radian mode. Calculate Arcsec(–2.005) using identity:

$\text{Sec}^{-1}x = \text{Cos}^{-1}(1/x) = \text{Cos}^{-1}(1/(-2.005))$

Enter $(1/(-2.005))$ and press the $\cos^{-1}$ key. The result is 2.093.

6.6 Chapter 6 Summary and Highlights

• *Inverse functions* are functions that result in the same value of x after the operations of the two functions are performed. Notation for inverse functions is $f^{-1}(x)$, (where $^{-1}$ is *not* an exponent). An inverse of a function has its domain and range equal to the range and domain, respectively, of the original function. If $f(x) = y$, then $f^{-1}(y) = x$.

• *Inverses of the six trigonometric functions* $y = \sin x, y = \cos x,$ $y = \tan x, y = \sec x, y = \csc x,$ and $y = \cot x$ are respectively $x = \text{Arcsin } y, x = \text{Arccos } y, x = \text{Arctan } y, x = \text{Arcsec } y, x = \text{Arccsc } y,$ and $x = \text{Arccot } y$, and they exist within defined intervals. Inverse functions are also written using $^{-1}$ as, for example, $\text{Sin}^{-1}x$ or $\text{Cos}^{-1}x$. It is necessary to find an inverse of a trigonometric function when the value of y in a function such as $y = \sin x$ is known, but the value of x, which may be an angle or a real number, is not known. To determine x, the inverse of the function must be calculated. In their natural form the six trigonometric functions are not true one-to-one functions and must be limited to an interval where there is a one-to-one correspondence between the domain (x) and range (y) values in order to define an inverse function. The defined intervals of *domain* and *range* values of the inverse trigonometric functions are part of their definitions. The domain and range values of cosine, sine, and tangent functions and their inverses are:

Function	Domain	Range
Sin x	$-\pi/2 \leq x \leq \pi/2$	$-1 \leq y \leq 1$
Sin^{-1}x	$-1 \leq x \leq 1$	$-\pi/2 \leq y \leq \pi/2$
Cos x	$0 \leq x \leq \pi$	$-1 \leq y \leq 1$
Cos^{-1}x	$-1 \leq x \leq 1$	$0 \leq y \leq \pi$
Tan x	$-\pi/2 < x < \pi/2$	$-\infty < y < \infty$
Tan^{-1}x	$-\infty < x < \infty$	$-\pi/2 < y < \pi/2$

• The graph of an inverse trigonometric function is a reflection of the graph of the corresponding trigonometric function over its restricted domain across the line y = x (dashed). For example,

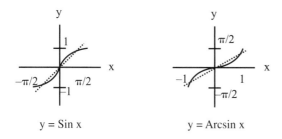

y = Sin x   y = Arcsin x

• Graphs of inverse trigonometric functions can be obtained by using a calculator in radian mode and selecting domain values from –1 to 1, and then plotting the resulting domain/range pairs. A graph can most easily be obtained using a graphing calculator or graphing software and entering the equation for the inverse function. It is also possible to graph, for example, y = Arcsin x, by choosing the coordinates of a few points on the graph of the restricted sine function y = Sin x, reversing their order, then using the points to sketch the graph of the inverse sine function.

• To calculate values for any of the inverse trigonometric functions, such as the value of x in an equation y = tan x, the value of x = arctan y can be found using a calculator with inverse trigonometric function keys, graphs and reference triangles, on-line resources, or by working backward and interpolating values in a trigonometric table. To determine the inverse functions $\csc^{-1} x$, $\sec^{-1} x$, and $\cot^{-1} x$ using a calculator that does not have these keys, the $\sin^{-1}$, $\cos^{-1}$, and $\tan^{-1}$ keys can be used instead with the following *inverse identities*:

$\mathrm{Cot}^{-1} x = \mathrm{Tan}^{-1}(1/x)$ for x > 0
$\mathrm{Cot}^{-1} x = \pi + \mathrm{Tan}^{-1}(1/x)$ for x < 0
$\mathrm{Sec}^{-1} x = \mathrm{Cos}^{-1}(1/x)$ for x ≥ 1 or x ≤ –1
$\mathrm{Csc}^{-1} x = \mathrm{Sin}^{-1}(1/x)$ for x ≥ 1 or x ≤ –1

• Notation for the inverse sine of x is written y = arcsin x or y = $\sin^{-1}x$, and for the inverse cosine of x, y = arccos x or y = $\cos^{-1}x$. The –1 represents the inverse function and is NOT an exponent. Capital letters in the trigonometric functions and inverse trigonometric functions are often used to signify that the interval over which the function is considered is limited to a one-to-one correspondence between domain and range values.

Chapter 7

Trigonometric Identities

7.1 Summary of Identities

• Identities are used to solve problems involving trigonometric or circular functions by expressing one trigonometric function in terms of another. Using trigonometric identities to make substitutions is often required to solve problems involving trigonometry. Trigonometric identities are equations that are true for all possible values of angles or real numbers. The *quotient and reciprocal identities*, the *Pythagorean identities*, the *negative angle/number identities*, and the *sum and difference of angles/numbers identities* are especially important and should be memorized because they can be used to derive other identities. The advent of calculators and computers has to some extent reduced the use of identities. Identities discussed in this chapter include the following:

217

Quotient Identities and Reciprocal Identities:

$\tan A = \sin A / \cos A$ $\cot A = \cos A / \sin A$

$\sin A = 1 / \csc A$ $\cos A = 1 / \sec A$ $\tan A = 1 / \cot A$

$\csc A = 1 / \sin A$ $\sec A = 1 / \cos A$ $\cot A = 1 / \tan A$

Pythagorean Identities:

$\sin^2(A) + \cos^2(A) = 1$

$\sin^2(A) = 1 - \cos^2(A)$

$\cos^2(A) = 1 - \sin^2(A)$

$1 + \tan^2(A) = \sec^2(A)$

$1 + \cot^2(A) = \csc^2(A)$

Negative Angle (Number) Identities:

$\sin(-A) = -\sin A$ $\cos(-A) = \cos A$ $\tan(-A) = -\tan A$

$\csc(-A) = -\csc A$ $\sec(-A) = \sec A$ $\cot(-A) = -\cot A$

Sum and Difference of Angles/Numbers Identities, Also Called Addition and Subtraction Identities:

$\sin(A + B) = \sin(A)\cos(B) + \sin(B)\cos(A)$

$\cos(A + B) = \cos(A)\cos(B) - \sin(A)\sin(B)$

$$\tan(A+B) = \left(\frac{\tan(A)+\tan(B)}{1-\tan(A)\tan(B)} \right)$$

$\sin(A - B) = \sin(A)\cos(B) - \sin(B)\cos(A)$

$\cos(A - B) = \cos(A)\cos(B) + \sin(A)\sin(B)$

$$\tan(A-B) = \left(\frac{\tan(A)-\tan(B)}{1+\tan(A)\tan(B)} \right)$$

Cofunction Identities:

$\sin A = \cos(\pi/2 - A)$ $\cos A = \sin(\pi/2 - A)$

$\tan A = \cot(\pi/2 - A)$ $\cot A = \tan(\pi/2 - A)$

$\sec A = \csc(\pi/2 - A)$ $\csc A = \sec(\pi/2 - A)$

These identities can be written using 90° because $\pi/2$ radians = 90°.

Supplementary Angle Relations:

$\sin(\pi - A) = \sin A$

$\cos(\pi - A) = -\cos A$

$\tan(\pi - A) = -\tan A$

Remember: Supplementary angles sum to π radians or 180°.

Double-Angle/Number Identities:

$\sin(2A) = 2 \sin(A) \cos(A)$

$\cos(2A) = \cos^2(A) - \sin^2(A)$

$\qquad = 1 - 2 \sin^2(A) = 2 \cos^2(A) - 1$

$$\tan(2A) = \left(\frac{2\tan(A)}{1 - \tan^2(A)}\right)$$

These identities are special cases of the sum/addition identities.

Half-Angle Identities:

$$\sin\left(\frac{A}{2}\right) = \pm\sqrt{\frac{1 - \cos(A)}{2}}$$

$$\cos\left(\frac{A}{2}\right) = \pm\sqrt{\frac{1 + \cos(A)}{2}}$$

$$\tan\left(\frac{A}{2}\right) = \sqrt{\frac{1 - \cos(A)}{1 + \cos(A)}} = \frac{\sin(A)}{1 + \cos(A)} = \frac{1 - \cos(A)}{\sin(A)}$$

These identities are derived from the double-angle identities.

Product to Sum Identities:

$\sin(A) \cos(B) = [1/2][\sin(A + B) + \sin(A - B)]$

$\cos(A) \sin(B) = [1/2][\sin(A + B) - \sin(A - B)]$

$\sin(A) \sin(B) = [1/2][\cos(A - B) - \cos(A + B)]$

$\cos(A) \cos(B) = [1/2][\cos(A + B) + \cos(A - B)]$

Sum/Difference to Product Identities:

$$\sin x + \sin y = 2 \sin\left(\frac{x+y}{2}\right) \cos\left(\frac{x-y}{2}\right)$$

$$\sin x - \sin y = 2 \sin\left(\frac{x-y}{2}\right) \cos\left(\frac{x+y}{2}\right)$$

$$\cos x + \cos y = 2 \cos\left(\frac{x+y}{2}\right) \cos\left(\frac{x-y}{2}\right)$$

$$\cos x - \cos y = -2 \sin\left(\frac{x+y}{2}\right) \sin\left(\frac{x-y}{2}\right)$$

Squared Formulas:

$$\sin^2 A = (1/2)(1 - \cos(2A))$$
$$\cos^2 A = (1/2)(1 + \cos(2A))$$

• In the following sections, we will use the fact that in a triangle, sine = opposite / hypotenuse, cosine = adjacent / hypotenuse, and tangent = opposite / adjacent. In addition, A, B, x, and y represent real numbers or angles in degrees or radians.

7.2 Quotient Identities and Reciprocal Identities

• The *quotient and reciprocal identities* are especially important to remember and used when solving problems and deriving other identities.

The *quotient identities* and *reciprocal identities* are:

$\tan A = \sin A / \cos A$	$\cot A = \cos A / \sin A$	
$\sin A = \cos A \tan A$	$\cot A = \cos A \csc A$	
$\sin A = 1 / \csc A$	$\cos A = 1 / \sec A$	$\tan A = 1 / \cot A$
$\csc A = 1 / \sin A$	$\sec A = 1 / \cos A$	$\cot A = 1 / \tan A$

These identities are easily derived using the definitions for sine, cosine, tangent, secant, cosecant, and cotangent.

7.3 Pythagorean Identities

• The Pythagorean identities are particularly important to remember and used when solving problems and deriving other identities.

The **Pythagorean identities** are:

$$\sin^2(A) + \cos^2(A) = 1$$
$$\sin^2(A) = 1 - \cos^2(A)$$
$$\cos^2(A) = 1 - \sin^2(A)$$
$$1 + \tan^2(A) = \sec^2(A)$$
$$1 + \cot^2(A) = \csc^2(A)$$

• Applications when the Pythagorean identities are often useful include making substitutions when solving problems that contain trigonometric functions. For example, suppose you are given the parametric equations (discussed in Chapter 10) $x = 2 \cos t$ and $y = 4 \sin t$ specified in the interval $0 < t \le 2\pi$, and you need to convert these parametric equations into rectangular form so that you can plot x and y pairs on a Cartesian coordinate system. To transform parametric equations into rectangular form we would normally isolate t and write the equations in terms of x and y only. Because they contain trigonometric functions it is difficult to isolate t, however we can isolate cos t and sin t, then substitute the Pythagorean identity in terms of t: $\sin^2 t + \cos^2 t = 1$.

Therefore, given $x = 2 \cos t$ and $y = 4 \sin t$, isolate cos t and sin t:

$$x/2 = \cos t \text{ and } y/4 = \sin t$$

Substitute the Pythagorean identity in terms of t, $\sin^2 t + \cos^2 t = 1$:

$$(x/2)^2 + (y/4)^2 = 1, \text{ or equivalently,}$$
$$x^2/4 + y^2/16 = 1$$

which is the equation for $x = 2 \cos t$ and $y = 4 \sin t$ in rectangular form.

However, in order to graph this equation on a rectangular coordinate system it must be solved for y.

$$y^2/16 = 1 - x^2/4$$
$$y^2 = 16 - 4x^2$$
$$y = \pm[16 - 4x^2]^{1/2}$$

See Section 10.4 for a graph of the rectangular and parametric equations, which forms an ellipse, generated by entering both equations into a graphing utility.

• To derive the *Pythagorean identities*, the **Pythagorean Theorem** $x^2 + y^2 = r^2$, can be used.

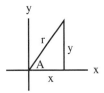

Begin by dividing the Pythagorean formula by r^2:

$(x/r)^2 + (y/r)^2 = 1$

Because $(x/r) = \cos A$ and $(y/r) = \sin A$, then:

$(x/r)^2 = \cos^2 A$ and $(y/r)^2 = \sin^2 A$

Substituting back in to $(x/r)^2 + (y/r)^2 = 1$ results in the first **Pythagorean identity**: $\cos^2 A + \sin^2 A = 1$

Note the equivalent forms obtained by algebra:

$\sin^2 A = 1 - \cos^2 A$ and $\cos^2 A = 1 - \sin^2 A$

To derive the next Pythagorean identity, begin by dividing the Pythagorean Theorem by x^2:

$1 + (y/x)^2 = (r/x)^2$

Substituting $(y/x)^2 = \tan^2 A$ and $(r/x)^2 = \sec^2 A$ results in the second **Pythagorean identity**: $1 + \tan^2 A = \sec^2 A$

To derive the next Pythagorean identity, begin by dividing the Pythagorean Theorem by y^2:

$(x/y)^2 + 1 = (r/y)^2$

Substituting $(x/y)^2 = \cot^2 A$ and $(r/y)^2 = \csc^2 A$ results in the third **Pythagorean identity**: $\cot^2 A + 1 = \csc^2 A$

7.4 Negative Number/Angle Identities

• The negative number/angle identities are important to remember and used when solving problems and deriving other identities.

The **negative angle (number) identities** are:

$\sin(-A) = -\sin A$ $\cos(-A) = \cos A$ $\tan(-A) = -\tan A$

$\csc(-A) = -\csc A$ $\sec(-A) = \sec A$ $\cot(-A) = -\cot A$

where sine is an odd function, cosine is an even function, and tangent is an odd function.

• The negative number/angle identities can be verified by observing a 30:60:90 triangles in a coordinate system.

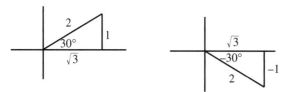

First consider sin(–A) = – sin A. From the figure it is clear that the values of the functions are:

   sin 30° = 1/2 and sin(–30°) = –1/2

The sine of –30° has the same magnitude but the opposite sign as the sine of 30°. Therefore,

   sin(–30°) = –sin 30°

For identity cos(–A) = cos A, from the figure it is clear that the values of the functions are:

   cos 30° = $\sqrt{3}$/2 and cos(–30°) = $\sqrt{3}$/2

The cosine of –30° has the same magnitude and the same sign as the cosine of 30°. Therefore,

   cos(–30°) = cos 30°

Remember: sin A = opposite/hypotenuse and cos A = adjacent/hypotenuse.

• These identities can be obtained using the *difference identities* discussed in Section 7.6:

   cos(A – B) = cos A cos B + sin A sin B
   sin(A – B) = sin A cos B – cos A sin B
   tan(A – B) = [tan A – tan B] / [1 + tan A tan B]

When angle A = 0 and angle B = θ:

   cos(0 – θ) = cos 0 cos θ + sin 0 sin θ = cos θ
   sin(0 – θ) = sin 0 cos θ – cos 0 sin θ = – sin θ
   tan(0 – θ) = [tan 0 – tan θ] / [1 + tan 0 tan θ] = – tan θ

• The other negative number/angle identities can be verified using the quotient and reciprocal trigonometric identities.

Because tan A = sin A / cos A, sin(–A) = – sin A, and cos(–A) = cos A, then –sin A / cos A = – tan A, or tan(–A) = – tan A

This can also be observed in the graph of tangent.

Because the cosecant, secant, and cotangent are reciprocal functions, csc A = 1 / sin A, sec A = 1 / cos A, and cot A = 1 / tan A, their sign relationships correspond to the respective sine, cosine, and tangent functions:

$$\sin(-A) = -\sin A \qquad \cos(-A) = \cos A \qquad \tan(-A) = -\tan A$$
$$\csc(-A) = -\csc A \qquad \sec(-A) = \sec A \qquad \cot(-A) = -\cot A$$

• The negative angle identities can also be verified using graphs of the cosine and sine functions.

Sine

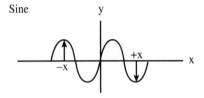

For a distance +x to the right of 0, the vertical distance (arrow) from the axis to the curve represents sin x. For an equivalent distance to the left of 0, the vertical distance (arrow) from the axis to the curve represents the sin(–x). The vertical distances have the same magnitudes but are measured in opposite directions. Therefore, it is clear from the figure that sin x = the opposite of sin(–x), or, sin x = –sin(–x).

Cosine

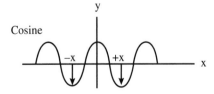

For a distance +x to the right of 0, the vertical distance (arrow) from the axis to the curve represents cos x. For an equivalent distance to the left of 0, the vertical distance (arrow) from the axis to the curve represents the cos(–x). The vertical distances have the same magnitudes and the same directions. Therefore, it is clear from the figure that cos x = cos(–x).

7.5 Verifying Trigonometric Identities

• It may be necessary to verify that a trigonometric relationship is also an identity, which means that the relationship is true for all possible values of angles or real numbers. Trigonometric relationships can be verified as identities using algebra and substitutions. In general, if you are given an equation that is not a known identity, it is possible to verify that the two sides are equivalent by using algebraic manipulations and substitutions, or using a graphing utility, such as a graphing calculator or graphing software. To verify using graphing, graph the expressions on each side of the equal sign on one graph and see if they overlap for all values of x.

Verifying a trigonometric identity is different from solving an equation. Solving an equation involves using properties of equality, such as adding or multiplying the same quantity to each side. These operations are not appropriate in the process of verifying that both sides of an equation are equal and therefore the equation is an identity. When verifying an identity, the idea is to begin with the expression on one side of the equal sign and manipulate it using algebra and substitutions of known identities to convert that expression into the same form as the expression on the other side. There are a number of general guidelines that can be observed when verifying identities, and they include the following:

1. In general, begin with the more complicated side of the identity and transform it into the less complicated side. The less complicated side represents the goal.
2. Use algebraic substitution and simplification operations on each side of the equation independently. In some cases, transforming both sides of the equation into the same simpler form may be helpful. Do not use standard equation-solving algebraic techniques in which the same operation is performed on both sides of the equation simultaneously.
3. Become familiar with the basic identities and their equivalent forms, and use them for substitutions.
4. Use algebraic operations such as factoring, multiplying, and operations on fractions, such as adding, subtracting, combining, splitting, reducing, transforming into equivalent forms, and multiplying the numerator and denominator of a fraction by the conjugate of either.
5. It is sometimes helpful to use substitutions to express all functions in the equation in terms of sine and cosine.

• **Example:** Suppose one side of a relation is tan A + cot A and it is necessary to express it in terms of sin A and cos A. Express tan A + cot A in terms of sin A and cos A, simplify, and use a graphing utility to verify that the expression in terms of sin A and cos A is equal to tan A + cot A.

Begin by substituting the quotient identities:

$$\tan A + \cot A = \sin A / \cos A + \cos A / \sin A$$

Simplify by adding the two fractions on the right side, using the common denominator cos A sin A (multiply the numerator and denominator, simplify, and substitute the Pythagorean identity $1 = \sin^2 A + \cos^2 A$):

$$\tan A + \cot A = \frac{\sin A \cos A \sin A}{\cos A \cos A \sin A} + \frac{\cos A \cos A \sin A}{\sin A \cos A \sin A}$$

$$= \frac{\sin^2 A}{\cos A \sin A} + \frac{\cos^2 A}{\cos A \sin A} = \frac{\sin^2 A + \cos^2 A}{\cos A \sin A} = \frac{1}{\cos A \sin A}$$

Therefore, tan A + cot A = 1 / cos A sin A.

Use a graphing utility to graph each side of the equation on one graph to verify that they are equal. Graph $y_1 = \tan x + \cot x$ (which may be entered as tan x +1/tan x if necessary) and $y_2 = 1 / \cos x \sin x$:

Graph of $y_1 = \tan x + \cot x$ and $y_2 = 1 / \cos x \sin x$

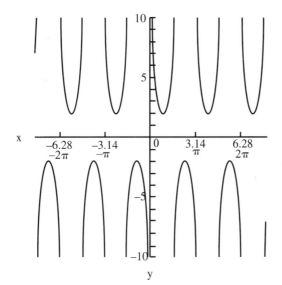

The perfect overlap of these two curves verifies that
$\tan A + \cot A = 1 / \cos A \sin A$.

• **Example:** Verify identity $\tan x + \cot x = \csc x / \cos x$.

In this case we can convert both sides independently.

Begin with $\tan x + \cot x$ and use quotient identities:

$$\tan x + \cot x = \sin x/\cos x + \cos x/\sin x$$

Combine fractions using a common denominator $\cos x \sin x$ and substitute the Pythagorean identity as in the previous example:

$$= \frac{\sin^2 x}{\cos x \sin x} + \frac{\cos^2 x}{\cos x \sin x} = \frac{\sin^2 x + \cos^2 x}{\cos x \sin x} = \frac{1}{\cos x \sin x}$$

Convert the right side $\csc x / \cos x$ to match the result from the left side using the reciprocal identity $\csc x = 1/\sin x$:

$$\frac{\csc x}{\cos x} = \frac{\csc x}{1} \times \frac{1}{\cos x} = \frac{1}{\sin x} \times \frac{1}{\cos x} = \frac{1}{\cos x \sin x}$$

Therefore, $\tan x + \cot x = \csc x / \cos x$.

• **Example:** Suppose one side of a relation is $\tan x \cot x$ and it is necessary to simplify it to match the other side to verify that the equation is an identity.

Use reciprocal identity $\cot x = 1/\tan x$ to simplify $\tan x \cot x$:

$$\tan x \cot x = (\tan x)(1/\tan x) = 1$$

Therefore, $\tan x \cot x = 1$.

• **Example:** Verify $(1 + \sin x)/(\cos x) = (\cos x)/(1 - \sin x)$.

Begin with the left side $(1 + \sin x)/(\cos x)$ and work toward the right side:

$$\frac{1 + \sin x}{\cos x} = \frac{(1 + \sin x)\cos x}{\cos x \cos x} = \frac{(1 + \sin x)\cos x}{\cos^2 x} = \frac{(1 + \sin x)\cos x}{1 - \sin^2 x}$$

$$= \frac{(1 + \sin x)\cos x}{(1 + \sin x)(1 - \sin x)} = \frac{\cos x}{1 - \sin x}$$

Therefore, $(1 + \sin x)/(\cos x) = (\cos x)/(1 - \sin x)$.

7.6 Sum and Difference of Angles/Numbers Identities, Also Called Addition and Subtraction Identities

• The *sum and difference identities* can be used to derive other trigonometric identities, such as the *cofunction identities*, the *double-angle/number identities* (which are used to derive the *half-angle identities*), the *product to sum identities*, and the *sum to product identities*. Sum and difference identities are also often used for substitutions when solving problems involving trigonometric and circular functions. These identities can be used to verify more complicated identities or to find exact values of trigonometric functions. These identities are true for all **real numbers** and **angles** in radian or degree measure. The identities for sine and cosine are worth remembering.

• The **sum and difference of angles/numbers identities** are:

$$\sin(A + B) = \sin(A)\cos(B) + \sin(B)\cos(A)$$
$$\cos(A + B) = \cos(A)\cos(B) - \sin(A)\sin(B)$$

$$\tan(A + B) = \left(\frac{\tan(A) + \tan(B)}{1 - \tan(A)\tan(B)} \right)$$

$$\sin(A - B) = \sin(A)\cos(B) - \sin(B)\cos(A)$$
$$\cos(A - B) = \cos(A)\cos(B) + \sin(A)\sin(B)$$

$$\tan(A - B) = \left(\frac{\tan(A) - \tan(B)}{1 + \tan(A)\tan(B)} \right)$$

These identities can be derived algebraically using properties from geometry and other basic trigonometric identities. (See the following paragraphs.)

• **Example:** Suppose we know the coordinates of Point 1 and Point 2 as $(2, 4)$ and $(3, 2)$ respectively, but need to find the angle β.

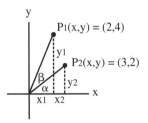

We can use the subtraction identity for tangent if we choose the angles as α and $(\alpha + \beta)$ (that form two right triangles), where the difference between α and $(\alpha + \beta)$ is β, or $\beta = (\alpha + \beta) - \alpha$

Therefore, $\tan(A - B)$ in the identity is $\tan((\alpha + \beta) - \alpha)$

First calculate tangents:

$$\tan(\alpha + \beta) = y_1/x_1 = 4/2 = 2$$
$$\tan(\alpha) = y_2/x_2 = 2/3$$

Substitute into the identity:

$$\tan \beta = \tan((\alpha + \beta) - \alpha)$$

$$= \left(\frac{\tan(\alpha + \beta) - \tan(\alpha)}{1 + \tan(\alpha + \beta)\tan(\alpha)} \right) = \frac{2 - 2/3}{1 + (2)(2/3)} = \frac{4/3}{7/3} = \frac{4}{7}$$

Therefore, $\beta = \arctan(4/7) \approx 0.519$ rad. $\approx 29.7°$

- The *addition and subtraction identities for cosine,*

$$\cos(A - B) = \cos(A)\cos(B) + \sin(A)\sin(B) \text{ and}$$
$$\cos(A + B) = \cos(A)\cos(B) - \sin(A)\sin(B)$$

can be derived using the fact that the distance between two points on a circle is the same whether a triangle between the two points is rotated or not. The distance between two points (x_1, y_1) and (x_2, y_2) is given by the distance formula:

$$d((x_1, y_1)(x_2, y_2)) = \sqrt{(x_2 - x_1)^2 + (y_2 - y_1)^2}$$

First, recall that on a unit circle (a circle with a radius of one) the coordinate of each point P can be described by $x = \cos A$ and $y = \sin A$, where A is a standard position angle. The coordinates of a point on a unit circle can therefore be described by $(x, y) = (\cos A, \sin A)$:

P(cosA, sinA)

On a unit circle two points can be positioned on the points of a triangle that is located at the origin. When the triangle is rotated about the origin the distance between the two points remains the same, as can be observed in the figure:

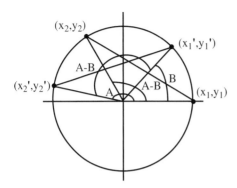

The x and y coordinates represent the original triangle location and the x'
and y' coordinates represent the rotated triangle position.

Angle B spans from (x_1, y_1) to (x_1', y_1').

Angle A spans from (x_1, y_1) to (x_2', y_2').

Angle A–B spans from (x_1', y_1') to (x_2', y_2') and an equivalent angle
from (x_1, y_1) to (x_2, y_2).

Coordinates of point (x_2', y_2') equals $(\cos A, \sin A)$.

Coordinates of point (x_1', y_1') equals $(\cos B, \sin B)$.

Coordinates of point (x_2, y_2) equals $(\cos(A-B), \sin(A-B))$.

Coordinates of point (x_1, y_1) equals $(\cos 0, \sin 0) = (1, 0)$.

The distance between points (x_1', y_1') and (x_2', y_2') is equal to the distance between points (x_1, y_1) and (x_2, y_2). Therefore,

$$d((x_1, y_1)(x_2, y_2)) = d((x_1', y_1')(x_2', y_2'))$$

$$= \sqrt{(x_2 - x_1)^2 + (y_2 - y_1)^2} = \sqrt{(x_2' - x_1')^2 + (y_2' - y_1')^2}$$

Substituting with the coordinates of points listed above:

$$= \sqrt{(\cos(A\text{-}B) - 1)^2 + (\sin(A\text{-}B) - 0)^2}$$

$$= \sqrt{(\cos(A) - \cos(B))^2 + (\sin(A) - \sin(B))^2}$$

Squaring both sides and using algebra:

$$(\cos(A-B) - 1)^2 + (\sin(A-B))^2 = (\cos A - \cos B)^2 + (\sin A - \sin B)^2$$

Simplifying the left side:

$(\cos(A–B) – 1)(\cos(A–B) – 1) + \sin^2(A–B)$
$= \cos^2(A–B) – 2\cos(A–B) + 1 + \sin^2(A–B)$

Simplifying the right side:

$= (\cos A – \cos B)(\cos A – \cos B) + (\sin A – \sin B)(\sin A – \sin B)$
$= \cos^2 A – 2 \cos A \cos B + \cos^2 B + \sin^2 A – 2 \sin A \sin B + \sin^2 B$

Therefore combining both sides:

$\cos^2(A–B) – 2 \cos(A–B) + 1 + \sin^2(A–B)$
$= \cos^2 A – 2 \cos A \cos B + \cos^2 B + \sin^2 A – 2 \sin A \sin B + \sin^2 B$

Substitute the identity $\cos^2 x + \sin^2 x = 1$ where it applies:

$2 – 2 \cos(A–B) = 2 – 2 \cos A \cos B – 2 \sin A \sin B$

Simplifying gives the **subtraction identity for cosine**:

$\cos(A–B) = \cos A \cos B + \sin A \sin B$

This formula applies to any value of the angles depicted above and also to all real numbers and angles in radian or degree measure.

• To obtain the identity for the *sum/addition identity for cosine* substitute –B for B in the *difference/subtraction identity for cosine*:

$\cos(A–(–B)) = \cos(A+B) = \cos A \cos(–B) + \sin A \sin(–B)$

Substituting the *negative number identities* $\cos(–x) = \cos x$ and $\sin(–x) = – \sin x$, results in the **sum/addition identity for cosine**:

$\cos(A+B) = \cos A \cos B – \sin A \sin B$

• The *sum/addition and difference/subtraction identities for sine* can be derived using right triangle cofunction relationships

$\sin A = \cos(\pi/2 – A)$ and $\cos A = \sin(\pi/2 – A)$

(which are discussed in the next section).

Right triangles have complimentary angles that can be measured by $90° – \phi$ or $\pi/2 – \phi$.

Using the complimentary angle relations, $\cos \phi = \sin(\pi/2 - \phi)$ and $\sin \phi = \cos(\pi/2 - \phi)$, and replacing angle ϕ with angle $(A + B)$, we have:

$$\sin(A + B) = \cos(\pi/2 - (A+B)) = \cos(\pi/2 - A - B) = \cos((\pi/2 - A) - B)$$

Substituting $\cos((\pi/2 - A) - B)$ into the *subtraction identity for cosine* $\cos(A - B) = \cos A \cos B + \sin A \sin B$, where A in this identity is represented by angle $(\pi/2 - A)$ gives:

$$\cos((\pi/2 - A) - B) = \cos(\pi/2 - A) \cos B + \sin(\pi/2 - A) \sin B$$

Substituting the cofunction identity $\sin A = \cos(\pi/2 - A)$, this equation becomes the **sum/addition identity for sine**:

$$\sin(A + B) = \sin A \cos B + \cos A \sin B$$

• To obtain the *difference identity for sine* substitute $-B$ for B in the *addition identity for sine* derived above:

$$\sin(A + (-B)) = \sin A \cos(-B) + \cos A \sin(-B)$$

Then substitute the *negative number identities* $\cos(-x) = \cos x$ and $\sin(-x) = -\sin x$, which results in the **difference/subtraction identity for sine**:

$$\sin(A - B) = \sin A \cos B - \cos A \sin B$$

• To obtain the *sum and difference identities for tangent*

$$\tan(A + B) = \left(\frac{\tan A + \tan B}{1 - \tan A \tan B} \right)$$

$$\tan(A - B) = \left(\frac{\tan A - \tan B}{1 + \tan A \tan B} \right)$$

use the sum and difference identities for sine and cosine and the identity $\tan x = \sin x / \cos x$:

$$\tan(A + B) = \frac{\sin(A + B)}{\cos(A + B)} = \frac{\sin A \cos B + \cos A \sin B}{\cos A \cos B - \sin A \sin B}$$

Multiply numerator and denominator by $1 / \cos A \cos B$, simplify, and substitute $\tan x = \sin x / \cos x$:

$$\frac{\dfrac{\sin A \cos B}{\cos A \cos B} + \dfrac{\cos A \sin B}{\cos A \cos B}}{\dfrac{\cos A \cos B}{\cos A \cos B} - \dfrac{\sin A \sin B}{\cos A \cos B}} = \frac{\dfrac{\sin A}{\cos A} + \dfrac{\sin B}{\cos B}}{1 - \dfrac{\sin A}{\cos A} \dfrac{\sin B}{\cos B}} = \frac{\tan A + \tan B}{1 - \tan A \tan B}$$

This results in the *sum/addition identity for tangent*:

$$\tan(A + B) = \left(\frac{\tan A + \tan B}{1 - \tan A \tan B} \right)$$

• To obtain the difference identity for tangent begin with the sum identity, substitute –B for B, and use the negative number identity for tangent $\tan(-x) = -\tan x$:

$$\tan(A + (-B)) = \left(\frac{\tan A + \tan(-B)}{1 - \tan A \tan(-B)} \right) = \left(\frac{\tan A - \tan B}{1 + \tan A \tan B} \right)$$

This results in the *difference/subtraction identity for tangent*:

$$\tan(A - B) = \left(\frac{\tan A - \tan B}{1 + \tan A \tan B} \right)$$

• **Example:** Simplify $\cos(x + \pi/2)$.

Substitute $x = A$ and $\pi/2 = B$ into the addition identity for cosine, $\cos(A + B)$, and simplify:

$$\cos(A + B) = \cos A \cos B - \sin A \sin B$$
$$\cos(x + \pi/2) = \cos x \cos \pi/2 - \sin x \sin \pi/2$$

Because $\cos \pi/2 = 0$ and $\sin \pi/2 = 1$:

$$\cos(x + \pi/2) = (\cos x)(0) - (\sin x)(1) = -\sin x$$

Therefore, $\cos(x + \pi/2) = -\sin x$.

This can be observed by superimposing the graphs of $y = \sin x$ and $y = \cos x$. The $+\pi/2$ in $\cos(x + \pi/2)$ causes a *phase shift* of $\pi/2$ to the *left* in the $y = \cos x$ graph, which result in an upside down graph of $y = \sin x$.

$y = \cos(x+\pi/2)$ (solid gray)
$y = \cos x$ (dashed)
$y = \sin x$ (solid black)

7.7 Cofunction Identities

• The *cofunction identities* are:

$$\sin A = \cos(\pi/2 - A) \qquad \cos A = \sin(\pi/2 - A)$$
$$\tan A = \cot(\pi/2 - A) \qquad \cot A = \tan(\pi/2 - A)$$
$$\sec A = \csc(\pi/2 - A) \qquad \csc A = \sec(\pi/2 - A)$$

These identities can be written using $90°$ because $\pi/2$ radians $= 90°$.

• Cofunction identities represent complementary angles, which can be viewed in the following figure:

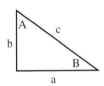

Angle A and angle B are *complementary angles* and have the property:

$$A + B = 90°, \text{ or equivalently, } A + B = \pi/2.$$

• **Example:** $\cos 30° = \sin 60°$ and $\sin 30° = \cos 60°$.

• The sum of the angles in any planar triangle is $180°$. In a right triangle, the measure of the right angle is $90°$, so the other two angles must sum to $90°$. Therefore, if one non-$90°$ angle is A, then the other non-$90°$ angle is $(90° - A)$.

In the right triangle depicted above,

$$\sin A = a/c \text{ and } \cos B = \cos(90° - A) = a/c$$
$$\cos A = b/c \text{ and } \sin B = \sin(90° - A) = b/c$$
$$\tan A = a/b \text{ and } \cot B = \cot(90° - A) = a/b$$
$$\cot A = b/a \text{ and } \tan B = \tan(90° - A) = b/a$$
$$\sec A = c/b \text{ and } \csc B = \csc(90° - A) = c/b$$
$$\csc A = c/a \text{ and } \sec B = \sec(90° - A) = c/a$$

The trigonometric functions of an angle A have the same values as the trigonometric cofunctions of $(\pi/2 - A)$, or $(90° - A)$. The values of $(\pi/2 - A)$, or $(90° - A)$, denote *the other angle*. The words *cosine*, *cotangent*, and *cosecant* are abbreviations for the sine of the complementary angle (the other angle), the tangent of the complementary angle (the other angle), and the secant of the complementary angle (the other angle), respectively.

• The *cofunction identities for sine and cosine* can be derived using the *difference identity for cosine*:

$$\cos(A - B) = \cos A \cos B + \sin A \sin B$$

When $A = \pi/2$ this becomes:

$$\cos(\pi/2 - B) = \cos \pi/2 \cos B + \sin \pi/2 \sin B$$
$$= 0 \cos B + 1 \sin B$$
$$= \sin B$$

resulting in the **cofunction identity for cosine** for any angle in radian measure or any real number:

$$\cos(\pi/2 - B) = \sin B$$

The cofunction identity in terms of degrees uses $90°$ rather than $\pi/2$:

$$\cos(90° - B) = \sin B$$

• To obtain this identity for the other angle A, substitute $B = (\pi/2 - A)$ in the identity $\cos(\pi/2 - B) = \sin B$:

$$\cos(\pi/2 - (\pi/2 - A)) = \sin(\pi/2 - A)$$

Simplifying the left side:

$$\cos A = \sin(\pi/2 - A)$$

resulting in the **cofunction identity for sine** for any angle in radian measure or any real number:

$$\sin(\pi/2 - A) = \cos A$$

The cofunction identity in terms of degrees uses $90°$ rather than $\pi/2$:

$$\sin(90° - A) = \cos A$$

• To obtain the *cofunction identity for tangent and cotangent*, begin with the quotient identity:

$\tan(\pi/2 - A) = [\sin(\pi/2 - A)] / [\cos(\pi/2 - A)]$

Substituting the cofunction identities $\cos(\pi/2 - A) = \sin A$ and $\sin(\pi/2 - A) = \cos A$ gives:

$\tan(\pi/2 - A) = \cos A / \sin A = \cot A$

which results in the **cofunction identity for tangent**:

$\tan(\pi/2 - A) = \cot A$

The cofunction identity in terms of degrees uses 90° rather than $\pi/2$:

$\tan(90° - A) = \cot A$

• To obtain this identity for the other angle A, substitute $A = (\pi/2 - B)$ in the identity $\tan(\pi/2 - A) = \cot A$:

$\tan(\pi/2 - (\pi/2 - B)) = \cot(\pi/2 - B)$
$\tan(B) = \cot(\pi/2 - B)$

which is the **cofunction identity for cotangent**:

$\cot(\pi/2 - B) = \tan B$

• To obtain the *cofunction identities for secant and cosecant*, begin with the reciprocal identity:

$\sec(\pi/2 - A) = 1 / [\cos(\pi/2 - A)]$

Substituting the cofunction identity $\cos(\pi/2 - A) = \sin A$:

$\sec(\pi/2 - A) = 1 / \sin A = \csc A$

which results in the **cofunction identity for secant**:

$\sec(\pi/2 - A) = \csc A$

• To obtain this identity for the other angle A, substitute $A = (\pi/2 - B)$ in the identity $\sec(\pi/2 - A) = \csc A$:

$\sec(\pi/2 - (\pi/2 - B)) = \csc(\pi/2 - B)$
$\sec(B) = \csc(\pi/2 - B)$

which is the **cofunction identity for cosecant**:

$\csc(\pi/2 - B) = \sec B$

• Combining cofunction and negative angle identities, where $\pi/2$ radians can be substituted with $90°$:

$$\sin (\pi/2 - A) = \cos A, \quad \sin (\pi/2 + A) = \cos A$$
$$\cos (\pi/2 - A) = \sin A, \quad \cos (\pi/2 + A) = -\sin A$$
$$\sin (90° - A) = \cos A, \quad \sin (90° + A) = \cos A$$
$$\cos (90° - A) = \sin A, \quad \cos (90° + A) = -\sin A$$
$$\tan (90° - A) = \cot A, \quad \tan (90° + A) = -\cot A$$
$$\sec (90° - A) = \csc A, \quad \sec (90° + A) = -\csc A$$
$$\csc (90° - A) = \sec A, \quad \csc (90° + A) = \sec A$$
$$\cot (90° - A) = \tan A, \quad \cot (90° + A) = -\tan A$$

Note that these identities can be obtained from observing their graphs.

7.8 Supplementary Angle Relations

• The *supplementary angle relations* are:

$$\sin(\pi - A) = \sin A$$
$$\cos(\pi - A) = -\cos A$$
$$\tan(\pi - A) = -\tan A$$

Remember: Supplementary angles sum to π radians or $180°$.

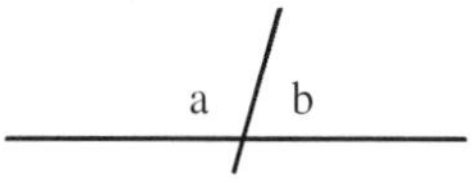

Angles a and b are supplementary angles and have the property:

$$a + b = 180°, \text{ or equivalently, } a + b = \pi$$

The trigonometric functions of an angle A have the same values as the trigonometric supplementary relations $(\pi - A)$, or $(180° - A)$.

• **Example:** $\sin(\pi - 2) = \sin 2$ and $\cos(\pi - 3) = -\cos 3$.

- A summary of supplementary angle relationships is:

$$\sin(180° - A) = \sin A, \qquad \sin(180° + A) = -\sin A$$
$$\cos(180° - A) = -\cos A, \qquad \cos(180° + A) = -\cos A$$
$$\tan(180° - A) = -\tan A, \qquad \tan(180° + A) = \tan A$$
$$\sec(180° - A) = -\sec A, \qquad \sec(180° + A) = -\sec A$$
$$\csc(180° - A) = \csc A, \qquad \csc(180° + A) = -\csc A$$
$$\cot(180° - A) = -\cot A, \qquad \cot(180° + A) = \cot A$$

These identities can be obtained by observing their graphs. For example, for $\sin(\pi - A) = \sin A$ and $\sin(\pi + A) = -\sin A$:

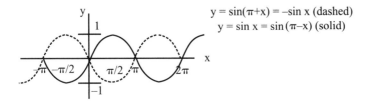

$y = \sin(\pi + x) = -\sin x$ (dashed)
$y = \sin x = \sin(\pi - x)$ (solid)

7.9 Double-Angle/Number Identities

- The *double-angle/number identities* are:

$$\sin(2A) = 2\sin(A)\cos(A)$$
$$\cos(2A) = \cos^2(A) - \sin^2(A)$$
$$= 1 - 2\sin^2(A)$$
$$= 2\cos^2(A) - 1$$

$$\tan(2A) = \left(\frac{2\tan(A)}{1 - \tan^2(A)}\right)$$

- Double-angle/number identities are special cases of the *sum/addition identities*. Double-angle/number identities are true for angles as well as real numbers and can be used to find exact values and when verifying more complicated identities or solving equations.

- Applications in which the double-angle/number identities are often useful include making substitutions when solving problems that contain trigonometric functions. For example, suppose you are given the polar equation (discussed in Chapter 10), and you need to convert it to rectangular form and plot x and y pairs on a Cartesian coordinate system. First, you will look up the conversion relationships for converting between rectangular and polar coordinates and equations (discussed in section 10.2) which are $x = r \cos \theta$, $y = r \sin \theta$, $\tan \theta = y/x$, and $r^2 = x^2 + y^2$.

Given the equation $r^2 \cos 2\theta = 2$ that we need to convert to rectangular coordinates, we notice that it contains a double angle. To convert this equation, begin by substituting the double angle formula for cosine $\cos 2\theta = \cos^2 \theta - \sin^2 \theta$ into $r^2 \cos 2\theta = 2$, and then transform it into a form in which the conversion relationships can be used:

$$r^2 (\cos^2 \theta - \sin^2 \theta) = 2$$
$$r^2 \cos^2 \theta - r^2 \sin^2 \theta = 2$$
$$(r \cos \theta)^2 - (r \sin \theta)^2 = 2$$

Then substitute $x = r \cos \theta$ and $y = r \sin \theta$ resulting in the rectangular form of $r^2 \cos 2\theta = 2$:

$$x^2 - y^2 = 2$$

See the end of Section 10.2 for a graph of both forms of the equation.

• Double-number identities can be derived using the sum/addition identities. The double-angle/number identities represent the case where $A = B$ in the *sum/addition identities*:

$$\sin(A + B) = \sin A \cos B + \cos A \sin B$$
$$\cos(A + B) = \cos A \cos B - \sin A \sin B$$

• To develop the *identity for sin 2A*, begin with the identity for $\sin(A + B)$:

$$\sin(A + B) = \sin A \cos B + \cos A \sin B$$

When angle B is equal to angle A this becomes:

$$\sin(A + A) = \sin A \cos A + \cos A \sin A$$

On the left, $\sin(A + A) = \sin 2A$, and both terms on the right are the product of sin A and cos A, or 2 sin A cos A. This results in the ***double-angle/number identity for sine***:

$$\sin 2A = 2 \sin A \cos A$$

• To develop the *identity for cos 2A* begin with the identity for $\cos(A + B)$:

$$\cos(A + B) = \cos A \cos B - \sin A \sin B$$

When angle B is equal to angle A this becomes:

$$\cos(A + A) = \cos A \cos A - \sin A \sin A$$

which simplifies to the **double-angle/number identity for cosine**:

$$\cos 2A = (\cos A)^2 - (\sin A)^2 = \cos^2 A - \sin^2 A$$

• There are two other useful forms of the *double-angle/number identity for cosine* that can be obtained by substituting the Pythagorean identities:

$$\sin^2 A + \cos^2 A = 1, \cos^2 A = 1 - \sin^2 A, \text{ and } \sin^2 A = 1 - \cos^2 A$$

These identities can be substituted into the first form of the double angle identity for cosine to get two other identities.

Begin with the cosine double-angle identity: $\cos 2A = \cos^2 A - \sin^2 A$

Use identity $\cos^2 A = 1 - \sin^2 A$ and replace $\cos^2 A$ with $1 - \sin^2 A$:

$$\cos 2A = (1 - \sin^2 A) - \sin^2 A = 1 - 2 \sin^2 A$$

which is another form of the **double-angle/number identity for cosine**:

$$\cos 2A = 1 - 2 \sin^2 A$$

A third form of the double angle/number identity for cosine, $\cos 2A$ can be obtained by using the identity $\sin^2 A = 1 - \cos^2 A$ and replacing $\sin^2 A$ with $1 - \cos^2 A$:

Begin with the first double-angle identity: $\cos 2A = \cos^2 A - \sin^2 A$

Replace $\sin^2 A$ with $1 - \cos^2 A$:

$$\cos 2A = \cos^2 A - (1 - \cos^2 A) = 2 \cos^2 A - 1$$

which is a third form of the **double-angle/number identity for cosine**:

$$\cos 2A = 2 \cos^2 A - 1$$

• The *double-number identity for tangent, tan 2A* can be developed using the double angle/number identities for sine and cosine:

$$\sin 2A = 2 \sin A \cos A \text{ and } \cos 2A = \cos^2 A - \sin^2 A$$

Because tangent equals sine divided by cosine, then tan 2A equals sin 2A divided by cos 2A:

$$\tan 2A = \frac{\sin 2A}{\cos 2A} = \frac{2 \sin A \cos A}{\cos^2 A - \sin^2 A}$$

To write this in terms of tan A, divide every term in the numerator and denominator by $\cos^2 A$ or equivalently, (cos A)(cos A):

$$\tan 2A = \frac{[2 \sin A \cos A] / [\cos A \cos A]}{[\cos^2 A] / [\cos^2 A] - [\sin^2 A] / [\cos^2 A]} = \frac{2 \tan A}{1 - \tan^2 A}$$

which results in the **double-angle/number identity for tangent**:

$$\tan 2A = \frac{2 \tan A}{1 - \tan^2 A}$$

• The *double-number identity for tangent* can also be derived directly from the sum identity for tangent:

$$\tan(A + B) = \left(\frac{\tan A + \tan B}{1 - \tan A \tan B} \right)$$

When A = B this becomes:

$$\tan(A + A) = \left(\frac{\tan A + \tan A}{1 - \tan A \tan A} \right) = \tan 2A = \frac{2 \tan A}{1 - \tan^2 A}$$

• **Example:** You launched a satellite over the ocean and while tracking it you notice that part of its payload dropped off at 6 miles above the ocean. How far must you travel from where you made your observation, at the beach, to recover part of your satellite remains? In other words, for a known distance above the ground (ocean) of 6 miles, find the distance across the ocean x (and also find the angles A) as shown in the diagram.

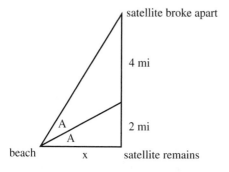

satellite broke apart

4 mi

2 mi

beach   x   satellite remains

In the lower triangle tan A = opposite/adjacent = 2/x, and in the entire triangle tan 2A = opposite/adjacent = 6/x. Therefore,

   tan A = 2/x and tan 2A = 6/x.

Substituting into the double-angle identity for tangent

$$\tan 2A = \frac{2\tan A}{1 - \tan^2 A}$$

$$= 6/x = \frac{2(2/x)}{1 - (2/x)^2} = \frac{4/x}{1 - 4/x^2} = \frac{(x^2)(4/x)}{(x^2)(1) - (x^2)(4/x^2)} = \frac{4x}{x^2 - 4}$$

Therefore, the double-angle identity for tangent for these values becomes:

$$\frac{6}{x} = \frac{4x}{x^2 - 4}$$

Solving for x:

   $(6)(x^2 - 4) = (4x)(x)$
   $6x^2 - 24 = 4x^2$
   $6x^2 - 4x^2 = 24$
   $2x^2 = 24$
   $x^2 = 12$
   $x = \sqrt{12} = \sqrt{(3)(2)(2)} = 2\sqrt{3}$ mi

or approximately 3.464 miles, which is how far you must travel across the ocean to recover part of your satellite remains.

To find angles A substitute x = 3.464 into tan A = 2/x from the figure:

   tan A = 2/3.464 = 0.5774
   A = arctan(0.5774) ≈ 30° (or 0.5236 rad)

Note that we could have also used tan A = $2/x = 2/(2\sqrt{3}) = 1/\sqrt{3}$ or A = arctan($1/\sqrt{3}$) = 30°

Therefore, the distance x is approximately 3.464 miles, and the angle A is 30° with the value of the double angle 2A as 60°.

Note that in the large right triangle, tan 60° = 6 / 3.464.

7.10 Half-Angle Identities

• The *half-angle identities* are:

$$\sin\left(\frac{A}{2}\right) = \pm\sqrt{\frac{1-\cos(A)}{2}}$$

$$\cos\left(\frac{A}{2}\right) = \pm\sqrt{\frac{1+\cos(A)}{2}}$$

$$\tan\left(\frac{A}{2}\right) = \sqrt{\frac{1-\cos(A)}{1+\cos(A)}} = \frac{\sin(A)}{1+\cos(A)} = \frac{1-\cos(A)}{\sin(A)}$$

• *Half-angle, or half-number, identities* are *double-angle identities* in an alternative form. Half-angle identities can be derived using double-angle identities. These identities can be used to find exact values for functions, such as $\sin(\pi/8)$, using the fact that $\pi/8$ is half of $\pi/4$. These identities can also be used in verifying more complicated identities and for substitutions when solving problems involving trigonometric functions.

• **Example:** If $\cos x = 1/3$, what is $\cos x/2$ in the interval between 0 and $\pi/2$?

Substitute into the identity:

$$\cos x/2 = \pm [(1 + \cos x)/2]^{1/2}$$
$$\cos x/2 = \pm [(1 + 1/3)/2]^{1/2}$$
$$\cos x/2 = \pm [(4/3)/2]^{1/2}$$
$$\cos x/2 = \pm [(2/3)]^{1/2}$$

In the interval $0 < x < \pi/2$, or $0 < x/2 < \pi/4$, then $x/2$ is between 0 and $\pi/4$ in the first quadrant and therefore has a positive cosine. (Remember the quadrants and signs from chapter 4):

Quad II	Quad I
cos α is −	cos α is +
sin α is +	sin α is +
tan α is −	tan α is +
Quad III	Quad IV
cos α is −	cos α is +
sin α is −	sin α is −
tan α is +	tan α is −

Therefore, if $\cos x = 1/3$, then $\cos x/2 = \sqrt{2/3}$

• The *half-angle/number identities sin x/2 and cos x/2* can be developed using the *double-angle identities*. In particular, development of the half-angle identities requires two of the forms of the cos 2A double-angle/number identity:

$$\cos 2A = 1 - 2 \sin^2 A$$
$$\cos 2A = 2 \cos^2 A - 1$$

• To obtain the *half-angle/number identity for sine*, begin with the *double-angle identity* for cosine in the form:

$$\cos 2A = 1 - 2 \sin^2 A$$

When A = x/2, this becomes:

$$\cos x = 1 - 2 \sin^2 x/2$$

Solving for sin(x/2):

$$2 \sin^2 x/2 = 1 - \cos x$$
$$\sin^2 x/2 = [1 - \cos x] / 2$$

Taking the square root of both sides, results in the **half-angle/number identity for sine**:

$$\sin x/2 = \pm\sqrt{\frac{1 - \cos x}{2}}$$

In this identity, the choice of the sign is determined by the quadrant of a coordinate system in which x/2 lies.

• To obtain the *half-angle/number identity for cosine*, begin with the *double-angle identity* for cosine in the form:

$$\cos 2A = 2 \cos^2 A - 1$$

When A = x/2, this becomes:

$$\cos x = 2 \cos^2 x/2 - 1$$

Solving for cos x/2:

$$2 \cos^2 x/2 = 1 + \cos x$$
$$\cos^2 x/2 = [1 + \cos x] / 2$$

Taking the square root of both sides, results in the ***half-angle/number identity for cosine***:

$$\cos x/2 = \pm\sqrt{\frac{1+\cos x}{2}}$$

In this identity, the choice of the sign is again determined by the quadrant of a coordinate system in which x/2 lies.

• To obtain a *half-angle/number identity for tangent*, the quotient identity and the half-angle formulas for sine and cosine can be used:

$$\tan x/2 = \frac{\sin x/2}{\cos x/2} = \frac{\pm\sqrt{\frac{1-\cos x}{2}}}{\pm\sqrt{\frac{1+\cos x}{2}}} = \pm\sqrt{\frac{1-\cos x}{1+\cos x}}$$

Therefore the ***half-angle/number identity for tangent*** is:

$$\tan x/2 = \pm\sqrt{\frac{1-\cos x}{1+\cos x}}$$

where the sign is determined by the quadrant of a coordinate system in which x/2 lies.

This form of the tangent identity can be expanded to obtain two other forms. Beginning with:

$$\tan x/2 = \pm\sqrt{\frac{1-\cos x}{1+\cos x}}$$

Multiply the numerator and denominator by $(1 + \cos x)$:

$$= \pm\sqrt{\frac{(1-\cos x)(1+\cos x)}{(1+\cos x)(1+\cos x)}} = \pm\sqrt{\frac{(1+\cos x - \cos x - \cos^2 x)}{(1+\cos x)^2}}$$

$$= \pm\sqrt{\frac{1-\cos^2 x}{(1+\cos x)^2}} = \pm\sqrt{\frac{\sin^2 x}{(1+\cos x)^2}}$$

resulting in a second form of the ***half-angle/number identity for tangent***:

$$\tan x/2 = \frac{\sin x}{1+\cos x}$$

To obtain a third form of the tangent half-angle/number identity begin with:

$$\tan x/2 = \pm\sqrt{\frac{1 - \cos x}{1 + \cos x}}$$

Multiply the numerator and denominator by $(1 - \cos x)$:

$$= \pm\sqrt{\frac{(1 - \cos x)(1 - \cos x)}{(1 + \cos x)(1 - \cos x)}} = \pm\sqrt{\frac{(1 - \cos x)^2}{1 - \cos^2 x}} = \pm\sqrt{\frac{(1 - \cos x)^2}{\sin^2 x}}$$

resulting in a third form of the *half-angle/number identity for tangent*:

$$\tan x/2 = \frac{1 - \cos x}{\sin x}$$

• **Example:** Find the exact value of $\sin(\pi/8)$.

Use the half-angle/number identity for sine

$$\sin x/2 = \pm\sqrt{\frac{1 - \cos x}{2}} \quad \text{and the fact that } \pi/8 \text{ is half of } \pi/4.$$

Therefore if $x/2 = \pi/8$, then $x = \pi/4$:

$$\sin \pi/8 = \sqrt{\frac{1 - \cos \pi/4}{2}} = \sqrt{\frac{1 - \sqrt{2}/2}{2}} = \sqrt{\frac{2 - \sqrt{2}}{4}} = \frac{1}{2}\sqrt{2 - \sqrt{2}}$$

To check this result we can use a calculator to estimate whether

$$\sin(\pi/8) = \frac{1}{2}\sqrt{2 - \sqrt{2}}:$$

$$\sin(\pi/8) \approx 0.38268, \text{ and } \frac{1}{2}\sqrt{2 - \sqrt{2}} \approx 0.38268.$$

7.11 Product-To-Sum Identities

• The *product-to-sum identities* are:

$$\sin A \cos B = [1/2][\sin(A + B) + \sin(A - B)]$$
$$\cos A \sin B = [1/2][\sin(A + B) - \sin(A - B)]$$
$$\sin A \sin B = [1/2][\cos(A - B) - \cos(A + B)]$$
$$\cos A \cos B = [1/2][\cos(A + B) + \cos(A - B)]$$

These identities involve the products of sines and cosines. The product-to-sum identities can be used when solving problems that involve a conversion of a product into a sum. They also have applications in sound and music. These identities can be developed using the four *sum and difference identities* for the sine and cosine.

• To obtain one of the product-to-sum identities, it is helpful to look at four *sum and difference identities* in order to identify which one can be used to derive the desired product-sum identity:

$$\sin(A + B) = \sin A \cos B + \cos A \sin B$$
$$\sin(A - B) = \sin A \cos B - \cos A \sin B$$
$$\cos(A + B) = \cos A \cos B - \sin A \sin B$$
$$\cos(A - B) = \cos A \cos B + \sin A \sin B$$

• To develop the *identity for sin A cos B* notice that in the sum and difference identities for sine, the sin A cos B term appears in both of these identities. Adding the sum and difference identities for sine:

$$\sin(A + B) = \sin A \cos B + \cos A \sin B$$
$$+ \ \sin(A - B) = \sin A \cos B - \cos A \sin B$$
$$\overline{\sin(A + B) + \sin(A - B) = 2 \sin A \cos B}$$

The cos A sin B terms had opposite signs and added to zero. To solve for term sin A cos B, multiply both sides by 1/2, resulting in the **product-to-sum identity for sin A cos B**:

$$\sin A \cos B = (1/2)[\sin (A + B) + \sin (A - B)]$$

• To develop the *identity for cos A sin B* notice that in the sum and difference identities for sine, the cos A sin B term appears in both of these identities. Subtracting the sum and difference identities for sine:

$$\sin(A + B) = \sin A \cos B + \cos A \sin B$$
$$- \ \sin(A - B) = \sin A \cos B - \cos A \sin B$$

$$\sin(A + B) - \sin(A - B)$$
$$= (\sin A \cos B + \cos A \sin B) - (\sin A \cos B - \cos A \sin B)$$

Simplifying:

$$\sin(A + B) - \sin(A - B) = 2 \cos A \sin B$$

The sin A cos B terms cancelled to zero. To solve for term cos A sin B, multiply both sides by 1/2, resulting in the ***product-to-sum identity for cos A sin B***:

cos A sin B = (1/2)[sin (A + B) − sin (A − B)]

• To develop the *identity for cos A cos B* notice that in the sum and difference identities for cosine, the cos A cos B term appears in both of these identities. Add the sum and difference identities for cosine:

cos (A + B) = cos A cos B − sin A sin B
+ cos (A − B) = cos A cos B + sin A sin B

cos (A + B) + cos (A − B) = 2 cos A cos B

The sin A sin B terms had opposite signs and added to zero. To solve for term cos A cos B, multiply both sides by 1/2, resulting in the ***product-to-sum identity for cos A cos B***:

cos A cos B = (1/2)[cos(A + B) + cos(A − B)]

• To develop the *identity for sin A sin B* notice that in the sum and difference identities for cosine, the sin A sin B term appears in both of these identities. Subtracting the sum and difference identities for cosine:

cos(A + B) = cos A cos B − sin A sin B
− cos(A − B) = cos A cos B + sin A sin B

cos(A + B) − cos(A − B) = − 2 sin A sin B

The cos A cos B terms cancelled to zero. To solve for term sin A sin B, multiply both sides by −1/2, resulting in the ***product-to-sum identity for sin A sin B***:

sin A sin B = (1/2)[cos(A − B) − cos(A + B)]

7.12 Sum/Difference-To-Product Identities

• The ***sum/difference-to-product identities*** are:

$$\sin x + \sin y = 2 \sin\left(\frac{x+y}{2}\right) \cos\left(\frac{x-y}{2}\right)$$

$$\sin x - \sin y = 2 \sin\left(\frac{x-y}{2}\right) \cos\left(\frac{x+y}{2}\right)$$

$$\cos x + \cos y = 2 \cos\left(\frac{x+y}{2}\right) \cos\left(\frac{x-y}{2}\right)$$

$$\cos x - \cos y = -2 \sin\left(\frac{x+y}{2}\right) \sin\left(\frac{x-y}{2}\right)$$

• Like the product-to-sum identities, the sum/difference-to-product identities can be used to express sums and differences involving sines and cosines. These identities can be derived from the *sum and difference identities* for cosine and sine or directly from the product-to-sum identities.

• To develop these identities, begin by changing variables:

Let x = A + B and y = A – B.

Next, add these two equations to solve for A and B:

$$x = A + B$$
$$+ y = A - B$$
$$\overline{}$$
$$x + y = 2 A$$

Solving for A: A = (x + y)/2

To obtain B subtract the two equations:

$$x = A + B$$
$$- y = A - B$$
$$\overline{}$$
$$x - y = 2 B$$

Solving for B: B = (x – y)/2

The *changed variables* are:

x = A + B, y = A – B, A = (x + y)/2, and B = (x – y)/2.

• To obtain the product-sum identities use the changed variables and the *sum and difference identities for sine and cosine*:

$\sin(A + B) = \sin A \cos B + \cos A \sin B$

$\sin(A - B) = \sin A \cos B - \cos A \sin B$

$\cos(A + B) = \cos A \cos B - \sin A \sin B$

$\cos(A - B) = \cos A \cos B + \sin A \sin B$

Now substitute the changed variables into these identity equations:

$\sin x = \sin((x + y)/2) \cos((x - y)/2) + \cos((x + y)/2) \sin((x - y)/2)$

$\sin y = \sin((x + y)/2) \cos((x - y)/2) - \cos((x + y)/2) \sin((x - y)/2)$

$\cos x = \cos((x + y)/2) \cos((x - y)/2) - \sin((x + y)/2) \sin((x - y)/2)$

$\cos y = \cos((x + y)/2) \cos((x - y)/2) + \sin((x + y)/2) \sin((x - y)/2)$

• To derive the identity for *sin x + sin y*, add the two changed-variable sum/difference identities for sin x and sin y:

$\sin x = \sin((x + y)/2) \cos((x - y)/2) + \cos((x + y)/2) \sin((x - y)/2)$

$+ \sin y = \sin((x + y)/2) \cos((x - y)/2) - \cos((x + y)/2) \sin((x - y)/2)$

_____

$\sin x + \sin y = 2 \sin((x + y)/2) \cos((x - y)/2)$

which is the **sum-to-product identity for sin x + sin y**.

• To derive the identity for *sin x − sin y*, subtract the two changed-variable sum/difference identities for sin x and sin y:

$\sin x = \sin((x + y)/2) \cos((x - y)/2) + \cos((x + y)/2) \sin((x - y)/2)$

$-\sin y = \sin((x + y)/2) \cos((x - y)/2) - \cos((x + y)/2) \sin((x - y)/2)$

_____

$\sin x - \sin y = 2 \cos((x + y)/2) \sin((x - y)/2)$

which is the **difference-to-product identity for sin x − sin y**.

• To derive the identity for *cos x + cos y*, add the two changed-variable sum/difference identities for cos x and cos y:

$\cos x = \cos((x + y)/2) \cos((x - y)/2) - \sin((x + y)/2) \sin((x - y)/2)$

$+ \cos y = \cos((x + y)/2) \cos((x - y)/2) + \sin((x + y)/2) \sin((x - y)/2)$

_____

$\cos x + \cos y = 2 \cos((x + y)/2) \cos((x - y)/2)$

which is the **sum-to-product identity for cos x + cos y**.

• To derive the identity for *cos x – cos y*, subtract the two changed-variable sum/difference identities for cos x and cos y:

$$\cos x = \cos((x + y)/2)\cos((x - y)/2) - \sin((x + y)/2)\sin((x - y)/2)$$
$$-\cos y = \cos((x + y)/2)\cos((x - y)/2) + \sin((x + y)/2)\sin((x - y)/2)$$

$$\overline{\cos x - \cos y = -2\sin((x + y)/2)\sin((x - y)/2)}$$

which is the ***difference-to-product identity for cos x – cos y***.

• The *sum/difference-to-product to identities* can also be derived directly from the *product-to-sum identities* and using the changed variables:

$$\sin A \cos B = [1/2][\sin(A + B) + \sin(A - B)]$$
$$\cos A \sin B = [1/2][\sin(A + B) - \sin(A - B)]$$
$$\sin A \sin B = [1/2][\cos(A - B) - \cos(A + B)]$$
$$\cos A \cos B = [1/2][\cos(A + B) + \cos(A - B)]$$

Rearranging:

$$2\sin A \cos B = [\sin(A + B) + \sin(A - B)]$$
$$2\cos A \sin B = [\sin(A + B) - \sin(A - B)]$$
$$2\sin A \sin B = [\cos(A - B) - \cos(A + B)]$$
$$2\cos A \cos B = [\cos(A + B) + \cos(A - B)]$$

Substitute the changed variables

$$x = A + B, y = A - B, A = (x + y)/2, \text{ and } B = (x - y)/2:$$
$$2\sin((x + y)/2)\cos((x - y)/2) = \sin x + \sin y$$
$$2\cos((x + y)/2)\sin((x - y)/2) = \sin x - \sin y$$
$$2\sin((x + y)/2)\sin((x - y)/2) = \cos y - \cos x$$
$$2\cos((x + y)/2)\cos((x - y)/2) = \cos x + \cos y$$

Rearranging gives the four ***sum/difference-to-product identities***:

$$\sin x + \sin y = 2\sin((x + y)/2)\cos((x - y)/2)$$
$$\sin x - \sin y = 2\cos((x + y)/2)\sin((x - y)/2)$$
$$\cos y - \cos x = 2\sin((x + y)/2)\sin((x - y)/2)$$
$$\cos x + \cos y = 2\cos((x + y)/2)\cos((x - y)/2)$$

Note that cos y − cos x = 2 sin((x + y)/2) sin((x − y)/2) is also written:

cos x − cos y = − 2 sin((x + y)/2) sin((x − y)/2)

• **Example:** Write the following as a product: cos 6A + cos 4A.

Using the sum to product identity for cos x + cos y:

cos x + cos y = 2 cos((x + y)/2) cos((x − y)/2)

Substitute x = 6A and y = 4A:

cos 6A + cos 4A = 2 cos((6A + 4A)/2) cos((6A − 4A)/2)
$$= 2 \cos(10A/2) \cos(2A/2)$$
$$= 2 \cos 5A \cos A$$

To check the result, choose A = 30°. Does cos 6A + cos 4A equal 2 cos 5A cos A?

cos 6(30°) + cos 4(30°) = −1.5
2 cos 5(30°) cos(30°) = −1.5

Using this rule would be helpful in a situation where, for example, 6A and 4A are angles where cosine is not easily known, whereas cos 5A and cos A are known.

7.13 Squared Formulas

• The *squared formulas* are:

$\sin^2 A = (1/2)(1 − \cos(2A))$
$\cos^2 A = (1/2)(1 + \cos(2A))$

• The *squared formulas* $\sin^2 A$ *and* $\cos^2 A$ can be developed using the *double-angle/number identities*. In particular, development of the squared formulas uses two of the forms of the cos 2A double-angle/number identity:

cos 2A = 1 − 2 sin² A
cos 2A = 2 cos² A − 1

To obtain the formula for $\sin^2 A$ solve the double-number equation:

cos 2A = 1 − 2 sin² A

Rearranging:

$$\cos 2A - 1 = -2 \sin^2 A$$
$$1 - \cos 2A = 2 \sin^2 A$$
$$\sin^2 A = (1/2)(1 - \cos 2A)$$

which is the *squared formula $\sin^2 A$*.

To obtain the formula for $\cos^2 A$ solve the double-number equation:

$$\cos 2A = 2 \cos^2 A - 1$$

Rearranging:

$$\cos 2A + 1 = 2 \cos^2 A$$
$$2 \cos^2 A = 1 + \cos 2A$$
$$\cos^2 A = (1/2)(1 + \cos 2A)$$

which is the *squared formula $\cos^2 A$*.

7.14 Chapter 7 Summary and Highlights

• *Identities* are used to solve problems involving trigonometric or circular functions by expressing one trigonometric function in terms of another. Using trigonometric identities to make substitutions is often required to solve problems involving trigonometry. Trigonometric identities are equations that are true for all possible values of angles or real numbers.

• The quotient and reciprocal identities, the Pythagorean identities, the negative angle/number identities, and the sum and difference of angles/ numbers identities are particularly important because they can be used to derive other identities. In particular, the sum and difference of angles/numbers identities can be used to derive the product-to-sum identities, the sum/difference-to-product identities, and the double-angle identities, which are used to derive the half-angle identities.

• The following are important identities to remember.

Quotient Identities and Reciprocal Identities:

$$\tan A = \sin A/\cos A, \csc A = 1/\sin A, \sec A = 1/\cos A, \text{ and}$$
$$\cot A = 1/\tan A = \cos A/\sin A$$

Pythagorean Identities:

$\sin^2(A) + \cos^2(A) = 1$, $\sin^2(A) = 1 - \cos^2(A)$, and
$\cos^2(A) = 1 - \sin^2(A)$

(These are derived from the Pythagorean formula $x^2 + y^2 = r^2$.)

Negative Angle/Number Identities:

$\sin(-A) = -\sin A$ and $\cos(-A) = \cos A$

Sum and Difference of Angles/Numbers Identities, also called Addition and Subtraction Identities:

$\sin(A + B) = \sin(A) \cos(B) + \sin(B) \cos(A)$
$\cos(A + B) = \cos(A) \cos(B) - \sin(A) \sin(B)$
$\sin(A - B) = \sin(A) \cos(B) - \sin(B) \cos(A)$
$\cos(A - B) = \cos(A) \cos(B) + \sin(A) \sin(B)$

- Identities that can be derived from the above include:

Cofunction Identities:

$\sin A = \cos(\pi/2 - A)$ and $\cos A = \sin(\pi/2 - A)$

Double-Angle/Number Identities:

$\sin(2A) = 2 \sin(A) \cos(A)$ and $\cos(2A) = \cos^2(A) - \sin^2(A)$

(These identities are special cases of the sum/addition identities.)

Half-Angle Identities:

$$\sin x/2 = \pm\sqrt{\frac{1 - \cos x}{2}} \text{ and } \cos x/2 = \pm\sqrt{\frac{1 + \cos x}{2}}$$

(These are derived from the double-angle identities.)

Product-to-Sum Identities:

$\sin(A) \cos(B) = [1/2][\sin(A + B) + \sin(A - B)]$
$\cos(A) \sin(B) = [1/2][\sin(A + B) - \sin(A - B)]$
$\sin(A) \sin(B) = [1/2][\cos(A - B) - \cos(A + B)]$
$\cos(A) \cos(B) = [1/2][\cos(A + B) + \cos(A - B)]$

Sum/Difference-to-Product Identities:

$$\sin x + \sin y = 2 \sin((x + y)/2) \cos((x - y)/2)$$
$$\sin x - \sin y = 2 \cos((x + y)/2) \sin((x - y)/2)$$
$$\cos y - \cos x = 2 \sin((x + y)/2) \sin((x - y)/2)$$
$$\cos x + \cos y = 2 \cos((x + y)/2) \cos((x - y)/2)$$

• The Supplementary Angle Relations, $\sin(\pi - A) = \sin A$ and $\cos(\pi - A) = -\cos A$ are also important.

• It may be necessary to verify that a trigonometric relationship is also an identity, which means that the relationship is true for all possible values of angles or real numbers. Trigonometric relationships can be verified as identities using algebra and substitutions. If you are given an equation that is not a known identity, it is possible to verify that the two sides are equivalent for all values by using algebraic manipulations and substitutions or by using a graphing calculator or graphing software and graphing the expressions on each side of the equal sign to see if they overlap for all values of x.

Chapter 8

Trigonometric Functions in Equations and Inequalities

8.1 Review of Solving Algebraic Equations

• This section provides a brief summary of a number of the important algebraic techniques that are applicable to solving equations and inequalities involving trigonometric and circular functions. Please see *Master Math: Algebra* for a comprehensive treatment of algebra that includes detailed explanations and examples of the information in this section. In an *equation* there are expressions on each side of an equal sign that are equivalent. In order to solve an equation, a value or values must be determined that make the equation true. The simplest algebraic equation is a *linear equation*, which contains no powers of the variable and has the form $ax + b = 0$, where a and b are constants and a does not equal zero. More complicated algebraic equations include *non-linear equations*, such as the *quadratic equation*, in which the variable is squared, and it has the standard form $ax^2 + bx + c = 0$, where a and b are constants and a does not equal zero. The variables in a nonlinear equation may also be raised to powers greater than 2. The solution(s) of polynomial equations are called ***roots***.

• The *argument* of an equation is represented by each *input* number to the function and is the first member of each ordered pair of the function. For example, in the functions

$$y = \log(x + 2), y = \sin(x - \pi), \text{ and } y = \cos x,$$

the arguments are

$$(x + 2), (x - \pi), \text{ and } x, \text{ respectively.}$$

If the argument has more than one term, it is generally enclosed in parentheses.

• To *solve an equation* for an unknown variable, the variable must be isolated to one side of the equal sign so that:

unknown variable = known numbers or values.

In the process of solving an equation, if a number is added, subtracted, multiplied, or divided to or from one side of the equal sign, the same operation must be executed on the other side of the equal sign. For example, to solve $4x = 32$, in order to isolate x we must divide both sides of the equation by 4:

$$4x/4 = 32/4$$

Therefore, $x = 8$.

Equations Containing Fractions

• In equations that contain fractions it is helpful to remove the fractions by reducing each term, identifying the lowest common denominator, multiplying each term by the lowest common denominator, and solving the equation by isolating the unknown. It may be beneficial to factor and reduce algebraic fractions by factoring the numerator and the denominator separately, then reducing by comparing the numerator to the denominator and canceling common factors. For example, to solve the simple equation $2x/3 - 2x/4 = 4$ for x:

First, reduce second term:

$2x/3 - x/2 = 4/1$

Identify the lowest common denominator as 6.

Multiply each term by 6:

$(6)2x/3 - (6)x/2 = (6)4/1$

The denominators will cancel with the common denominator resulting in no fractions in the equation:

$4x - 3x = 24$

Combine like terms:

$1x = 24$, or

$x = 24$

To verify an answer, substitute the result back into the original equation:

$2(24)/3 - 2(24)/4 = 4$

$2(8) - 2(6) = 4$

$16 - 12 = 4$

$4 = 4$

• When solving an equation, it may be necessary to multiply, divide, add, or subtract fractions within the equation. The following paragraphs describe each of these operations.

• To *multiply algebraic fractions*, factor each numerator and denominator, reduce each fraction by canceling factors common to the numerator and denominator, multiply the numerators with each other and the denominators with each other, and reduce the resulting fraction by canceling factors common to the numerator and denominator. For example, multiply $(6y + 6z)/(4x^2 + 4x)$ by $(4x)/(2y + 2z)$.

$$\frac{6y + 6z}{4x^2 + 4x} \times \frac{4x}{2y + 2z} = \frac{6(y + z)}{4x(x + 1)} \times \frac{4x}{2(y + z)} = \frac{3(y + z)}{2x(x + 1)} \times \frac{2x}{(y + z)}$$

$$= \frac{3(y + z)(2x)}{(2x)(x + 1)(y + z)} = \frac{3}{(x + 1)}$$

• To *divide algebraic fractions*, change the division format into a multiplication format by multiplying the first fraction by the reciprocal of the second fraction. Using the reciprocal of the second fraction provides a multiplication format. Therefore, to multiply the first fraction by the reciprocal of the second fraction, use the same steps as multiplying fractions listed in the previous paragraph.

• To *add or subtract fractions with common denominators*, add or subtract the numerators, place the result over the common denominator, and reduce the resulting fraction by factoring and canceling factors common to the numerator and denominator. For example, subtract:

$$\frac{2x}{3(x-2)} - \frac{4}{3(x-2)} = \frac{2x-4}{3(x-2)} = \frac{2(x-2)}{3(x-2)} = \frac{2}{3}$$

• To *add or subtract fractions with different denominators*, first form fractions that are equivalent to the original fractions but have a common denominator, then add or subtract the numerators. The procedure is as follows:

1. Find the lowest common denominator contained in each term. The lowest common denominator will contain each different factor to the highest power it occurs in any of the denominators. For example, if (x + 1) occurs twice in one of the original denominators, it must occur twice in the common denominator. To find a common denominator, factor each denominator and then write multiples of each denominator until a common multiple in found.

2. Multiply each fraction by a different fraction (with its numerator equal to its denominator) to create new fractions with common denominators that are equivalent to the original fractions. (By having the numerator equal to the denominator, the value of each of the fractions remains unchanged (for example, 1/2 = 2/4)). To determine what each multiplying fraction needs to be, compare the new common denominator with the denominators of each of the original fractions and create new fractions that contain the factors that are in the common denominator but not in each original denominator.

3. After new equivalent fractions with common denominators have been created, add or subtract the numerators, place the result over the common denominator, and reduce the resulting fraction by factoring and canceling factors common to the numerator and denominator.

For example, add the following fractions:

$$\frac{(x+2)}{2(x-2)} + \frac{4(x-2)}{x(x+2)}$$

The lowest common denominator is $(2)(x)(x-2)(x+2)$.

The fractions become:

$$= \frac{(x+2)(x)(x+2)}{(2)(x)(x-2)(x+2)} + \frac{(4)(x-2)(2)(x-2)}{(2)(x)(x-2)(x+2)}$$

$$= \frac{(x+2)(x)(x+2)+(4)(x-2)(2)(x-2)}{(2)(x)(x-2)(x+2)}$$

Multiplying and adding like terms results in:

$$\frac{x^3 + 12x^2 - 28x + 32}{2x^3 - 8x}$$

Equations with Square Roots

• In equations that contain *square roots* it is helpful to isolate the *radical* term on one side of the equal sign, combine like terms, square each side of the equation to eliminate the radical sign, and solve the equation for the unknown variable. Note that if the equation contains a cubed root or fourth root, remove the radical sign by raising both sides of the equation to a power of three or four. In addition, remember that the square root, cubed root, or fourth root can be represented as $\sqrt{x}$, $\sqrt[3]{x}$, or $\sqrt[4]{x}$, or equivalently as $(x)^{1/2}$, $(x)^{1/3}$, or $(x)^{1/4}$, respectively.

Example: Solve $2(x-1)^{1/2} = 8$.

Isolate radical term:

$$(x-1)^{1/2} = 8/2$$
$$(x-1)^{1/2} = 4$$

Square both sides and solve for x:

$$(x-1)^{(1/2)(2)} = 4^2$$
$$x - 1 = 16$$
$$x = 17$$

8.2 Review of Solving Algebraic Quadratic Equations

• *Quadratic equations* are nonlinear equations with a second-degree term. If x is the unknown variable in a quadratic equation, its highest power is 2 and the variable would be expressed as x^2. Quadratic equations have the form $ax^2 + bx + c = 0$. In a quadratic equation the coefficient a can never be zero, but coefficient b or c can be zero. When quadratic equations are written, the coefficients for the second-degree term, the first-degree term, and the constant term are represented by a, b, and c, respectively. Examples of quadratic equations include:

$$ax^2 + bx + c = 0$$
$$3x^2 + 2x + 5 = 0$$
$$x^2 + 2 = 0$$

• There are several methods that are used to solve quadratic equations with one unknown variable. Certain methods are more applicable to particular forms of the equation, and they include:

1. *Factoring* is useful for solving quadratic equations with two or three terms.
2. The *quadratic formula* is useful for solving any quadratic equation, particularly those with three terms.
3. The *square-root method* is useful for solving quadratic equations with two terms if the b coefficient is zero, resulting in the absence of the first-degree term.
4. The *method of completing the square* is useful for solving quadratic equations, particularly those with a non-zero b coefficient. This is an alternative method to using the quadratic formula.

Factoring

• The method of *factoring* is useful for solving many quadratic equations with one unknown variable that have the form of a binomial or a trinomial. The factored form of a polynomial is an expression of the polynomial as a product of monomials or polynomials that, when multiplied together, equal the polynomial.

• When solving equations or simplifying expressions, check to see if there is a *common monomial factor* in each term. When factoring a polynomial with a common monomial factor, factor out the greatest common factor. For example, factoring $2x^2 + 4x^2y$ results in:

$$2x^2 + 4x^2y = 2x^2(1 + 2y)$$

where $2x^2$ and $(1 + 2y)$ are factors and $2x^2$ is the greatest common factor.

• When solving equations or simplifying expressions containing *trinomials* in the form $ax^2 + bx + c$, it is generally beneficial to factor the trinomial. Factoring a trinomial in the form $ax^2 + bx + c$ results in two *binomials*. Factoring a trinomial is the reverse of multiplying two binomials. Recall the steps involved in multiplying binomials.

For example, multiply the binomials $(x + 2)(x + 3)$:

$$(x + 2)(x + 3) = x^2 + 3x + 2x + (2)(3) = x^2 + 5x + 6$$

In this case the trinomial is $x^2 + 5x + 6$, and the factored form is $(x + 2)(x + 3)$.

• The factored binomial form and the trinomial form of a simple expression can be illustrated as:

$(x + m)(x + n)$ which is the *factored form*
$x^2 + (m+n)x + mn$ which is the *trinomial form*

where m and n represent numbers.

Similarly for a more complicated expression, the factored binomial form and the trinomial form can be illustrated as:

$(px + m)(qx + n)$, which is the *factored form*
$= pqx^2 + pnx + qmx + mn$
$pqx^2 + (pn+qm)x + mn$, which is the *trinomial form*,

where p, q, m, and n represent numbers.

• The following procedure can be used to factor a trinomial:

1. Write the factored format ( )( ).
2. Find sets of two values that, when multiplied together, equal the first term of the trinomial.
3. Find sets of two values that, when multiplied together, equal the last term of the trinomial.
4. Choose the sets such that the sum of the outer product and the inner product of the binomial is equal to the second term (the first degree term) of the trinomial. (Remember to be careful of negative signs.)

5. Multiply the resulting binomials to check that the original trinomial is
   obtained.

- **Example:** Factor the trinomial $x^2 + 5x + 6$.

Write the factored format ( )( )

Find sets of two values that, when multiplied together, equal the first term
of the trinomial. The only set is: x and x.

  (x)(x)

Find sets of two values that, when multiplied together, equal the last term
of the trinomial.

The possible sets are 2 and 3 or 1 and 6

  Set 1: (2)(3)
  Set 2: (1)(6)

Choose these sets such that the sum of the outer product and the inner
product of the binomial is equal to the second term of the trinomial.

The second term of the trinomial is 5x.

Therefore: Outer product + inner product must equal 5x.

  Set 1: 3x + 2x = 5x
  Set 2: 6x + –1x = 5x

Because there are no negative signs in the original trinomial, Set 2 is
eliminated. Therefore, the Set 1 binomial, $(x + 2)(x + 3)$, must be the
factored binomial.

Multiply the chosen binomial set to check that it produces the original
trinomial.

  $(x + 2)(x + 3) = x^2 + 3x + 2x + 6 = x^2 + 5x + 6$

Therefore, the factored form of $x^2 + 5x + 6$ is $(x + 2)(x + 3)$.

- There are special binomial products that are worth remembering:

1. The *difference of two squares* $x^2 - y^2 = (x + y)(x - y)$, because the
   sum of the inner and outer products will always equal zero.

2. The *sum of two squares* $x^2 + y^2$ cannot be further factored.
3. The *binomial squared* has the two following forms:

   Sum squared $(x + y)^2 = x^2 + 2xy + y^2$,
   because $(x + y)(x + y) = x^2 + xy + xy + y^2 = x^2 + 2xy + y^2$
   Difference squared $(x - y)^2 = x^2 - 2xy + y^2$,
   because $(x - y)(x - y) = x^2 - xy - xy + y^2 = x^2 - 2xy + y^2$

- To solve a quadratic equation using the method of *factoring*:

1. Express the equation in the form of a quadratic equation,
   $ax^2 + bx + c = 0$.
2. Factor the quadratic expression, $ax^2 + bx + c$.
3. Set each factor equal to zero and solve each resulting equation for the
   unknown variable.
4. Check the solutions by substituting into the original equation.

- **Example:** Solve $x^2 - 2x = 3$.

First express in the form, $ax^2 + bx + c = 0$:

$$x^2 - 2x - 3 = 0$$

Factor the quadratic expression $(x^2 - 2x - 3)$ by finding sets of two values
that, when multiplied, equal the first term of the trinomial. Find sets of
two values that when multiplied equal the last term of the trinomial.
Possible binomial sets are:

$(x + 1)(x - 3)$
$(x - 1)(x + 3)$

Choose these sets where the sum of the outer product and the inner
product of the binomial is equal to the second term (the first-degree term)
of the trinomial.

$(x + 1)(x - 3)$ outer product $= -3x$, inner product $= 1x$, sum $= -2x$
$(x - 1)(x + 3)$ outer product $= 3x$, inner product $= -1x$, sum $= 2x$

The second term of the trinomial is $-2x$.

Therefore, choose $(x + 1)(x - 3)$.

Check factors by multiplying the resulting binomials:

$(x + 1)(x - 3) = x^2 - 3x + 1x - 3 = x^2 - 2x - 3$.

Therefore, the factors are $(x + 1)$ and $(x - 3)$.

Set each factor equal to zero.

$(x + 1) = 0$

$(x - 3) = 0$

Solve for the unknown variable in each equation.

$x + 1 = 0$

$x = -1$

$x - 3 = 0$

$x = 3$

Therefore, the solutions for $x^2 - 2x = 3$, are $x = -1$ and $x = 3$.

Check these solutions by substituting each one into the original equation $x^2 - 2x = 3$.

Substitute $x = -1$: $1 + 2 = 3$, or $3 = 3$

Substitute $x = 3$: $9 - 6 = 3$, or $3 = 3$

(See Section 8.4 where this example is solved using graphing.)

Quadratic Formula

• The *quadratic formula* can be used to find the solution to any quadratic equation, particularly equations that have the form of a trinomial. The quadratic formula is worth memorizing. The quadratic formula is:

$$x = \frac{-b \pm \sqrt{b^2 - 4ac}}{2a}$$

The expression inside the square root is called the *discriminant* and can be used to determine if the equation will have *real or imaginary roots*. Remember that the roots of an equation are the solutions, or values of x, of the equation and correspond to where the graph of the equation crosses the X-axis (where $y = 0$).

If $b^2 - 4ac > 0$, the equation has 2 distinct real roots.

If $b^2 - 4ac < 0$, the equation has 2 imaginary roots.

If $b^2 - 4ac = 0$, the equation has 1 real root.

Note that the quadratic formula can be derived using the method of completing the square.

- To solve a quadratic equation using the quadratic formula:

1. Express the equation in the form of a quadratic equation, $ax^2 + bx + c = 0$.
2. Identify the values for the coefficients a, b, and c.
3. Substitute the values for a, b, and c into the quadratic formula.
4. Reduce the resulting equation by performing the indicated arithmetic operations and simplify the radical.
5. Check the solutions by substituting into the original equation.

- **Example:** Solve $3x^2 + 2 = -5x$.

Express in the form of a quadratic equation $ax^2 + bx + c = 0$:

$$3x^2 + 5x + 2 = 0$$

Identify the values for the coefficients a, b, and c:

$$a = 3, b = 5, \text{ and } c = 2$$

Substitute the values for a, b, and c into the quadratic formula.

$$x = \frac{-5 \pm \sqrt{5^2 - (4)(3)(2)}}{2(3)} = \frac{-5 \pm \sqrt{1}}{6} = \frac{-5 \pm 1}{6}$$

Because of the $\pm$ sign, both $+$ and $-$ must be accounted for:

$$x = (-5 + 1)/6 \quad \text{and} \quad x = (-5 - 1)/6$$

Reduce each equation:

$$x = -2/3 \quad \text{and} \quad x = -1$$

Therefore, the solutions for $3x^2 + 2 = -5x$ are $x = -2/3$ and $x = -1$.

The solutions can be verified by substituting into the original equation:

$$3(-2/3)^2 + 2 = -5(-2/3) \text{ and } 3(-1)^2 + 2 = -5(-1).$$

Square-Root Method

• The *square-root method* is useful for solving *quadratic equations* with two terms if the b coefficient is zero, resulting in the absence of the first-degree term. When the b coefficient is zero, the form of the quadratic equation changes from $ax^2 + bx + c = 0$ to $ax^2 + c = 0$.

• To solve a quadratic equation using the *square-root method*:

1. Isolate the x^2 variable on one side of the equal sign (not the whole term).
2. Take the square root of both sides of the equation.
   This transforms x^2 into x, because $\sqrt{x^2} = \pm x$.
3. Simplify.
4. Check the solutions by substituting into the original equation.

• **Example:** Solve $3x^2 + 4 = 31$.

Subtract 4 from both sides of the equation:

$$3x^2 = 27$$

Isolate the x^2 variable onto one side of the equal sign by dividing both sides by 3.

$$x^2 = 9$$

Take the square root of both sides of the equation.

$$\sqrt{x^2} = \sqrt{9}$$
$$x = \pm 3$$

Therefore, the solutions to $3x^2 + 4 = 31$ are $x = 3$ and $x = -3$.

The solutions can be verified by substituting them into the original equation: $3(3)^2 + 4 = 31$ and $3(-3)^2 + 4 = 31$.

Method of Completing the Square

• The *method of completing the square* is an extension of the square-root method that is useful for solving *quadratic equations*, particularly equations with a non-zero b coefficient. It is an alternate method to the quadratic formula described previously.

• To solve a quadratic equation using the method of completing the square:

1. Express the equation in the form $x^2 + bx = c$. To obtain this form it may be necessary to divide each term by the coefficient a:

    $ax^2/a + bx/a = c/a.$

2. Complete the square by finding one-half of the coefficient b, or (b/2). Then square one-half of coefficient b, $(b/2)^2$, and add the result to each side of the equation.
3. Factor the resulting perfect square trinomial expression into a binomial-squared, and combine like terms.
4. Solve using the square-root method, by isolating the x^2 variable on one side of the equal sign, taking the square root of both sides of the equation, and simplifying the radicals.
5. Check the solutions by substituting into the original equation.

(Please see *Master Math: Algebra* for a detailed discussion with examples.)

8.3 Review of Solving Algebraic Inequalities

• Instead of solving an algebraic equation with an equal sign, it may be necessary to solve an algebraic *inequality*. Inequalities are identified by the symbols $>$, $<$, $\geq$, and $\leq$. *Inequalities* are represented by the symbols for greater than $>$, less than $<$, greater than or equal to $\geq$, and less than or equal to $\leq$, and describe expressions in which the value of the expression on one side of the symbol is greater than or greater than or equal to the value of the expression on the other side of the symbol.

• When solving inequalities involving adding, subtracting, multiplying, and dividing positive and negative numbers, the following rules apply:

1. If a number is *added to or subtracted from* both sides of the inequality, the inequality sign *remains unchanged*.
2. If a *positive* number is *multiplied to or divided into* both sides of the inequality, the inequality sign *remains unchanged*.
3. If a *negative* number is *multiplied to or divided into* both sides of the inequality, the inequality sign *reverses*.

• Except for adjusting the inequality sign as described in these rules, inequalities are solved using the same techniques that are used to solve equations.

For example, solve $3 - (x/3) \le -5$.

Subtract 3 from both sides.

$-(x/3) \le -8$

Multiply both sides by -3. (Inequality sign reverses.)

$x \ge (-8)(-3)$
$x \ge 24$

8.4 Solving Algebraic Equations and Inequalities Using Graphing

• An equation can be solved graphically using two methods. In one method the equation is plotted and where the curve crosses the X-axis corresponds to the roots, or solutions, to the equation. This is called the *x-intercept method.* In the second method the expressions on the two sides of the equal sign are each plotted on the same graph, and the solutions correspond to the x-coordinates of the points where the two curves or lines intersect. This is called *intersection-of-graphs.*

• To solve an equation graphically using the *x-intercept method* where the solution(s) are the points where the curve crosses the X-axis:

1. Simplify the equation to the form $f(x) = 0$ with all the terms on one side of the equal sign, and if it is a quadratic equation put it in standard form $ax^2 + bx + c = 0$, and replace 0 with y.
2. Graph the equation by hand or using a graphing utility.
3. Determine the solutions, or roots, for x by estimating the points where the curve crosses the X-axis (at values of $y = 0$). Any number that satisfies the equation is an x-intercept of the graph.

• To solve an *inequality* graphically using the *x-intercept method*, the solution set for $f(x) < 0$ consists of the x-values of the points on the graph of $f(x)$ that lie below the X-axis. The solution set for $f(x) > 0$ consists of the x-values of points on the graph of $f(x)$ that lie above the X-axis. If the inequality involves $\le$ or $\ge$, the solutions include the solution of the equation $f(x) = 0$ as well as the solution set for $f(x) < 0$ or $f(x) > 0$.

• To solve an equation using the *intersection-of-graphs* method where the solutions correspond to the x-coordinates of the points where the two curves or lines intersect:

1. Begin with the equation in the form $f(x) = g(x)$.
2. Plot the expressions on each side of the equal sign as $y_1 = f(x)$ and $y_2 = g(x)$.
3. The solutions correspond to the x-coordinate(s) of the point(s) where the graphs of the two curves or lines intersect each other.

• To solve an ***inequality*** using the ***intersection-of-graphs*** method, the solution set for $f(x) < g(x)$ is the set of real numbers x where the curve or line representing $f(x)$ is below the curve or line representing $g(x)$. In other words, the solution set of $f(x) < g(x)$ consists of all x-values (that correspond to points on the X-axis), where the graph of $f(x)$ lies below the graph of $g(x)$. Conversely, the solution set for $f(x) > g(x)$ is the set of real numbers x where the curve or line representing $f(x)$ is above the curve or line representing $g(x)$. In other words, the solution set of $f(x) > g(x)$ consists of all x-values (that correspond to points on the X-axis), where the graph of $f(x)$ lies above the graph of $g(x)$. If the inequality involves $\leq$ or $\geq$, the solutions include the solution of the equation $f(x) = g(x)$ as well as the solution set for $f(x) < g(x)$ or $f(x) > g(x)$.

• Note that the *intersection-of-graphs* method can also be applied to *solving two equations*. Two equations can be solved graphically by plotting both equations. The solutions correspond to the point(s) where the graphs of the two equations intersect each other.

• Also note that if two sides of an *identity* are graphed the curves will overlap for all points, and if two sides of an equation are graphed the curves will overlap at points that represent the solutions of the equation.

• **Example:** In the discussion on the factoring method in the previous section, 8.2, the equation $x^2 - 2x = 3$ was solved and the solutions, or roots, were found to be $x = -1$ and $x = 3$. Solve this same equation using the ***intersection-of-graphs*** method.

Plot the expressions on each side of the equal sign $y = x^2 - 2x$ and $y = 3$:

Graph of y = x² – 2x (curve) and y = 3 (line)

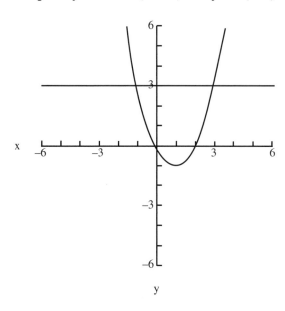

The solutions correspond to the x-coordinate(s) of the point(s) where the graphs of the two curves or lines intersect each other. In this graph the solutions correspond to $x = -1$ and $x = 3$, which are the same solutions obtained using factoring.

• **Example:** Solve this same equation, $x^2 - 2x = 3$, that was solved using the factoring method and the intersection of graphs method in the previous example using the ***x-intercept method***.

Simplify the equation to the form $f(x) = 0$ and replace 0 with y:

$y = x^2 - 2x - 3.$

Graph the equation using a graphing utility.

Graph of y = x² – 2x – 3

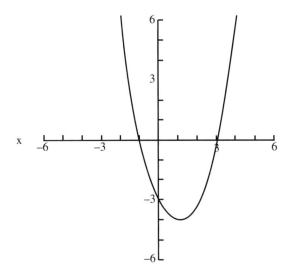

The solutions, or roots, for x correspond to the points where the curve crosses the X-axis (at values of y = 0). In this graph the solutions correspond to x = –1 and x = 3, which are the same solutions obtained using factoring and the intersection-of-graphs method.

8.5 Introduction to Solving Trigonometric Equations and Inequalities

• Equations that involve trigonometric and circular functions of an unknown angle can be *identities* and therefore are true for all angles or real numbers, or *conditional equations* and therefore are true for certain values of angles or real numbers. For example, sin x = 0 is a conditional equation because it is not true for all values of x, such as π/2 (or 90°). This equation is true for x = 0 and x = π (or 180°), or x = nπ where n is an integer.

• *Equations* involving trigonometric or circular functions can be solved for the unknown variable using algebraic techniques such as factoring and substituting into the quadratic equation, substituting identities, reference triangles, location of a point on a unit circle, and using graphical techniques such as the x-intercept and intersection-of-graphs methods. Trigonometric or circular *inequalities* can be solved using the same algebraic techniques or by observing where the graph of the corresponding equation lies above or below the X-axis. When solving these equations and inequalities, substitutions using trigonometric identities, such as the double-number and half-number identities are often helpful. When an equation or inequality cannot be solved analytically using algebra, it may be solved (or the solutions estimated) graphically. Algebraic techniques may result in exact solutions or approximate solutions; however, graphing techniques represent approximate solutions. In addition, many calculators and computer software packages have *solver* programs that can be used to solve equations.

• Because trigonometric and circular functions are periodic, and numerous coterminal angles can exist (multiples of 2π or $360°$ for any angle in standard position), multiple solutions often occur. Therefore, solutions to equations involving trigonometric or circular functions are specified within a **defined interval**, usually $0 \le x < 2\pi$ or $0 \le x < 360°$.

• The remainder of this chapter provides examples of solving equations and inequalities that involve trigonometric functions using various techniques.

8.6 Solving Simple Trigonometric Equations Using Standard Position Angles, Reference Triangles, and Identities

• There is often more than one method that can be used to solve a given trigonometric equation. Simple trigonometric equations can, for example, be solved using standard position angles, reference triangles, and substitutions of identities.

• **Example:** Solve $\sin A = 1/(2)^{1/2}$ for $0 \le A \le 360°$.
(Remember: $(x)^{1/2}$ denotes the square root of x.)

To solve this equation, find the angle A whose sine is $1/(2)^{1/2}$.

Remember that the sine = opposite/adjacent and that a 45:45:90 triangle has sides proportional to:

In a coordinate system for an angle in standard position within the range of $0 \leq A \leq 360°$, a 45:45:90 *reference triangle* depicts two possible *standard position angles* for A when $\sin A = 1/\sqrt{2}$, which are 45° and 135°.

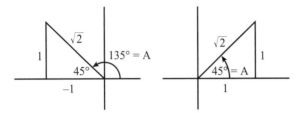

Therefore, the solution to $\sin A = 1/(2)^{1/2}$ for $0 \leq A \leq 360°$ in degrees are the angles 45° and 135°.

To check these results calculate:

$1/(2)^{1/2} \approx 0.7071$

$\sin 45° \approx 0.7071$

$\sin 135° \approx 0.7071$

We can also see that $A = \text{Arcsin } 1/(2)^{1/2} = 45°$, and because there are two possible cases within the defined interval that a 45° reference triangle can be drawn, $180° - 45° = 135°$.

• **Example:** Solve $3 \csc A = -3(2)^{1/2}$ for $0 \leq A \leq 360°$.
(Remember: $(x)^{1/2}$ denotes the square root of x.)

To solve this equation, isolate csc A and find the angle A.

$3 \csc A = -3(2)^{1/2}$

Divide both sides by 3 and use square root symbol:

$\csc A = -\sqrt{2}$

Use the *reciprocal identity* $\csc x = 1/\sin x$, or $\sin x = 1/\csc x$, which gives the equation:

$\sin A = 1/\csc A = 1/(-\sqrt{2})$ or $\sin A = -1/\sqrt{2}$

Similar to the previous example, $\sin A = -1/\sqrt{2}$ corresponds to a 45:45:90 triangle, except with a negative sign:

In a coordinate system for an angle in standard position within the range of $0 \leq A \leq 360°$, a 45:45:90 reference triangle depicts two possible *standard position angles* for A when $\sin A = -1/\sqrt{2}$, which are 225° and 315°.

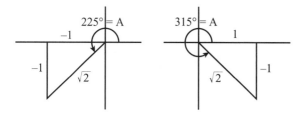

Therefore, the solution to $3 \csc A = -3(2)^{1/2}$ for $0 \leq A \leq 360°$ in degrees are the angles 225° and 315°.

To check these results calculate:

$$-1/(2)^{1/2} \approx -0.7071$$
$$\sin 225° \approx -0.7071$$
$$\sin 315° \approx -0.7071$$

We can also see that $A = \text{Arcsin } -1/(2)^{1/2} = -45°$, and because there are two possible cases within the defined interval that a $-45°$ reference triangle can be drawn, $180° + 45° = 225°$ and $360° - 45° = 315°$.

• By observing the graph of $y = \sin x$, we can also see that $\sin x$ is negative left of the Y-axis for $-\pi < x < 0$, or right of the Y-axis for $\pi < x < 2\pi$.

8.7 Solving Trigonometric Equations Involving Powers Using Factoring, a Unit Circle, and Identities

• More than one method can usually be used to solve a given trigonometric equation. Trigonometric equations involving powers can, for example, be solved using factoring, the location of a point on a unit circle, and substitution of identities.

• Factoring is often used when solving trigonometric equations containing powers. The following two examples demonstrate factoring of trigonometric functions.

• **Example:** Factor the trigonometric equation $\sin^2 x = 1$. First, rearrange equation $\sin^2 x = 1$ as $\sin^2 x - 1 = 0$.

Written in factored form this becomes:

$(\sin x - 1)(\sin x + 1) = 0$

We can check this factored result by multiplying:

$0 = (\sin x - 1)(\sin x + 1) = \sin^2 x + \sin x - \sin x - 1 = \sin^2 x - 1$

Therefore, $0 = \sin^2 x - 1$

Rearranging gives: $\sin^2 x = 1$

• **Example:** Factor the trigonometric equation $\tan^2 x + \tan x = 0$.

This equation can be written in factored form as:

$\tan x (\tan x + 1) = 0$

This result can be verified by multiplying through $\tan x$.

• The following three examples are solved using factoring, the location of a point on a unit circle, and substitution of identities.

• **Example:** Solve the equation $\cos^2 A - 1 = 0$ for $0 \leq A \leq 2\pi$.

This equation factors to:

$(\cos A - 1)(\cos A + 1) = 0$

To solve set each factor equal to zero and solve the simple equations:

$\cos A - 1 = 0$ and $\cos A + 1 = 0$
$\cos A = 1$ $\cos A = -1$

Find angle A using the unit circle and the fact that the coordinates on a unit circle can be represented as $(\cos x, \sin x)$.

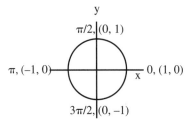

When cos A = 1, the angle in radians (or degrees) is 0.

When cos A = −1, the angle in radians is π (or in degrees is 180°).

Therefore, the solutions to this equation $\cos^2 A − 1 = 0$ for $0 \leq A \leq 2\pi$ are: A = 0 and A = π (or 180°).

Check by substituting 0 and π back into the original factored equation:

$$(\cos 0 − 1)(\cos 0 + 1) = 0$$
$$(\cos \pi − 1)(\cos \pi + 1) = 0$$

- **Example:** Solve sin x + cos x = 1 for $0 \leq x < 2\pi$.

This example can be solved using factoring, a unit circle, and an identity.

Remember the Pythagorean identity $\sin^2 x + \cos^2 x = 1$. Rearranging gives $\sin^2 x = 1 − \cos^2 x$ or, equivalently, $\sin x = \pm\sqrt{1 − \cos^2 x}$.

Substituting for sin x into the original equation and isolating the radical:

$$\pm\sqrt{1 − \cos^2 x} + \cos x = 1$$

$$\pm\sqrt{1 − \cos^2 x} = 1 − \cos x$$

Squaring both sides and using the + radical (see below for − radical):

$$1 − \cos^2 x = (1 − \cos x)^2$$
$$1 − \cos^2 x = (1 − \cos x)(1 − \cos x)$$
$$1 − \cos^2 x = 1 − 2 \cos x + \cos^2 x$$
$$2 \cos x − 2 \cos^2 x = 0$$

Factoring out 2 cos x:

$$(2 \cos x)(1 − \cos x) = 0$$

Set the two expressions equal to zero, solve each equations for cos x and locate the points on the unit circle:

2 cos x = 0   and   1 – cos x = 0

cos x = 0   cos x = 1

when cos x = 0   when cos x = 1

x = π/2 and 3π/2   x = 0

Remember: On the unit circle the coordinates of a point can be represented as (cos x, sin x). Observe particular points on the circle, or on a graph of cosine:

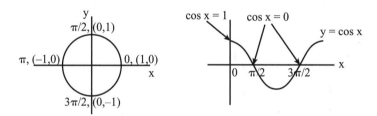

cos π/2 = 0, cos 3π/2 = 0, and cos 0 = 1

Finally, consider the negative (–) radical of:

$$\pm\sqrt{1-\cos^2 x} = 1 - \cos x$$

Squaring both sides and using the – radical:

$$-(1 - \cos^2 x) = (1 - \cos x)^2$$
$$-1 + \cos^2 x = (1 - \cos x)(1 - \cos x)$$
$$-1 + \cos^2 x = 1 - 2 \cos x + \cos^2 x$$

Rearranging:

$$-1 - 1 + \cos^2 x - \cos^2 x = -2 \cos x$$
$$-2 = -2 \cos x$$
$$-2/(-2) = \cos x$$
$$\cos x = 1$$

On the unit circle we can see that when cos x = 1, x = 0. Also, cos 0 = 1.

Therefore, the solutions to sin x + cos x = 1 for 0 ≤ x < 2π are:

x = 0, x = π/2, and x = 3π/2.

Because of the squaring involved in this solution, it is a good idea to check these values by substituting into the original equation:

$\sin x + \cos x = 1$

$\sin 0 + \cos 0 = 1$

$0 + 1 = 1$

$\sin(\pi/2) + \cos(\pi/2) = 1$

$1 + 0 = 1$

$\sin(3\pi/2) + \cos(3\pi/2) = 1$

$-1 + 0 = -1$, which is NOT 1

This solution was obtained due to squaring and is called an extraneous solution.

- **Example:** Solve $\sin^2(x/2) = 1/4$ for $0 \le x < 2\pi$.

Because this equation contains a half-angle, consider the half-angle identity for sine:

$$\sin(x/2) = \pm\sqrt{\frac{1- \cos x}{2}}$$

For $\sin^2(x/2)$ this becomes $\sin^2(x/2) = (1 - \cos x)/2$.

Substituting $\sin^2(x/2) = (1 - \cos x)/2$ into the original equation $\sin^2(x/2) = 1/4$:

$(1 - \cos x)/2 = 1/4$

$1 - \cos x = 1/2$

$2/2 - 1/2 = \cos x$

$\cos x = 1/2$

When $\cos x = 1/2$, the solutions for x can be observed on the unit circle depicted below and are: $x = \pi/3$ and $5\pi/3$. Remember: The coordinates on a unit circle can be expressed as $(\cos x, \sin x)$, where x is the angle or arc length.

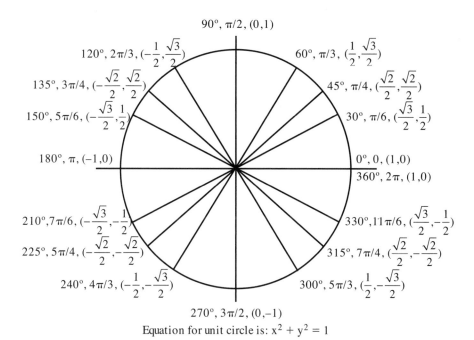

90°, π/2, (0,1)

120°, 2π/3, $(-\frac{1}{2},\frac{\sqrt{3}}{2})$

135°, 3π/4, $(-\frac{\sqrt{2}}{2},\frac{\sqrt{2}}{2})$

150°, 5π/6, $(-\frac{\sqrt{3}}{2},\frac{1}{2})$

180°, π, (−1,0)

210°,7π/6, $(-\frac{\sqrt{3}}{2},-\frac{1}{2})$

225°, 5π/4, $(-\frac{\sqrt{2}}{2},-\frac{\sqrt{2}}{2})$

240°, 4π/3, $(-\frac{1}{2},-\frac{\sqrt{3}}{2})$

60°, π/3, $(\frac{1}{2},\frac{\sqrt{3}}{2})$

45°, π/4, $(\frac{\sqrt{2}}{2},\frac{\sqrt{2}}{2})$

30°, π/6, $(\frac{\sqrt{3}}{2},\frac{1}{2})$

0°, 0, (1,0)

360°, 2π, (1,0)

330°,11π/6, $(\frac{\sqrt{3}}{2},-\frac{1}{2})$

315°, 7π/4, $(\frac{\sqrt{2}}{2},-\frac{\sqrt{2}}{2})$

300°, 5π/3, $(\frac{1}{2},-\frac{\sqrt{3}}{2})$

270°, 3π/2, (0,−1)

Equation for unit circle is: $x^2 + y^2 = 1$

Therefore the solutions to $\sin^2(x/2) = 1/4$ for $0 \leq x < 2\pi$ are:

$x = \pi/3$ and $5\pi/3$.

To check these results substitute x into $\sin^2(x/2) = 1/4$ or the equivalent equation cos x = 1/2: cos π/3 = 1/2 and cos 5π/3 = 1/2.

8.8 Solving Trigonometric Equations and Inequalities Using the Quadratic Formula, Identities, Unit Circles, Factoring, and Graphing

• Trigonometric equations and inequalities can be solved using a combination of techniques including substituting into the quadratic formula, substituting identities, the location of a point on a unit circle, the method of factoring, and graphing methods.

• **Example:** Solve (cot x + 3)(cot x) = 1 for $0 \leq x < 2\pi$ and verify results using x-intercept graphing solution.

To solve this equation multiply the factors and put it in the standard form $ax^2 + bx + c = 0$:

$\cot^2 x + 3 \cot x - 1 = 0$

The *quadratic formula* can be used to solve for cot x:

$$x = \frac{-b \pm \sqrt{b^2 - 4ac}}{2a}$$

Substitute the coefficients a = 1, b = 3, and c = −1 into the quadratic equation:

$$\cot x = \frac{-3 \pm \sqrt{3^2 - 4(1)(-1)}}{2(1)} = \frac{-3 \pm \sqrt{9 + 4}}{2} = \frac{-3 \pm \sqrt{13}}{2}$$

Use a calculator to determine the right side of the equation:

cot x ≈ 0.3027756 for + radical, and

cot x ≈ −3.3027756 for − radical.

To find x we need the arccotangent, which may not be available on many calculators therefore, we can use the *reciprocal identity* tan x = 1/cot x:

tan x = 1/cot x ≈ 1/(0.3027756) = 3.302776

tan x = 1/cot x ≈ 1/(−3.3027756) = −0.3027756

Take the arctangent using a calculator:

x ≈ arctan(3.302776) ≈ 1.276795

x ≈ arctan(−0.3027756) ≈ −0.2940013

Rounding results in x = 1.27 and x = −0.294

The equation (cot x + 3)(cot x) = 1 defines the interval for the solutions as 0 ≤ x < 2π. Therefore the negative value x = −0.294 is not in the interval of the solutions. However, because of the periodic nature of cotangent and the fact that its natural period is π, other solutions may exist at multiples of π plus −0.294, which are in the interval 0 ≤ x < 2π.

π + (−0.294) ≈ 2.85

2π + (−0.294) ≈ 5.99

Because 2π ≈ 6.28, then 5.99 is the maximum in the interval.

The value of x = 1.28 is in the interval and is a solution to the equation. Again, because of the periodic nature of cotangent and the fact that its natural period is π, other solutions should exist at multiples of π plus 1.28 within the interval 0 ≤ x < 2π.

$\pi + 1.28 \approx 4.42$

$2\pi + 1.28 \approx 7.56$, which is outside the interval and not a solution.

Therefore, solutions to equation (cot x + 3)(cot x) = 1 in the interval $0 \le x < 2\pi$ are: x ≈ 2.85, x ≈ 5.99, x ≈ 1.28, x ≈ 4.42.

This problem is somewhat tricky, so it is a good idea to verify the results using a graphing technique. Graph (cot x + 3)(cot x) = 1 using a graphing utility, and the solutions according to the *x-intercept method* will be where the graph crosses the X-axis within the interval $0 \le x < 2\pi$.

Arrange (cot(x) + 3)(cot(x)) = 1 into standard form and enter it into a graphing utility, as $\cot^2 x + 3 \cot x - 1 = 0$:

Graph of $\cot^2 x + 3 \cot x - 1 = 0$

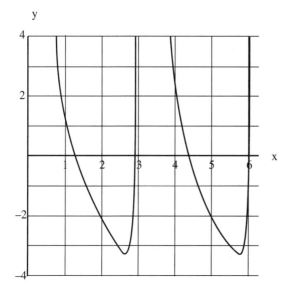

Observing the graph we can verify solutions where the curve crosses the X-axis at approximately x ≈ 2.85, x ≈ 5.99, x ≈ 1.28, and x ≈ 4.42 within the interval $0 \le x < 2\pi$. Many graphing utilities will also calculate roots.

• **Example:** Solve equation cos(2x) = cos x and inequality cos 2x > cos x for the interval $0 \le x < 2\pi$.

Because this equation involves a double number, consider the *double-number identity* for cosine. (See section 7.1 for a summary of the trigonometric identities.) There are a few forms of this identity:

$$\cos(2x) = 2\cos^2 x - 1$$
$$\cos(2x) = 1 - 2\sin^2 x$$
$$\cos(2x) = \cos^2 x - \sin^2 x$$

Substitute the first identity for $\cos(2x) = 2\cos^2 x - 1$ into the equation, which maintains the equation in terms of cos x:

$$\cos(2x) = \cos x$$
$$2\cos^2 x - 1 = \cos x$$

Put equation in standard form and *factor*:

$$2\cos^2 x - \cos x - 1 = 0$$
$$(2\cos x + 1)(\cos x - 1) = 0$$

which is the *factored equation*.

Check factoring by multiplying:

$$2\cos^2 x - 2\cos x + \cos x - 1 = 0$$
$$2\cos^2 x - \cos x - 1 = 0$$

Set each side of factored equation equal to zero and solve for x:

$$(2\cos x + 1) = 0 \quad \text{and} \quad (\cos x - 1) = 0$$
$$2\cos x = -1 \qquad\qquad\qquad \cos x = 1$$
$$\cos x = -1/2$$

When $\cos x = -1/2$, the solutions for x can be observed on the *unit circle* and are: $x = 2\pi/3$ and $4\pi/3$.

When $\cos x = 1$, the solution for x can be observed on the unit circle and is: $x = 0$. (Remember x can be an angle or arc.)

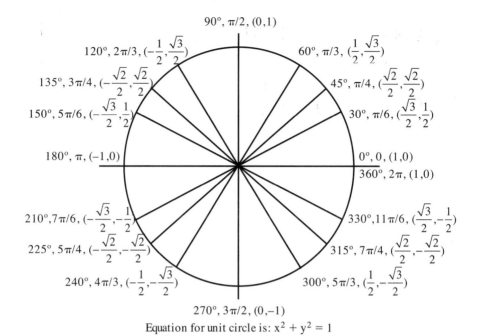

$90°, \pi/2, (0,1)$

$120°, 2\pi/3, (-\frac{1}{2}, \frac{\sqrt{3}}{2})$

$60°, \pi/3, (\frac{1}{2}, \frac{\sqrt{3}}{2})$

$135°, 3\pi/4, (-\frac{\sqrt{2}}{2}, \frac{\sqrt{2}}{2})$

$45°, \pi/4, (\frac{\sqrt{2}}{2}, \frac{\sqrt{2}}{2})$

$150°, 5\pi/6, (-\frac{\sqrt{3}}{2}, \frac{1}{2})$

$30°, \pi/6, (\frac{\sqrt{3}}{2}, \frac{1}{2})$

$180°, \pi, (-1,0)$

$0°, 0, (1,0)$
$360°, 2\pi, (1,0)$

$210°, 7\pi/6, (-\frac{\sqrt{3}}{2}, -\frac{1}{2})$

$330°, 11\pi/6, (\frac{\sqrt{3}}{2}, -\frac{1}{2})$

$225°, 5\pi/4, (-\frac{\sqrt{2}}{2}, -\frac{\sqrt{2}}{2})$

$315°, 7\pi/4, (\frac{\sqrt{2}}{2}, -\frac{\sqrt{2}}{2})$

$240°, 4\pi/3, (-\frac{1}{2}, -\frac{\sqrt{3}}{2})$

$300°, 5\pi/3, (\frac{1}{2}, -\frac{\sqrt{3}}{2})$

$270°, 3\pi/2, (0,-1)$

Equation for unit circle is: $x^2 + y^2 = 1$

The solutions $x = 0, x = 2\pi/3$, and $x = 4\pi/3$ can be verified by substituting into the original equation $\cos(2x) = \cos x$:

$\cos((2)(0)) = \cos(0)$, or $1 = 1$

$\cos(4\pi/3) = \cos(2\pi/3)$, or $-1/2 = -1/2$

$\cos(8\pi/3) = \cos(4\pi/3)$, or $-1/2 = -1/2$

The equation $\cos(2x) = \cos x$ can also be solved using the *intersection-of-graphs* or *x-intercept* graphing methods. We can verify our solutions for the equation found above, and also determine the *inequality* $\cos 2x > \cos x$ by graphing.

Graph the equation $\cos(2x) - \cos x = 0$ in the interval $0 \le x < 2\pi$.

Graph of y = cos(2x) − cos x

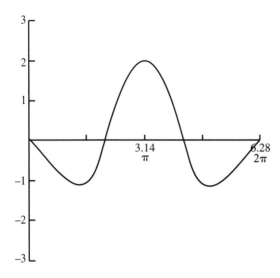

The solutions for the *equation* cos(2x) − cos x = 0 in the interval
0 ≤ x < 2π can be confirmed where the curve crosses the X-axis at:
x = 2π/3, 4π/3, and x = 0 from the graph, by estimating the values:

   x = 2π/3 ≈ 2.094395 and x = 4π/3 ≈ 4.188790.

These solutions are where the curve crosses the X-axis.

The solution set for the *inequality* is given by cos 2x > cos x rearranged
as cos 2x − cos x > 0 in the interval 0 ≤ x < 2π, and is the x-values
where the graph of y = cos(2x) − cos x is above the x-axis. This solution
set for the inequality can be observed on the graph as the x-values that
are above the X-axis in the open interval (x = 2π/3, x = 4π/3). On the
graph we can see the approximate values x = 2π/3 ≈ 2.094395 and
x = 4π/3 ≈ 4.188790 where the curve crosses *above* the X-axis within
the interval 0 ≤ x < 2π, so the solution is 2π/3 ≤ x ≤ 4π/3.

Note that the solution set for the opposite inequality cos 2x < cos x exists
where the graph of the curve is *below* the X-axis.

8.9 Estimating Solutions to Trigonometric Equations and Inequalities Using Graphing

• Solutions to trigonometric equations and inequalities can be estimated by graphing the equations or inequalities according to the *intersection-of-graphs* and *x-intercept methods*. The following two examples demonstrate these techniques.

• **Example:** Estimate the solution to $2x = \cos x$ for $0 \le x < 2\pi$ using the intersection-of-graphs and x-intercept methods.

For the *intersection-of-graphs method* plot each side of the equation as $y = 2x$ and $y = \cos x$ and observe the graph in the interval $0 \le x < 2\pi$.

Graph of $y = 2x$ (line) and $y = \cos x$ (curve)

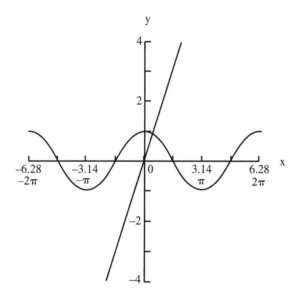

From the graph we can see that there is one intersection point with the solution for x between $x = 0$ and $x = 1.57$ ($\pi/2$) in the interval $0 \le x < 2\pi$. To observe this more carefully, expand the graph in this region.

From the expanded graph (below) the intersection value of x appears to be between $x = 0.4$ and $x = 0.5$.

Expanded graph of y = 2x (line) and y = cos x (curve)

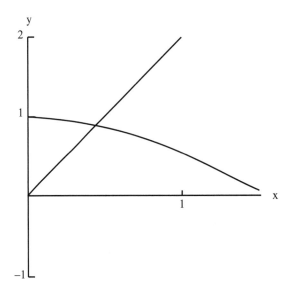

As a comparison, use the *x-intercept method*. Place the equation in standard form $2x - \cos x = 0$, and graph.

Graph of y = 2x − cos x

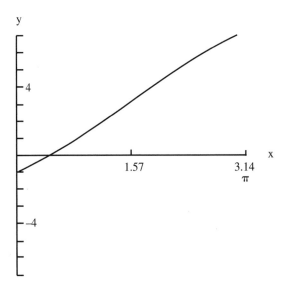

From the graph we can see that there is one point where the graph intersects the X-axis with the solution for x between x = 0 and x = 1.57 ($\pi/2$) in the interval $0 \le x < 2\pi$. To observe this more carefully, expand the graph in this region.

Expanded graph of y = 2x – cos x

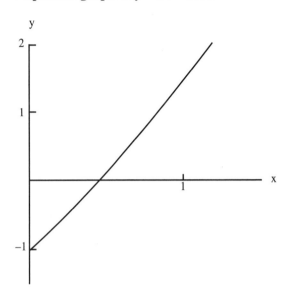

As with the intersection-of-graphs method, the x-intercept value appears to be between x = 0.4 and x = 0.5.

To test and pinpoint this estimate substitute the value between x = 0.4 and x = 0.5, or 0.45 into the original equation 2x = cos x, or

$$2x - \cos x = 0$$
$$2(0.45) - \cos(0.45) = ?$$
$$0.900 - 0.9004471 \approx 0$$

Therefore, x = 0.45 is a good estimate for this equation.

- **Example:** Find the *inequality* cos(sin x) > sin(cos x) for $0 \le x < 2\pi$ using the *intersection-of-graphs method*.

Graph the inequality as two equations y = cos(sin x) and y = sin(cos x) in the interval $0 \le x < 2\pi$.

Graph of y = cos(sin x) (top curve) and y = sin(cos x) (bottom curve)

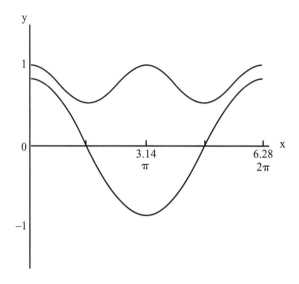

On this graph $y = \cos(\sin x)$ is the curve above $y = \sin(\cos x)$. Because in the interval $0 \leq x < 2\pi$ the curve of $\cos(\sin x)$ is above the curve of $\sin(\cos x)$ for all x, then the solution for the inequality $\cos(\sin x) > \sin(\cos x)$ exists for all values of x within this interval.

8.10 Chapter 8 Summary and Highlights

• This chapter includes a brief summary of important algebraic techniques that are applicable to solving equations and inequalities involving trigonometric and circular functions. Included are explanations of solving (1.) algebraic equations for an unknown variable, (2.) algebraic equations that contain fractions and combining fractions, (3.) algebraic equations that contain square roots, (4.) quadratic equations using the method of factoring, the quadratic formula, and the square-root method, (5.) algebraic inequalities, and (6.) algebraic equations and inequalities using graphing.

• *Equations* and *inequalities* containing trigonometric or circular functions can be solved using a number of techniques. There is often more than one method that can be used to solve a given trigonometric equation, and in addition, combining several techniques may be necessary for solving more complex equations and inequalities. Trigonometric equations and inequalities can be solved using techniques such as: factoring, substituting into the quadratic formula, substituting identities, standard position angles and reference triangles, the location of a point on a unit circle, and using graphical techniques such as the x-intercept and intersection-of-graphs methods. When an equation or inequality cannot be solved analytically using algebra, it may be solved (or the solutions estimated) graphically.

• Because trigonometric and circular functions are periodic, and numerous coterminal angles can exist (multiples of 2π or $360°$ for any angle in standard position), a number of solutions can exist. Therefore, solutions to equations involving trigonometric or circular functions are often specified within defined intervals, such as $0 \le x < 2\pi$, or $0 \le x < 360°$.

Chapter 9

Trigonometric Functions and Vectors

9.1 Definitions of Vectors

• Vectors are used to solve problems in various disciplines, including trigonometry, mathematics, physics, and engineering. Applications include calculations in velocity, force, navigation, area, and volume. This chapter provides a reference for the elementary principles of vectors.

• *Scalars* are quantities that represent magnitude and can be described by one number, either positive, negative, or zero. Scalars are real numbers and can be compared with each other when they have the same physical dimensions, or units. Examples of scalars include temperature, work, density, and mass.

• A *vector* represents a quantity that is described by both a numerical value for *magnitude* (or *length*) and a *direction*. A vector is depicted as a line segment with an initial point and a terminal point that has an arrow pointing in the direction of the terminal point.

The length of a vector represents the magnitude of the vector quantity. Examples of vectors include displacement, velocity, acceleration, electric field strength, force, and moment of force.

• A *displacement vector* represents the movement or displacement between two points in a *coordinate system*. The *length* of a displacement vector is the distance between the two points and the *direction* of a displacement vector is the direction it is pointing.

• A *velocity vector* describes an object in motion and has a magnitude representing the speed of the object and a direction representing the direction of motion. A *force vector* is a vector that represents the direction and magnitude of an applied force. If there are two forces acting on an object, then the sum of the two forces acts as a single force on the object. (See Section 9.4.)

• Vectors that point in the *same direction* and have the *same length* (or *magnitude*) are **equivalent vectors** even if they are not in the same location. A vector can be relocated and still be considered the same vector as long as its length and direction remain the same.

• The *zero vector* **0** has a length (or magnitude) of zero and no direction. Its initial and terminal points coincide.

• The **negative of a vector** is a vector with the same length but pointing in the *opposite* direction. If **A** is a vector, **−A** is its negative or opposite vector. The sum of vector **A** and its negative **−A** is a *zero vector*.

If vector **A** $= \overrightarrow{ab}$ and points from a to b and vector

B $= \overrightarrow{ba}$ and points from b to a, then **A** = **−B**.

a ⟺ b
    A
    B

• The *magnitude (or length)* of a vector is indicated with vertical bars as used to denote *absolute value*. Remember: The absolute value of a number n is represented by $|n|$, where $|1| = 1$ and $|-1| = 1$.

The following represent magnitudes of vectors: $|\mathbf{A}|$, $|\mathbf{B}|$, and $|\overrightarrow{AB}|$.

Sometimes double bars are used to represent magnitude: $||\mathbf{A}||$ or $||\mathbf{B}||$.

• A *unit vector* **u** has a length (or magnitude) of one. If unit vector **u** is pointing in the direction of vector **A**, and **A** is not a zero vector, then $\mathbf{u} = \mathbf{A}/|\mathbf{A}|$. Therefore, vector **A** divided by its *length* (magnitude) $|\mathbf{A}|$ results in a *unit vector* pointing in the *direction of vector* **A**. The direction of **A** is $\mathbf{u} = \mathbf{A}/|\mathbf{A}|$ and its length is $|\mathbf{A}|$. Therefore, length multiplied by direction gives **A** as $\mathbf{u}|\mathbf{A}| = \mathbf{A}$.

• *Notation for a vector* includes boldface single letters **A**, **a**, **B**, **b**, etc., two boldface letters, **AB**, etc., or one or two letters with an arrow $\vec{A}$, $\vec{B}$, $\vec{a}$, $\overrightarrow{AB}$, etc. When two letters are used to represent a vector, the first of the two letters represents the initial point and the second letter represents the terminal point.

9.2 Representing Vectors in Terms of Their Components in a Coordinate System

• A vector can be described by its horizontal and vertical components, which are also vectors.

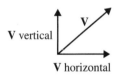

V vertical

V horizontal

Vectors **V**vertical and **V**horizontal are each *component vectors* of vector **V**. The horizontal and vertical components of a vector are useful in applications of vectors such as force, navigation, and surveying.

• It is often useful to resolve a vector into components that run along horizontal and perpendicular axes of a coordinate system. For example, in the drawing,

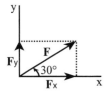

the force vector **F** can be resolved into horizontal and vertical component vectors using the principles of trigonometry. In this drawing the horizontal component is $\mathbf{F}_x = \mathbf{F}\cos 30°$, and the vertical component is $\mathbf{F}_y = \mathbf{F}\sin 30°$. In this drawing, **F** is the vector sum of the two vectors $\mathbf{F}_x$ and $\mathbf{F}_y$.

• In the above figure, suppose the magnitude of **F** is 50 and we want to find the horizontal and vertical component vectors $\mathbf{F}_x$ and $\mathbf{F}_y$ of **F**. Remember: The magnitude is the absolute value of a number n and is represented by $|n|$. To find the magnitude of $\mathbf{F}_x$ calculate

$$|\mathbf{F}_x| = \mathbf{F}\cos 30° = 50\cos 30° \approx 43$$

To find the magnitude of $\mathbf{F}_y$ calculate

$$|\mathbf{F}_y| = \mathbf{F}\sin 30° = 50\sin 30° = 25$$

• The magnitude or length of a vector **V** is also given by the square root of the sum of the squares of the x- and y-components, such that for:

$$|\mathbf{V}| = \sqrt{x^2 + y^2}$$

Applying this to our force problem above, we find that the length of the x-component 43 and the y-component 25 give the length of vector **F**:

$$|\mathbf{F}| = \sqrt{43^2 + 25^2} \approx 50 \quad \text{which was given initially.}$$

Remember: The magnitude itself is a real number and a vector is defined by both magnitude and direction.

• A vector with its initial point at the origin of a *coordinate system* is called a *position vector*. A position vector is a vector in *standard position* and is also called a *radius vector*. A position vector is defined according to the location or coordinates of its terminal point.

For example, if its terminal point is at B then vector $\overrightarrow{AB}$ is a position vector of point B.

• A *position vector* represents the position of a point with respect to the origin, and a **displacement vector** represents the change or displacement between two points in a *coordinate system*. The *length* of a displacement vector is the distance between the two points, and the *direction* of a displacement vector is the direction it is pointing.

• A vector in a coordinate system can be represented in terms of its **magnitude** and a **direction angle**, which is the angle it makes with the horizontal axis. The relationships of the horizontal and vertical components of a position vector to a direction angle and magnitude are important characteristics of vectors.

• In a coordinate system, a position vector is placed with its initial point at the origin, and the angle that the vector makes with the X-axis is called the *direction angle* θ. The direction angle is measured like a *standard position angle* from the X-axis.

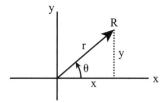

From trigonometric principles we know that in this diagram vector **R** has a magnitude of $|\mathbf{R}| = r$, so that the *magnitude* of the horizontal component can be written $x = r \cos \theta$, and *magnitude* of the vertical component can be written $y = r \sin \theta$. In addition, $x^2 + y^2 = r^2$ and $\tan \theta = y/x$, where x is not zero.

• In a coordinate system, a **position vector** with its initial point at the origin can also be described by the coordinates, or location, of its terminal end point. In other words, a position vector can be described by an *ordered pair of real numbers* in a coordinate system.

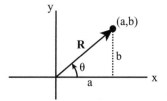

In this diagram, the position vector **R** can be called vector (a, b). There is a special notation for position vectors described by the coordinates of their endpoint, which uses brackets rather that parentheses. Vector (a, b) can therefore be written $\langle a,b \rangle$. In this diagram the real numbers a and b represent the x-component and y-component, respectively. The real numbers a and b are called scalar components of vector $\langle a,b \rangle$. The length, or magnitude, of vector **R** $= \langle a,b \rangle$ is given by $|\mathbf{R}| = \sqrt{a^2 + b^2}$.

• In the above figure, if the magnitude of **R** is represented by r and the direction angle is θ, then vector **R** can also be described by:

$$\mathbf{R} = \langle r \cos \theta, r \sin \theta \rangle$$

• Note that a position vector can be described in three-dimensional space or even higher dimensional spaces using its endpoint coordinates. For example, in a three-dimensional coordinate system a vector may be written $\langle a, b, c \rangle$, where c would correspond to the Z-axis. Vectors can be characterized and expressed in more than three dimensions or components. For example:

$$\mathbf{A} + \mathbf{B} = \langle a_1, a_2, a_3, a_4 \rangle + \langle b_1, b_2, b_3, b_4 \rangle$$

9.3 Representing Vectors in Terms of Their Components in a Coordinate System Using the Unit Vectors i, j, and k

• Vectors can be represented in terms of their *components in a coordinate system*. The vector components can be defined by their directions along the X,Y, and Z axes of a coordinate system using unit vectors denoted by *i, j,* and **k**. The *i, j, and k unit vectors* have *magnitudes* of one and *directions* pointing parallel to the X,Y, and Z axes, respectively, in a rectangular coordinate system.

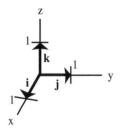

The coordinates of **i**, **j**, and **k** can be used to represent these vectors as

$\mathbf{i} = \langle 1, 0, 0 \rangle$, $\mathbf{j} = \langle 0, 1, 0 \rangle$, and $\mathbf{k} = \langle 0, 0, 1 \rangle$.

- Vector **A** can be written in three dimensions using the **i**, **j**, **k** unit vectors as: $\mathbf{A} = a_1\mathbf{i} + a_2\mathbf{j} + a_3\mathbf{k}$, where a_1, a_2 and a_3 are scalar quantities and $a_1\mathbf{i}$, $a_2\mathbf{j}$ and $a_3\mathbf{k}$ are the vector components of **A**. Therefore,

$$\mathbf{A} = a_1\mathbf{i} + a_2\mathbf{j} + a_3\mathbf{k} = \langle a_1, a_2, a_3 \rangle$$

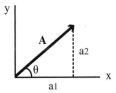

(**k** or z dimension not shown)

The *magnitude (or length)* of **A** is given by: $|\mathbf{A}| = \sqrt{a_1^2 + a_2^2 + a_3^2}$

- If a *position vector* described in two dimensions has its starting point at the origin and its terminal point at point P = (5, 6), then in two-dimensions vector **A** is written:

$\mathbf{A} = 5\mathbf{i} + 6\mathbf{j}$

It has length $|\mathbf{A}| = \sqrt{(5)^2 + (6)^2} = \sqrt{25 + 36} = \sqrt{61}$
and is depicted as:

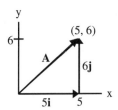

- The *direction of a vector in a coordinate system* can be represented by the angle it makes with the positive X-axis. For example, the direction of vector **A** can be written in terms of the *angle* θ that it makes with the positive X-axis. Vector $\mathbf{A} = a_1\mathbf{i} + a_2\mathbf{j}$ makes an angle $\theta = \tan^{-1}(a_2/a_1)$ with the X-axis and can be written as:

$\mathbf{A} = \mathbf{i}\,|\mathbf{A}|\cos\theta + \mathbf{j}\,|\mathbf{A}|\sin\theta$, where $a_1 = |\mathbf{A}|\cos\theta$ and $a_2 = |\mathbf{A}|\sin\theta$.

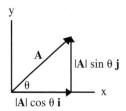

- Note that a **unit vector** **u** for vector **A** can be written in terms of **i**, **j**, and **k** as: $\mathbf{u} = \dfrac{\mathbf{A}}{|\mathbf{A}|} = \mathbf{i} \cos \theta + \mathbf{j} \sin \theta$ (in this case the coefficient $a_3 = 0$.)

Also, $|\mathbf{u}|^2 = \cos^2\theta + \sin^2\theta = 1$.

- Notation for vectors also includes writing them as *column vectors* and *row vectors*:

$$\mathbf{A} = \begin{bmatrix} a_1 \\ a_2 \end{bmatrix}, \mathbf{B} = [b_1 \ b_2], \mathbf{v} = \begin{bmatrix} v_1 \\ v_2 \end{bmatrix}, \mathbf{r} = [r_1 \ r_2].$$

where a_1 and a_2 are components of **A**, b_1 and b_2 are components of **B**, v_1 and v_2 are components of **v**, and r_1 and r_2 are components of **r**.

- A vector $\mathbf{A} = 5\mathbf{i} + 3\mathbf{j} - 6\mathbf{k}$ can be written in column vector format:

$$\mathbf{A} = 5\mathbf{i} + 3\mathbf{j} - 6\mathbf{k} = \begin{bmatrix} 5 \\ 3 \\ -6 \end{bmatrix}$$

- Unit vectors **i**, **j**, **k** can be represented as *column vectors*:

$$\mathbf{i} = \begin{bmatrix} 1 \\ 0 \\ 0 \end{bmatrix}, \mathbf{j} = \begin{bmatrix} 0 \\ 1 \\ 0 \end{bmatrix}, \mathbf{k} = \begin{bmatrix} 0 \\ 0 \\ 1 \end{bmatrix}$$

- The **zero vector** having zero length can be written in terms of **i**, **j**, **k**:

$$\mathbf{0} = 0\mathbf{i} + 0\mathbf{j} + 0\mathbf{k}$$

9.4 Addition and Subtraction of Vectors

- Two *vectors can be added or subtracted* if they have the same dimensions by adding or subtracting the corresponding components (or elements). For example, a two-dimensional vector can be added to another two-dimensional vector; however, a two-dimensional vector cannot be added to a three-dimensional vector.

• The *sum of two vectors* can be depicted by positioning the vector such that the initial point of the second vector is at the terminal point of the first vector. The sum of the two vectors is a third vector with its initial point at the initial point of the first vector and its final point at the final point of the second vector. In other words, the sum of two vectors **a** and **b** is the combined displacement from applying vector **a** then applying vector **b**.

Consider the figure below depicting the following two examples of adding vectors **a** and **b**:

To add vectors **a** and **b** in *Illustration 1* that follows, place the initial point of **b** at the final point of **a**. The sum is the vector joining the initial point of **a** to the final point of **b**, or vector **c**. The sum is also the diagonal of a *parallelogram* that can be constructed on **a** and **b**. Remember that the starting point of a vector can be moved as long as its length and direction stay the same. From the parallelogram we can see that **a** + **b** = **b** + **a**.

In *Illustration 2*, the initial point of **b** is already at the final point of **a**. The sum is the vector joining the initial point of **a** to the final point of **b**, which results in vector **c**.

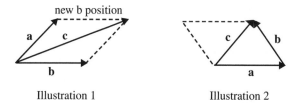

Illustration 1   Illustration 2

Both figures represent **a** + **b** = **c**.

Note that c is called the *resultant vector*. Illustration 1 represents what is sometimes called the *parallelogram rule* because a parallelogram is completed as the resultant vector **c** is determined.

• *Subtraction of two vectors* is equivalent to addition of the first vector with the negative of the second vector. The *negative of a vector* is a vector with the same length but pointing in the opposite direction.

• To subtract two vectors, reverse the direction of the second vector, then add the first vector with the negative of the second vector by positioning the vectors so that the initial point of the (negative) second vector is at the final point of the first vector. The sum of two vectors will be a third vector with its initial point at the initial point of the first vector and its final point at the final point of the second (negative) vector. This figure represents **a** – **b** = **c**:

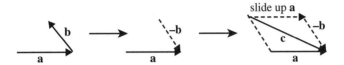

In a second example of vector subtraction, subtract two vectors **a** – **b** = **c**, where **a** – **b** can be represented using the negative of **b**, then slide –**b** up to place initial point of –**b** at terminal point of **a**. The sum is the vector joining the initial point of **a** to the final point of –**b**, which results in **c**. This figure represents **a** – **b** = **c**:

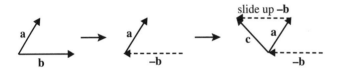

• The **sum of two vectors** $\overrightarrow{AB}$ and $\overrightarrow{CD}$ can be written:

$$\overrightarrow{AB} + \overrightarrow{CD} = \overrightarrow{AD}$$

• The sum of two vectors **A** and **B** can also be written using components or in column format:

$$\mathbf{A} + \mathbf{B} = \begin{bmatrix} a_1 \\ a_2 \end{bmatrix} + \begin{bmatrix} b_1 \\ b_2 \end{bmatrix} = \begin{bmatrix} a_1 + b_1 \\ a_2 + b_2 \end{bmatrix}$$

If $\mathbf{A} = \begin{bmatrix} 2 \\ 3 \end{bmatrix}$ and $\mathbf{B} = \begin{bmatrix} 3 \\ 3 \end{bmatrix}$, then $\mathbf{A} + \mathbf{B} = \begin{bmatrix} 5 \\ 6 \end{bmatrix}$

This is represented graphically as:

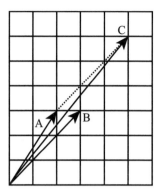

Vector **A** is the smallest vector on the left, vector **B** is the medium length vector pointing slightly to the right of **A**, and vector **C** is the resultant sum of vectors **A** and **B**. Vector **B** was slid up (dashed arrow) so that its initial point is at the terminal point of **A** to create vector **C**.

• Two *vectors can be added or subtracted* and expressed using unit vectors. If vector $\mathbf{A} = a_1\mathbf{i} + a_2\mathbf{j}$ and vector $\mathbf{B} = b_1\mathbf{i} + b_2\mathbf{j}$, then:

$$\mathbf{A} + \mathbf{B} = a_1\mathbf{i} + a_2\mathbf{j} + b_1\mathbf{i} + b_2\mathbf{j} = (a_1 + b_1)\mathbf{i} + (a_2 + b_2)\mathbf{j}$$
$$\mathbf{A} - \mathbf{B} = a_1\mathbf{i} + a_2\mathbf{j} - (b_1\mathbf{i} + b_2\mathbf{j}) = (a_1 - b_1)\mathbf{i} + (a_2 - b_2)\mathbf{j}$$

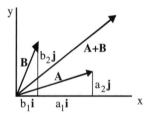

If $\mathbf{A} = 2\mathbf{i} + 3\mathbf{j}$ and $\mathbf{B} = 3\mathbf{i} + 4\mathbf{j}$, then $\mathbf{A} + \mathbf{B} = 5\mathbf{i} + 7\mathbf{j}$.

• Remember, the *direction of a vector in a coordinate system* is represented by the angle it makes with the positive X-axis. For example, the direction of vector **A** can be written in terms of the angle θ that it makes with the positive X-axis. Vector $\mathbf{A} = a_1\mathbf{i} + a_2\mathbf{j}$ makes an angle $\theta = \tan^{-1}(a_2/a_1)$ with the X-axis and can be written as:

$$\mathbf{A} = \mathbf{i}\,|\mathbf{A}|\,\cos\theta + \mathbf{j}\,|\mathbf{A}|\,\sin\theta$$

where $a_1 = |\mathbf{A}|\cos\theta$ and $a_2 = |\mathbf{A}|\sin\theta$.

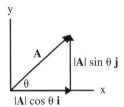

- Properties of adding vectors include:

    Vector addition is commutative $\mathbf{A} + \mathbf{B} = \mathbf{B} + \mathbf{A}$.

    Vector addition is associative $(\mathbf{A} + \mathbf{B}) + \mathbf{C} = \mathbf{A} + (\mathbf{B} + \mathbf{C})$.

    Vector with negative $\mathbf{A} + (-\mathbf{A}) = (-\mathbf{A}) + \mathbf{A} = \mathbf{0}$.

    Vector with zero vector $\mathbf{A} + \mathbf{0} = \mathbf{0} + \mathbf{A} = \mathbf{A}$.

9.5 Simple Vector Problems

- The following examples demonstrate the use of trigonometric functions and principles of vectors to solve problems.

- **Example:** Consider a ship initially moving east along the ocean at a velocity $\mathbf{v} = 15$ km/hr relative to the water, which has a current $\mathbf{c} = 2$ km/hr in the northeast direction. An angle $\theta = 45°$ exists between the direction the ship is traveling and the direction of the ocean current.

Use two different methods to calculate the actual speed of the ship relative to land and the angle that the ship is deviating from vector $\mathbf{v}$. In one method use *heading* in describing angle measurements.

Method 1:
The true velocity of the ship with respect to land is equal to the sum of the two vectors $\mathbf{v} + \mathbf{c}$.

To calculate the actual speed of the ship relative to land, set the velocity of the ship along the X-axis so that $\mathbf{v} = (15)\mathbf{i}$ with a zero $\mathbf{j}$ component for $\mathbf{v}$, and the magnitude of the ocean current $|\mathbf{c}| = 2$.

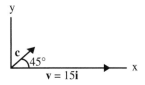

The angle **c** makes with the X-axis (**v**) is 45° and is the hypotenuse of the right triangle formed by **c**. Therefore, the **i** component of **c** is 2 cos 45° and the **j** component of **c** is 2 sin 45°.

Therefore:

$$\mathbf{c} = (2 \cos 45°)\mathbf{i} + (2 \sin 45°)\mathbf{j} \approx 1.4\mathbf{i} + 1.4\mathbf{j}$$

The actual velocity **S** of the ship relative to land is:

$$\mathbf{S} = \mathbf{v} + \mathbf{c} = 15\mathbf{i} + 1.4\mathbf{i} + 1.4\mathbf{j} = 16.4\mathbf{i} + 1.4\mathbf{j}$$

Therefore, the *magnitude of the speed of the ship* relative to land is:

$$|\mathbf{S}| = \sqrt{(16.4)^2 + (1.4)^2} \approx 16.46 \text{ km/hr.}$$

To find the angle θ that the ship is deviating from **v**, we can use the fact that for resultant vector **S**, tan θ is its **j** component 1.4 over its **i** component 16.4. Therefore, the *angle the ship is deviating from v* due to the current is:

$$\theta = \tan^{-1}(1.4/16.4) \approx 4.9° \approx 0.085 \text{ radians.}$$

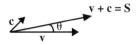

Method 2:
In this method we use the same information that a ship is moving along the ocean at a velocity **v** = 15 km/hr relative to the water, which has a current **c** = 2 km/hr. The ship is initially pointing east and therefore has a heading of 90°, and in addition, the 45° angle from **v** in the northeast direction gives the current a heading of 45°. We are asked to calculate the actual speed of the ship relative to land and the angle that the ship is deviating from its original heading, vector **v**. Remember that heading is measured from due north clockwise.

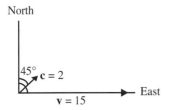

To find the actual speed $|\mathbf{S}|$ of the ship we can use the *Law of Cosines* for the oblique triangle formed as we slide the vector **c** over to the end of vector **v**. The Law of Cosines in terms of angle B is:

$$b^2 = a^2 + c^2 - 2ac \cos B$$

In an oblique triangle, a, b, and c represent the sides opposite the angles A, B, and C.

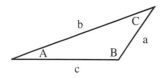

In the oblique triangle formed by sliding over the current vector **c**:

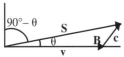

Angle θ represents how much the ship is deviating from **v** and the actual heading of the ship is $90° - \theta$. The Law of Cosines equation for this diagram has $b = |\mathbf{S}|$, $a = \mathbf{c} = 2$, $c = \mathbf{v} = 15$, and $B = B$:

$$|\mathbf{S}|^2 = 2^2 + 15^2 - 2(2)(15) \cos B$$

To find B, use the fact that we were given that vector **c** makes a 45° angle from **v**. Because angle B forms a supplementary angle with the 45° angle that vector **c** forms with **v**, angle B is:

$B = 180° - 45° = 135°$. Therefore,

$|\mathbf{S}|^2 = 2^2 + 15^2 - 2(2)(15) \cos 135° \approx 271$

Take the square root to obtain the *magnitude of the speed of the ship*:

$|\mathbf{S}| \approx 16.46$ km/hr

which is what we determined in Method 1!

Now find the angle the ship is deviating from vector $\mathbf{v}$ and also find the heading of the ship due to the current, which is $90° - \theta$. To do this we can use the *Law of Sines* for oblique triangles:

$(\sin A / a) = (\sin B / b)$.

In our triangle $b = |\mathbf{S}| = 16.46$, $a = \mathbf{c} = 2$, $B = 135°$, and $A = \theta$. Therefore,

$\sin \theta/2 = \sin 135°/16.46$

$\sin \theta = (2)(\sin 135°/16.46)$

$\theta = \arcsin [(2)(\sin 135°/16.46)] \approx 4.9°$

(Note that in Method 1 angle θ, which is the deviation from $\mathbf{v}$, was found to be $4.9°$.)

Therefore, the heading of the ship is $90° - \theta = 90° - 4.9° \approx 85°$.

• **Example:** Consider the navigation of an airplane. Basic definitions include:

The *heading* of an airplane, which is measured clockwise from north.

The *airspeed*, which is determined by an indicator in still air.

The *groundspeed*, which is the speed relative to the ground.

The *drift angle*, which is the difference between the heading and the actual course due to the wind and is a wind-corrected angle.

The *course*, which is the direction the plane moves relative to the ground and is measured clockwise from north.

Suppose an airplane can fly 250 mi/hr in still air. If a 40 mi/hr wind is blowing from the west, what drift angle, or wind-corrected angle, will be required for the course of the airplane to be due north? What will be the resulting groundspeed?

In other words, given that the airplane can fly 250 mi/hr airspeed, in order to fly due north, what heading do we need to fly in a 40 mi/hr wind from the west and what will be the groundspeed? Diagram what we know:

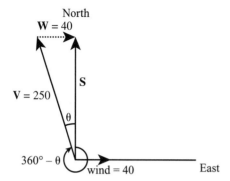

Note that the wind vector **W** was slid up to form a right triangle. This diagram depicts that the heading of the airplane will be $360° - \theta$, where θ is the drift angle. Vector **S** is the resultant vector and the actual groundspeed will be the magnitude of **S**, or $|S|$.

To find θ, use the trigonometric functions for a right triangle:

$\sin \theta = $ opposite/hypotenuse $= |40| \, / \, |250|$, or
$\theta = \arcsin(|40| \, / \, |250|) \approx 9.2°$

which is the drift angle or wind-corrected angle.

The *heading* the plane must fly will be

$360° - \theta = 360° - 9.2° = 350.8°$

The groundspeed is the magnitude of **S**, the resultant vector. To find the magnitude of side **S** use

$|V|^2 = |S|^2 + |W|^2$, or $|S|^2 = |V|^2 - |W|^2$

$|S| = \sqrt{250^2 - 40^2} \approx 247$ mi/hr which is the actual *groundspeed*.

Note that the groundspeed can also be calculated using:

$\cos \theta = $ adjacent/hypotenuse $= |S| \, / \, |V|$, or
$|S| = |V| \cos \theta = |250| \cos 9.2° \approx 247$ mi/hr

• **Example:** Incline plane problems generally have an object on a plane with an angle of inclination θ.

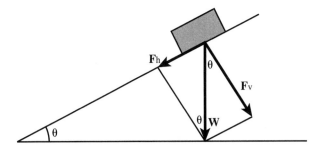

There is a force vector horizontal to the plane F_h and a force vector normal to the plane F_v, which are components of the weight W of the object. From the diagram,

$$\cos \theta = |F_v| / |W|, \text{ or } F_v = |W| \cos \theta$$
$$\sin \theta = |F_h| / |W|, \text{ or } F_h = |W| \sin \theta$$

If a 400-pound block is on a slick frictionless incline of 20°, what force is required to hold the block in place and what is the force of the block perpendicular to the incline plane?

F_h is the force along the plane, therefore a force equal and opposite to F_h is required to hold the block in place. To find this force calculate:

$$F_h = |W| \sin \theta = (400 \text{ pounds}) \sin 20° \approx 137 \text{ pounds}$$

The block exerts a force F_v normal to the incline plane and it is:

$$F_v = |W| \cos \theta = (400 \text{ pounds}) \cos 20° \approx 376 \text{ pounds}$$

9.6 Multiplying a Vector with a Scalar

• *Multiplying vector A with scalar c* results in a *vector* having a magnitude of $|c||A|$ and a direction of A, where $|c|$ represents the absolute value of the scalar c and $|A|$ represents the magnitude of vector A. When $c > 0$, the vector cA is parallel to A and pointing in the same direction as A. When $c < 0$, the vector cA is parallel to A but pointing in the opposite direction as A.

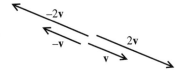

• Any vector **A** can be *multiplied by a scalar or constant* c. Using the **i**, **j**, and **k** unit vectors:

$$c\mathbf{A} = ca_1\mathbf{i} + ca_2\mathbf{j} + ca_3\mathbf{k}$$

• Using bracket notation for a vector defined according to its terminal point, scalar multiplication of vector $\mathbf{A} = \langle a, b \rangle$ with scalar c can be written:

$$c\mathbf{A} = c\langle a, b \rangle = \langle ca, cb \rangle$$

For example, if $\mathbf{A} = \langle -4, 3 \rangle$ and $c = -2$,

$$\text{then } c\mathbf{A} = -2\mathbf{A} = -2\langle -4, 3 \rangle = \langle 8, -6 \rangle$$

• Properties for multiplying scalars with vectors include the following (where **A** and **B** are vectors and c and d are scalars):

$$c(\mathbf{A} + \mathbf{B}) = c\mathbf{A} + c\mathbf{B}$$
$$\mathbf{A}(c + d) = c\mathbf{A} + d\mathbf{A}$$
$$c(d\mathbf{A}) = (cd)\mathbf{A}$$
$$1\mathbf{A} = \mathbf{A}$$

9.7 Dot or Scalar Products

• The *dot product* is used in applications where angles between vectors must be determined, and to solve problems in such fields as physics, engineering, geometry, and mathematics, such as calculating work done. The dot product of two vectors is a real number, or *scalar*, not a vector.

• The **dot product** (also called the **scalar product** or **inner product**) of two vectors is defined as: $\mathbf{A} \cdot \mathbf{B} = |\mathbf{A}||\mathbf{B}| \cos \theta$ where $|\mathbf{A}|$ and $|\mathbf{B}|$ represent the **magnitudes** (or *lengths*) of vectors **A** and **B** and θ is the angle between vectors **A** and **B**.

Rearranging, we find: $\cos \theta = (\mathbf{A} \cdot \mathbf{B})/(|\mathbf{A}||\mathbf{B}|)$

- The *dot product of a vector with itself*, $\mathbf{A} \bullet \mathbf{A}$, has $\theta = 0$, and because $\cos 0 = 1$, then $\mathbf{A} \bullet \mathbf{A} = |\mathbf{A}|^2 =$ length-squared.

- The dot product can be used to find the angle between two vectors. Remember: If the initial points do not coincide, a vector can be repositioned as long as the magnitude (length) and direction are unchanged. The dot product can be used to compute *angle θ* by first computing $\cos \theta$ and then θ, the angle between $\mathbf{A}$ and $\mathbf{B}$.

$$\cos \theta = \frac{\mathbf{A} \bullet \mathbf{B}}{|\mathbf{A}||\mathbf{B}|}, \text{ where } \theta = \arccos \frac{\mathbf{A} \bullet \mathbf{B}}{|\mathbf{A}||\mathbf{B}|}$$

- Vectors $\mathbf{A}$ and $\mathbf{B}$ are **perpendicular** if $\mathbf{A} \bullet \mathbf{B} = 0$, providing $\mathbf{A}$ or $\mathbf{B}$ does not equal zero. This is true because $\cos 90° = \cos(\pi/2) = 0$.

- Vectors $\mathbf{A}$ and $\mathbf{B}$ are **parallel** if $\mathbf{A} \bullet \mathbf{B} = |\mathbf{A}||\mathbf{B}|$, providing $\mathbf{A}$ or $\mathbf{B}$ does not equal zero. This is true because $\cos 0 = 1$.

- The dot product can be written in terms of unit vectors $\mathbf{i}, \mathbf{j}$, and $\mathbf{k}$ as defined in a rectangular coordinate system.

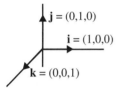

where the coordinates of the unit vectors are $\mathbf{i} = (1,0,0), \mathbf{j} = (0,1,0)$, and $\mathbf{k} = (0,0,1)$.

The dot product of *perpendicular* (or *orthogonal*) vectors $\mathbf{i} = (1,0,0)$ and $\mathbf{j} = (0,1,0)$ is $\mathbf{i} \bullet \mathbf{j} = 0$, because $\mathbf{i}$ and $\mathbf{j}$ are perpendicular (or *orthogonal*) to each other.

The dot product of *parallel* vectors $\mathbf{i} = (1,0,0)$ and $\mathbf{i} = (1,0,0)$ is $\mathbf{i} \bullet \mathbf{i} = 1$, because $\mathbf{i}$ and $\mathbf{i}$ are parallel to each other.

In summary, the dot products of *unit vectors combine* as follows:

$$\mathbf{i} \bullet \mathbf{i} = \mathbf{j} \bullet \mathbf{j} = \mathbf{k} \bullet \mathbf{k} = 1$$
$$\mathbf{i} \bullet \mathbf{j} = \mathbf{i} \bullet \mathbf{k} = \mathbf{j} \bullet \mathbf{i} = \mathbf{j} \bullet \mathbf{k} = \mathbf{k} \bullet \mathbf{i} = \mathbf{k} \bullet \mathbf{j} = 0$$

• The dot product written in the form of $|\mathbf{A}||\mathbf{B}|\cos\theta$ represents $\mathbf{A} \cdot \mathbf{B}$ without coordinates. The dot product can also be written in the form $[a_1b_1 + a_2b_2]$ that does involve coordinates.

• The dot product of vector $\mathbf{A} = a_1\mathbf{i} + a_2\mathbf{j}$ and vector $\mathbf{B} = b_1\mathbf{i} + b_2\mathbf{j}$ can be written as:

$$\mathbf{A} \cdot \mathbf{B} = \begin{bmatrix} a_1 \\ a_2 \end{bmatrix} \cdot \begin{bmatrix} b_1 \\ b_2 \end{bmatrix} = a_1b_1 + a_2b_2$$

which can be determined by:

$$\begin{aligned} \mathbf{A} \cdot \mathbf{B} &= (a_1\mathbf{i} + a_2\mathbf{j}) \cdot (b_1\mathbf{i} + b_2\mathbf{j}) \\ &= a_1b_1(\mathbf{i} \cdot \mathbf{i}) + a_1b_2(\mathbf{i} \cdot \mathbf{j}) + a_2b_1(\mathbf{j} \cdot \mathbf{i}) + a_2b_2(\mathbf{j} \cdot \mathbf{j}) \\ &= a_1b_1(1) + a_1b_2(0) + a_2b_1(0) + a_2b_2(1) \\ &= a_1b_1 + a_2b_2 \end{aligned}$$

Therefore:

$$\mathbf{A} \cdot \mathbf{B} = (a_1\mathbf{i} + a_2\mathbf{j}) \cdot (b_1\mathbf{i} + b_2\mathbf{j}) = a_1b_1 + a_2b_2$$

In three-dimensions the dot product of $\mathbf{A}$ and $\mathbf{B}$ is:

$$\mathbf{A} \cdot \mathbf{B} = a_1b_1 + a_2b_2 + a_3b_3$$

• The dot product of position vectors in a coordinate system can be written using bracket notation:

$$\mathbf{A} \cdot \mathbf{B} = \langle a_1, a_2 \rangle \cdot \langle b_1, b_2 \rangle = a_1b_1 + a_2b_2$$

For example, if $\mathbf{A} = \langle 1, 2 \rangle$ and $\mathbf{B} = \langle -1, 2 \rangle$, then

$$\mathbf{A} \cdot \mathbf{B} = \langle 1, 2 \rangle \cdot \langle -1, 2 \rangle = (1)(-1) + (2)(2) = -1 + 4 = 3$$

• The dot product of two vectors $\mathbf{A}$ and $\mathbf{B}$ can be written in numerous forms including:

$$\mathbf{A} \cdot \mathbf{B} = \langle a_1, a_2 \rangle \cdot \langle b_1, b_2 \rangle = (a_1\mathbf{i} + a_2\mathbf{j}) \cdot (b_1\mathbf{i} + b_2\mathbf{j}) = a_1b_1 + a_2b_2$$

$$\mathbf{A} \cdot \mathbf{B} = |\mathbf{A}||\mathbf{B}|\cos\theta = \begin{bmatrix} a_1 \\ a_2 \end{bmatrix} \cdot \begin{bmatrix} b_1 \\ b_2 \end{bmatrix}$$

- *Properties of the dot product* of vectors **A**, **B**, **C**, and scalar c include:

$$c(\mathbf{A} \bullet \mathbf{B}) = (c\mathbf{A}) \bullet \mathbf{B} = \mathbf{A} \bullet (c\mathbf{B})$$

$$\mathbf{A} \bullet (\mathbf{B} + \mathbf{C}) = (\mathbf{A} \bullet \mathbf{B}) + (\mathbf{A} \bullet \mathbf{C})$$

$$(\mathbf{A} + \mathbf{B}) \bullet \mathbf{C} = (\mathbf{A} \bullet \mathbf{C}) + (\mathbf{B} \bullet \mathbf{C})$$

$$\mathbf{A} \bullet \mathbf{B} = \mathbf{B} \bullet \mathbf{A}$$

$$\mathbf{A} \bullet \mathbf{A} = |\mathbf{A}|^2$$

$$\mathbf{A} \bullet \mathbf{0} = 0$$

- One application of the dot product is to find the *angle between two vectors*. Consider vectors **A** and **B** where $\mathbf{A} = \langle 2, 3 \rangle = (2\mathbf{i} + 3\mathbf{j})$ and $\mathbf{B} = \langle 3, 4 \rangle = (3\mathbf{i} + 4\mathbf{j})$. They can be depicted as:

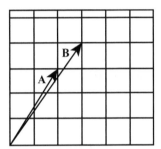

The angle θ between vectors **A** and **B** can be found using the dot product in the equation:

$$\cos \theta = \frac{\mathbf{A} \bullet \mathbf{B}}{|\mathbf{A}||\mathbf{B}|} = \frac{\langle 2, 3 \rangle \bullet \langle 3, 4 \rangle}{\sqrt{2^2 + 3^2}\sqrt{3^2 + 4^2}} = \frac{(2)(3) + (3)(4)}{\sqrt{13}\sqrt{25}} = \frac{18}{5\sqrt{13}}$$

Therefore, $\theta = \arccos \dfrac{18}{5\sqrt{13}} \approx 3.2°$

- Another application of the dot product is that the dot product of *force* **F** and *distance* **d** equals *work* done W:

$$\mathbf{F} \bullet \mathbf{d} = W$$

where **F** is acting on an object to displace it.

The dot product of force **F** and distance **d** can be written using the angle between the vectors, or $\cos \theta$:

$$W = \mathbf{F} \bullet \mathbf{d} = |\mathbf{F}||\mathbf{d}| \cos \theta$$

where **F** is acting on an object to displace it by distance **d**.

The work done by **F** in displacement is the *magnitude* $|\mathbf{F}|$ of the force multiplied by length $|\mathbf{d}|$ of the displacement multiplied with cosine of the angle θ between **F** and **d**. The work is zero if **F** and **d** are perpendicular to each other.

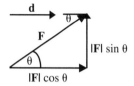

If $\theta = 45°$ then, $W = \mathbf{F} \cdot \mathbf{d} = |\mathbf{F}||\mathbf{d}| \cos 45° = |\mathbf{F}||\mathbf{d}| \sqrt{2}/2$

If $\theta = 90°$ then, $W = \mathbf{F} \cdot \mathbf{d} = |\mathbf{F}||\mathbf{d}| \cos 90° = |\mathbf{F}||\mathbf{d}|(0) = 0$

9.8 Vector or Cross Product

• The **vector product** or **cross product** is used in problems involving torque, volume, and area. The **vector product** or **cross product** of two vectors is defined as:

$\mathbf{A} \times \mathbf{B} = |\mathbf{A}||\mathbf{B}| \sin \theta$

where $|\mathbf{A}|$ and $|\mathbf{B}|$ represent the **magnitudes** (or *lengths*) of vectors **A** and **B**, and θ is the angle between vectors **A** and **B**. The product exists in three dimensions with **A** and **B** in a plane and $\mathbf{A} \times \mathbf{B}$ normal (or perpendicular) to the plane. The *cross product* of two vectors produces a third vector with a length of $|\mathbf{A}||\mathbf{B}| \sin \theta$ and a direction perpendicular to **A** and **B**. The length of $\mathbf{A} \times \mathbf{B}$ depends on $\sin \theta$ and is greatest when $\theta = 90°$ or $\sin \theta = 1$.

• The cross product of two vectors occurs geometrically according to what is referred to as the *right-hand screw rule*. This rule denotes that when taking the cross product $\mathbf{A} \times \mathbf{B}$ and moving from vector **A** to vector **B** through angle θ results in vector $\mathbf{A} \times \mathbf{B}$, which is perpendicular to both **A** and **B**. The right-hand rule can be visualized by curling the fingers of the right hand from **A** to **B**, where $\mathbf{A} \times \mathbf{B}$ points in the direction of the right thumb. Conversely, for the cross product $\mathbf{B} \times \mathbf{A}$, moving from vector **B** to vector **A** through angle θ results in a vector perpendicular to both **A** and **B** but pointing in the opposite direction of $\mathbf{A} \times \mathbf{B}$. Therefore, by the right-hand rule, $\mathbf{A} \times \mathbf{B}$ and $\mathbf{B} \times \mathbf{A}$ point in opposite directions but have the same magnitude.

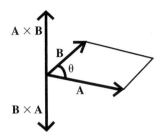

• The *vector or cross product* of two vectors written in terms of their components as vector $\mathbf{A} = a_1\mathbf{i} + a_2\mathbf{j} + a_3\mathbf{k}$ and vector $\mathbf{B} = b_1\mathbf{i} + b_2\mathbf{j} + b_3\mathbf{k}$ can be calculated using the same procedure as when calculating a *determinant*. (See *Master Math: Calculus* Section 5.6. for a summary of determinants.)

$$\mathbf{A} \times \mathbf{B} = (a_1\mathbf{i} + a_2\mathbf{j} + a_3\mathbf{k}) \times (b_1\mathbf{i} + b_2\mathbf{j} + b_3\mathbf{k}) = \begin{vmatrix} \mathbf{i} & \mathbf{j} & \mathbf{k} \\ a_1 & a_2 & a_3 \\ b_1 & b_2 & b_3 \end{vmatrix}$$

$$= (a_2b_3 - a_3b_2)\mathbf{i} + (a_3b_1 - a_1b_3)\mathbf{j} + (a_1b_2 - a_2b_1)\mathbf{k}$$

$$= (a_2b_3 - a_3b_2)\mathbf{i} - (a_1b_3 - a_3b_1)\mathbf{j} + (a_1b_2 - a_2b_1)\mathbf{k}$$

The terms with $\mathbf{i} \times \mathbf{i}$, $\mathbf{j} \times \mathbf{j}$, and $\mathbf{k} \times \mathbf{k}$ equal zero. The cross product of the $\mathbf{i}, \mathbf{j}, \mathbf{k}$ *unit vectors combine* as follows:

$\mathbf{i} \times \mathbf{i} = \mathbf{j} \times \mathbf{j} = \mathbf{k} \times \mathbf{k} = 0$

$\mathbf{i} \times \mathbf{j} = \mathbf{k}, \quad \mathbf{i} \times \mathbf{k} = -\mathbf{j}$

$\mathbf{j} \times \mathbf{i} = -\mathbf{k}, \quad \mathbf{j} \times \mathbf{k} = \mathbf{i}$

$\mathbf{k} \times \mathbf{i} = \mathbf{j}, \quad \mathbf{k} \times \mathbf{j} = -\mathbf{i}$

• Considering the nature of how the unit vectors combine, $\mathbf{A} \times \mathbf{B}$ can be written out as:

$(a_1\mathbf{i} + a_2\mathbf{j} + a_3\mathbf{k}) \times (b_1\mathbf{i} + b_2\mathbf{j} + b_3\mathbf{k})$

$= a_1b_1\mathbf{i} \times \mathbf{i} + a_1b_2\mathbf{i} \times \mathbf{j} + a_1b_3\mathbf{i} \times \mathbf{k}$

$+ a_2b_1\mathbf{j} \times \mathbf{i} + a_2b_2\mathbf{j} \times \mathbf{j} + a_2b_3\mathbf{j} \times \mathbf{k}$

$+ a_3b_1\mathbf{k} \times \mathbf{i} + a_3b_2\mathbf{k} \times \mathbf{j} + a_3b_3\mathbf{k} \times \mathbf{k}$

$= 0 + a_1b_2\mathbf{k} + a_1b_3(-\mathbf{j}) + a_2b_1(-\mathbf{k}) + 0 + a_2b_3\mathbf{i}$

$+ a_3b_1\mathbf{j} + a_3b_2(-\mathbf{i}) + 0$

$= (a_2b_3 - a_3b_2)\mathbf{i} + (a_3b_1 - a_1b_3)\mathbf{j} + (a_1b_2 - a_2b_1)\mathbf{k}$

• Because the unit vectors $\mathbf{i}, \mathbf{j}$, and $\mathbf{k}$ are *perpendicular* to each other, which was found using the determinant above, the angle between $\mathbf{i}$ and $\mathbf{j}$ is $\pi/2$ and by the right-hand rule the cross product of $\mathbf{i}$ and $\mathbf{j}$ is:

$\mathbf{i} \times \mathbf{j} = |\mathbf{i}||\mathbf{j}| \sin (\pi/2) = \mathbf{k}$

The cross product of $\mathbf{i}$ with itself is: $\mathbf{i} \times \mathbf{i} = |\mathbf{i}||\mathbf{i}| \sin 0 = \mathbf{0}$.

• The *maximum value of the cross product* of two vectors occurs when the angle θ is $\pi/2$ and $\sin \pi/2 = 1$, and therefore the two *vectors are perpendicular* to each other. Conversely, the *minimum value of the cross product* of two vectors occurs when the angle θ is 0 or π and $\sin 0 = \sin \pi = 0$, and therefore the two *vectors are parallel*.

• The magnitude of the cross product can represent the area of a parallelogram with sides $\mathbf{A}$ and $\mathbf{B}$, where the value resulting from $\mathbf{A} \times \mathbf{B}$ is both the *length of vector* $\mathbf{A} \times \mathbf{B}$ and the *area of the parallelogram*. The length of the cross product is the area, and the area of the parallelogram is $|\mathbf{A} \times \mathbf{B}|$, which is the magnitude of the area. A parallelogram with sides $\mathbf{A}$ and $\mathbf{B}$ has area $|a_1 b_2 - a_2 b_1|$. In an XY plane $\mathbf{A} \times \mathbf{B} = (a_1 b_2 - a_2 b_1)\mathbf{k}$.

$\mathbf{A} \times \mathbf{B}$

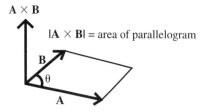

$|\mathbf{A} \times \mathbf{B}| =$ area of parallelogram

• Another important application of the *cross product* is *torque*, which is a force acting on an object to cause rotation. A force $\mathbf{F}$ can be applied to a lever arm, or a radius vector $\mathbf{r}$, which has its initial point located at the origin of rotation and causes the object to rotate. The torque is a vector having a magnitude that measures the force of the rotation and a direction of the axis of rotation. The cross product $\mathbf{F} \times \mathbf{r} = \mathbf{T}$ is the torque of the force about the origin for a force $\mathbf{F}$ acting at a point with position vector $\mathbf{r}$.

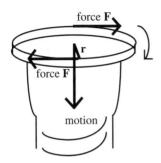

• An application of both the *cross product* and *dot product* together is the *volume of a parallelepiped*. A parallelepiped with sides given by vectors **A, B**, and **C** is represented by:

$$|\mathbf{A} \cdot \mathbf{B} \times \mathbf{C}| = (a_1\mathbf{i} + a_2\mathbf{j} + a_3\mathbf{k}) \cdot (b_1\mathbf{i} + b_2\mathbf{j} + b_3\mathbf{k}) \times (c_1\mathbf{i} + c_2\mathbf{j} + c_3\mathbf{k}) =$$

$$\begin{vmatrix} a_1 & a_2 & a_3 \\ b_1 & b_2 & b_3 \\ c_1 & c_2 & c_3 \end{vmatrix} = a_1(b_2c_3 - b_3c_2) + a_2(b_3c_1 - b_1c_3) + a_3(b_1c_2 - b_2c_1)$$

$= |\mathbf{B} \times \mathbf{C}|$ times $|\mathbf{A}|$ cos θ, where $|\mathbf{B} \times \mathbf{C}|$ is area of base and $|\mathbf{A}|$ cos θ is height. Volume $= |\mathbf{A} \cdot \mathbf{B} \times \mathbf{C}|$

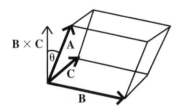

(Notice that taking the magnitude of the vector reduces it to a scalar quantity.)

• *Properties of the cross product* involving vectors **A, B, C**, and scalar c include:

$$\mathbf{A} \times \mathbf{B} = -(\mathbf{B} \times \mathbf{A})$$
$$c(\mathbf{A} \times \mathbf{B}) = (c\mathbf{A}) \times \mathbf{B} = \mathbf{A} \times (c\mathbf{B})$$
$$\mathbf{A} \times (\mathbf{B} + \mathbf{C}) = (\mathbf{A} \times \mathbf{B}) + (\mathbf{A} \times \mathbf{C})$$
$$\mathbf{A} \cdot \mathbf{B} \times \mathbf{C} = \mathbf{A} \times \mathbf{B} \cdot \mathbf{C} = \mathbf{B} \cdot \mathbf{C} \times \mathbf{A} = \mathbf{C} \cdot \mathbf{A} \times \mathbf{B}$$
$$|\mathbf{A} \cdot \mathbf{B}|^2 + |\mathbf{A} \times \mathbf{B}|^2 = |\mathbf{A}|^2|\mathbf{B}|^2 \cos^2\theta + |\mathbf{A}|^2|\mathbf{B}|^2 \sin^2\theta$$
$$= |\mathbf{A}|^2|\mathbf{B}|^2$$

9.9 Chapter 9 Summary and Highlights

• *Scalars* are quantities that represent magnitude and can be described by one number, which is positive, negative, or zero. Examples of scalars include temperature, work, density, and mass. A *vector* represents a quantity that is described by both a numerical value for *magnitude* (or *length*) and a *direction*. A vector is depicted as a line segment with an initial point and a terminal point that has an arrow pointing in the direction of the terminal point. The length of a vector represents the magnitude of the vector quantity. Examples of vectors include displacement, velocity, acceleration, electric field strength, and force. A vector can be relocated and still be considered the same vector as long as its length and direction remain the same.

• A vector can be described by its horizontal and vertical components, which are also vectors. A vector with its initial point at the origin of a *coordinate system* is called a *position vector*. A position vector is defined according to the location or coordinates of its terminal point.

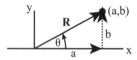

Vector **R** in the preceding figure can be called vector (a, b), where a and b represent the x-component and y-component, respectively. The length, or magnitude, of vector **R** is given by $|\mathbf{R}| = \sqrt{a^2 + b^2}$. If the magnitude of **R** is represented by r and the direction angle is θ, then vector **R** can also be described by: $\mathbf{R} = \langle r\cos\theta, r\sin\theta \rangle$, where the direction angle is the angle it makes with the horizontal axis.

• The vector components can also be defined by their directions along the X, Y, and Z axes of a coordinate system using the **i**, **j**, and **k** unit vectors. The direction of vector **A** can be written in terms of the *angle* θ that it makes with the positive X-axis. Vector $\mathbf{A} = a_1\mathbf{i} + a_2\mathbf{j}$ makes an angle $\theta = \tan^{-1}(a_2/a_1)$ with the X-axis and can be written as:

$$\mathbf{A} = \mathbf{i}\,|\mathbf{A}|\cos\theta + \mathbf{j}\,|\mathbf{A}|\sin\theta, \text{ where } a_1 = |\mathbf{A}|\cos\theta \text{ and } a_2 = |\mathbf{A}|\sin\theta.$$

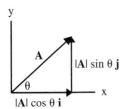

• Two vectors can be added or subtracted if they have the same dimensions by adding or subtracting the corresponding components. The *sum of two vectors* can be depicted by positioning the vectors such that the initial point of the second vector is at the terminal point of the first vector. The sum of the two vectors is a third vector with its initial point at the initial point of the first vector and its final point at the final point of the second vector. Multiplying vector **A** with scalar c results in a vector having a magnitude of $|c||\mathbf{A}|$ and a direction of **A**.

• The *dot product* can be used to find angles between vectors and is a scalar, not a vector. The dot product of two vectors is defined as: $\mathbf{A} \cdot \mathbf{B} = |\mathbf{A}||\mathbf{B}|\cos\theta$, where $|\mathbf{A}|$ and $|\mathbf{B}|$ represent the *magnitudes* (or *lengths*) of vectors **A** and **B** and θ is the angle between vectors **A** and **B**.

The angle θ between **A** and **B** is: $\theta = \arccos\dfrac{\mathbf{A} \cdot \mathbf{B}}{|\mathbf{A}||\mathbf{B}|}$

• The *vector product* or *cross product* of two vectors is defined as: $\mathbf{A} \times \mathbf{B} = |\mathbf{A}||\mathbf{B}|\sin\theta$, where $|\mathbf{A}|$ and $|\mathbf{B}|$ represent the *magnitudes* (or *lengths*) of vectors **A** and **B**, and θ is the angle between the vectors. The product exists in three dimensions with **A** and **B** in a plane and $\mathbf{A} \times \mathbf{B}$ normal (or perpendicular) to the plane.

Chapter 10

Trigonometric Functions in Polar Coordinates and Equations, and Parametric Equations

10.1 Polar Coordinates Defined

• Using *rectangular,* or *Cartesian, coordinates*, it is possible to locate a point on a plane with two numbers. The first number indicates the distance of the point from the origin along the horizontal X-axis, and the second number indicates the distance of the point from the origin along the vertical Y-axis. The numbers are written as ordered pairs enclosed in parentheses as (x, y) or using unit vectors $\mathbf{i}$ and $\mathbf{j}$, which designate directions along the X and Y axes. The unit vectors $\mathbf{i}$ and $\mathbf{j}$ indicate directions and have a length of one. For example, the location of a point 3 units to the right of the origin and 4 units below the origin can be described by using either one of the notations: $(3, -4)$ or $3\mathbf{i} - 4\mathbf{j}$. The notation $(3, -4)$ represents an ordered pair (x, y), and the notation $3\mathbf{i} - 4\mathbf{j}$ represents vector notation such that $+3$ is in the x direction and -4 is in the y direction.

• Although rectangular coordinates are most often used to map points in a plane, ***polar coordinates*** are also used frequently and are particularly applicable for graphing numerous relations. *Polar coordinates* describe points in a plane or in space, similar to rectangular coordinates. The difference is that in polar coordinates, there is an r-coordinate that maps the distance of a point from the origin of the coordinate system, and there is a θ-coordinate that measures the angle the r-ray makes from the horizontal positive X-axis called the *polar axis*.

• The *polar coordinate system* is based on a point, called the **pole,** and a *ray*, called the **polar axis**. The polar axis is usually drawn in the direction of the positive X-axis:

The polar axis lies on the positive side of the horizontal X-axis beginning at the origin of a rectangular coordinate system, so that the polar axis coincides with the positive X-axis.

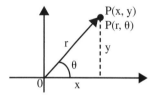

Point P has coordinates (x, y) in the rectangular coordinate system, and point P designates the ordered pair (r, θ) in polar coordinates. Point P is located by giving the angle θ from the positive X-axis to ray OP and the distance r from the pole to point P.

• The relationship between polar and *rectangular coordinates* can be visualized in the figure below using the *Pythagorean Theorem* for a right triangle where $r^2 = x^2 + y^2$. The r-coordinate is the *hypotenuse* and measures the distance from origin to the point of interest. The angle θ between r and the positive part of the X-axis can be described by $\tan \theta = y/x$. The relationships between rectangular and polar coordinate systems in two dimensions are:

$$r = \sqrt{x^2 + y^2}$$
$$\tan \theta = y/x \text{ or } \theta = \tan^{-1}(y/x)$$
$$x = r \cos \theta$$
$$y = r \sin \theta$$

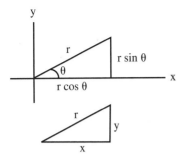

A point in a plane can be described in a polar coordinate system using a *direction angle* θ and the *magnitude r* of the corresponding ray or vector. The ordered pair (r, θ) designates the **polar coordinates** of the point. As we will see, there are numerous (r, θ) coordinates, which can describe any single point in a polar coordinate system. To identify a point in a plane using a polar coordinate system, choose a point to represent the origin, which is the pole, then draw a horizontal line to the right, which forms the polar axis. The 0 degree direction generally points directly right from the origin. Any point in the plane can be identified by the two coordinates r and θ, where

r = distance from the origin to the point and

θ = angle between the 0 degree line and the line drawn from the origin to the point.

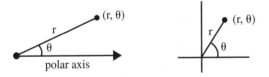

• To locate any point P in a plane, the polar coordinates (r, θ) are identified beginning with the polar axis as the initial side of an angle and rotating the terminal side until it (or its extension in the negative direction through the pole) intersects the point. The angle θ can be measured in degrees or radians. Angle θ is *positive* if the rotation is counterclockwise and *negative* if the rotation is clockwise. The r-coordinate, which is the directed distance from the pole to the point P, is *positive* if measured from the pole along the terminal side of θ and *negative* if measured along the terminal side extended through the pole. (See figure below.) A point $(-r, \theta)$ with a negative r-value represents a point a distance r away from the origin in the opposite direction as a point (r, θ).

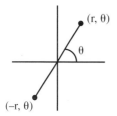

• A given point in a polar coordinate system can have numerous polar coordinates. For example, the following graphs of polar coordinates (r, θ) represent the same point:

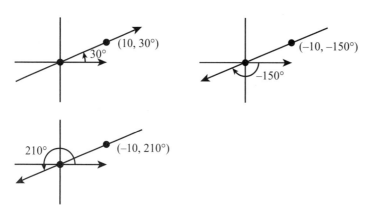

where (r, θ) = (10, 30°) can also be in radians (10, π/6),

(–10, –150°) can be in radians (–10, –5π/6), and

(–10, 210°) can be in radians (–10, 7π/6).

• *Positive angles* are measured counterclockwise from the positive X-axis, and *negative angles* are measured clockwise from the positive X-axis. If, for example, a point is a distance of 3 units from the origin at a positive angle of 50 degrees, its location can be described in terms of an ordered pair of r and θ as (3, 50°). There are many angles that can be used to locate this point, which include measuring counterclockwise 50° from the positive X-axis describing the point as (3, 50°) or measuring 310° in the clockwise (negative) direction describing the point as (3, –310°). It is also possible to rotate through either –130° or +230° and then back up 3 units in the negative direction of r, which results in the same point but is described by (–3, –130°) and (–3, 230°).

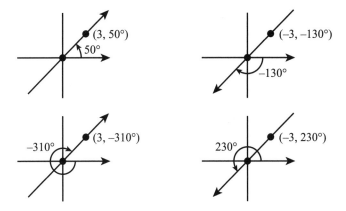

It may be helpful to turn through the angle first and then go in the indicated direction if r is a positive number and back up in the negative direction of r if r is a negative number.

• There is special polar graph paper that can be used for polar coordinates. For example, a graph of two points, P(2, 45°) and Q(–2, 45°), would be depicted as:

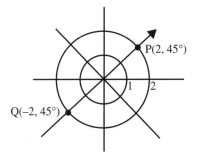

10.2 Converting Between Rectangular and Polar Coordinate Systems and Equations

• It is often useful to transform coordinates or equations from rectangular form into polar form, or from polar form into rectangular form. The relationships between point (x, y) in a two-dimensional rectangular coordinate system and point (r, θ) in a two-dimensional polar coordinate system are:

$r^2 = x^2 + y^2$, or $r = \sqrt{x^2 + y^2}$

$\tan \theta = y/x$, or $\theta = \tan^{-1}(y/x)$, where x is not 0

$x = r \cos \theta$, or $\cos \theta = x/r$

$y = r \sin \theta$, or $\sin \theta = y/r$

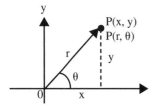

Note that the signs of x and y determine the quadrant for θ. The angle θ is usually chosen so that $-\pi < \theta \leq \pi$, or $-180° < \theta \leq 180°$.

• To convert rectangular coordinates (x, y) to polar coordinates (r, θ) in a plane, it is convenient to use:

$r = \sqrt{x^2 + y^2}$ and $\theta = \arctan(y/x)$

To convert polar coordinates (r, θ) to rectangular coordinates (x, y), it is convenient to use:

$x = r \cos \theta$ and $y = r \sin \theta$

• **Example:** Find x and y if r and θ are given as $r = 5$ and $\theta = \pi/2$.

Calculate $x = r \cos \theta$ and $y = r \sin \theta$:

$x = (5) \cos(\pi/2) = 0$

$y = (5) \sin(\pi/2) = 5$

Therefore, the point is at $(0, 5)$ in rectangular coordinates.

y
5 (x = 0, y = 5) and (r = 5, θ = π/2)
x

• **Example:** Find r and θ if x and y are given as $x = 2$ and $y = 3$.

Calculate $r = \sqrt{x^2 + y^2}$ and $\theta = \tan^{-1}(y/x)$:

$$r = \sqrt{2^2 + 3^2} = [13]^{1/2} \approx 3.6$$
$$\theta = \tan^{-1}(3/2) \approx 0.98 \text{ rad} \approx 56°$$

Therefore, the point is at $(3.6, 56°)$ in polar coordinates.

• **Example:** Convert rectangular coordinate $(-4, 3)$, or equivalently, $-4\mathbf{i} + 3\mathbf{j}$, to polar coordinates and express four different ways.

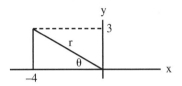

Determine the length of the hypotenuse r of the right triangle and the measure of the *polar angle* from the positive X-axis.

The length of r is $r = \sqrt{-4^2 + 3^2} = 5$
and $\theta = \tan^{-1}(3/-4) \approx -37°$

Therefore, r is 5 and the small angle is 37°. If we measure counterclockwise from the positive X-axis, the polar angle is 143°; and if we measure clockwise from the positive X-axis, the polar angle is –217°. If we use negative magnitudes of r, the corresponding polar angles are fourth-quadrant angles –37° and 323°. (Remember: Half of a circle formed in the coordinate system is 180°.)

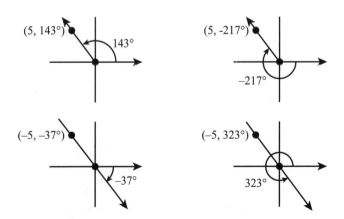

• **Example:** Convert polar coordinate $(-10, -245°)$ to a rectangular coordinate.

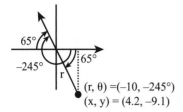

(r, θ) = (–10, –245°)
(x, y) = (4.2, –9.1)

In this diagram, we first measure –245° and then step back 10 units from the direction the arrow is pointing to locate point (–10, –245°). The angle of the triangle formed to r = –10 is 65°, (180° – 245°). We can then find the x value and the y value of the reference triangle:

$$x = r \cos \theta = 10 \cos(-65°) \approx 4.2$$
$$y = r \sin \theta = 10 \sin(-65°) \approx -9.1$$

We can also calculate x and y by using the standard position angle –245° and the negative r value:

$$x = r \cos \theta = -10 \cos(-245°) \approx 4.2$$
$$y = r \sin \theta = -10 \sin(-245°) \approx -9.1$$

Therefore, as depicted on the figure, the rectangular coordinates using either ordered pairs or the unit vectors **i** and **j** are:

$$(4.2, -9.1) \text{ or } 4.2\mathbf{i} - 9.1\mathbf{j}$$

• **Example:** Convert polar coordinate (–5, 120°) to a rectangular coordinate.

120°
60°
r
(r, θ) = (–5, 120°)
(x, y) = (2.5, –4.3)

In this diagram, we first measure +120° and then step back 5 units from the direction the arrow is pointing to locate point (–5, 120°). The angle of the reference triangle formed by drawing a line to the horizontal axis is 60°, (120° – 180°). We can then find the x value and the y value of the triangle.

$$x = r \cos \theta = 5 \cos(-60°) = 2.5$$
$$y = r \sin \theta = 5 \sin(-60°) \approx -4.3$$

We can also calculate x and y by using the standard angle 120° and the negative r value:

$$x = r \cos \theta = -5 \cos(120°) = 2.5$$
$$y = r \sin \theta = -5 \sin(120°) \approx -4.3$$

Therefore, the rectangular coordinate is (2.5, –4.3).

• **Example:** Convert rectangular coordinate (–2, 2) to two polar coordinates for that point.

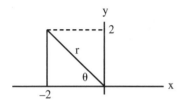

Determine the length of the hypotenuse r of the reference right triangle and then measure the *polar angle* from the positive X-axis.

The length of r is $r = \sqrt{-2^2 + 2^2} = \sqrt{8} = 2\sqrt{2} \approx 2.83$
$$\theta = \tan^{-1}(2/-2) = \tan^{-1}(-1) = -45°$$

For angle θ to be in the second quadrant –45° corresponds to 135°. A positive value of r can also have negative angle –225°. There are numerous possible polar coordinates for a single point.

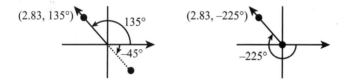

Therefore, two pairs of polar coordinates for the rectangular point (x, y) = (–2, 2) include (r, θ) = (2√2 , 135°) and (2√2 , –225°).

Converting Equations

• Rectangular and polar *equations* can be converted from one form to the other using the same relationships for x, y, r, and θ that are used to convert coordinates. For two-dimensions these relations are:

$r^2 = x^2 + y^2$ or $r = \sqrt{x^2 + y^2}$

$\tan \theta = y/x$, or $\theta = \tan^{-1}(y/x)$, where x is not 0

$x = r \cos \theta$, or $\cos \theta = x/r$

$y = r \sin \theta$, or $\sin \theta = y/r$

• **Example:** Transform equation $x^2 + y^2 = x$ into polar form and graph both rectangular and polar forms.

To transform this equation substitute $r^2 = x^2 + y^2$ and $x = r \cos \theta$:

$x^2 + y^2 = x$

$r^2 = r \cos \theta$

Rearrange:

$r^2 - r \cos \theta = 0$

Factor:

$r(r - \cos \theta) = 0$

$r = 0$ and $r - \cos \theta = 0$ or $r = \cos \theta$, are equations in terms of r and θ.

$r = 0$ represents the pole.

To plot the equations on one graph rearrange $x^2 + y^2 = x$ to $y = \pm\sqrt{x - x^2}$

Then, plot the positive and negative roots of $y = \pm\sqrt{x - x^2}$ and the polar equation $r = \cos \theta$ on the same graph.

Graph of $y = \pm\sqrt{x - x^2}$ and $r = \cos \theta$

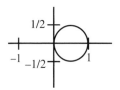

This graph is a complete overlap of the rectangular and polar graphs, with the positive root for y forming a semicircle above the X-axis, the negative root for y forming a semicircle below the X-axis, and the polar equation forming a complete circle.

• **Example:** Convert polar equation $r^2 \cos 2\theta = 2$ to a rectangular form of the equation, and graph both rectangular and polar forms of the equation on one graph.

To convert this equation, substitute the double angle formula for cosine $\cos 2x = \cos^2 x - \sin^2 x$ (see summary of identities in section 7.1):

$$r^2 \cos 2\theta = 2$$
$$r^2(\cos^2\theta - \sin^2\theta) = 2$$
$$r^2 \cos^2\theta - r^2 \sin^2\theta = 2$$
$$(r \cos \theta)^2 - (r \sin \theta)^2 = 2$$

Then substitute $x = r \cos \theta$ and $y = r \sin \theta$, resulting in:

$$x^2 - y^2 = 2$$

To plot the equations on one graph, rearrange $x^2 - y^2 = 2$:

$$y^2 = x^2 - 2, \text{ or } y = \pm\sqrt{x^2 - 2}$$

To plot $r^2 \cos 2\theta = 2$, rearrange:

$$r^2 = 2 / \cos 2\theta, \text{ or } r = \pm\sqrt{2/\cos 2\theta}$$

Graph of $y = \pm\sqrt{x^2 - 2}$ and $r = \pm\sqrt{2 / \cos 2\theta}$

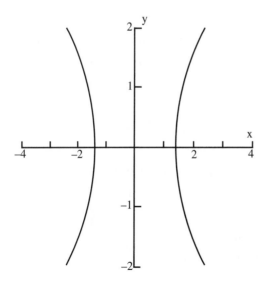

The graph forms an overlap of the rectangular and polar curves, with the positive root for y existing above the X-axis, the negative root for y existing below the X-axis, and the polar equation existing above and below the X-axis.

• Note that using a graphing calculator or graphing software to plot both rectangular and polar forms of an equation on a single graph will provide a means to verify conversions.

10.3 Graphing Polar Equations

• Various functions and their graphs will lend themselves to being represented in polar coordinates and others will be better represented in rectangular coordinates.

• *Polar equations* are expressed in terms of the variables r and θ and may be graphed using the same methods used for rectangular equations. To graph a polar equation, calculate r values for a range of θ values for any specified interval or until a pattern develops, and then plot the resulting ordered pairs of r and θ values (r, θ) in a coordinate system. An alternative to calculating r and θ values and plotting them is entering an equation into a graphing utility, such as a graphing calculator or graphing software. It is important to be consistent as far as the angle mode (degree or radian) that you are using for calculations and graphs. Also remember that the coordinates for r and θ correspond to the standard angle and the magnitude (or length) of r from the origin.

• There are *families of polar functions* that form important graphs. These functions are described in the following paragraphs and include:

Circles with equation forms $r = a \cos \theta$ and $r = a \sin \theta$,

Archimedes' spiral with equation form $r = a\theta$,

Cardioid with equation forms $r = a + a \cos \theta$ and $r = a + a \sin \theta$,

Roses with equation forms $r = a \sin n\theta$ and $r = a \cos n\theta$, including

Three-leaved rose with equation forms $r = a \cos 3\theta$ and $r = a \sin 3\theta$,

Four-leaved rose with equation forms $r = a \cos 2\theta$ and $r = a \sin 2\theta$,

Lemniscate with equation forms $r^2 = a^2 \cos 2\theta$ and $r^2 = a^2 \sin 2\theta$,

Vertical line with equation form $r = a / \cos \theta$,

Horizontal line with equation form $r = a / \sin \theta$, and

Radial line with equation form $\theta = a$,

where a is a constant.

• The graphs in this section illustrate that relatively small variations between the polar equations listed above result in diverse and characteristic shapes.

• A *circle* on a coordinate system can be represented by the equation $r = \cos \theta$ or $r = \sin \theta$, where substituting values of θ around the coordinate system will produce points on the circle.

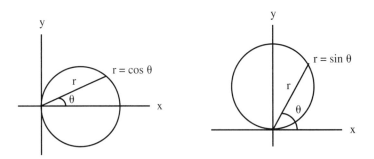

The following graph depicts $r = \cos \theta$, $r = \sin \theta$, $r = 2 \cos \theta$, and $r = 2 \sin \theta$, with the cosine curves lying to the right of the Y-axis and the sine curves lying above the X-axis.

Graph of $r = \cos \theta$ (right side small), $r = 2 \cos \theta$ (right side large), $r = \sin \theta$ (top small), and $r = 2 \sin \theta$ (top large)

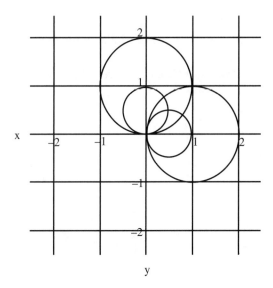

• **Example:** Make a table for r = 2 cos θ and graph it in a polar coordinate system, where θ is in degrees. Begin making calculations using multiples of 30°.

r = 2 cos θ

θ	r	θ	r
0°	2.00	210°	−1.73
30°	1.73	240°	−1.00
60°	1.00	270°	0.00
90°	0.00	300°	1.00
120°	−1.00	330°	1.73
150°	−1.73	360°	2.00
180°	−2.00		

Remember: Negative r values correspond to the negative end of r.

Graph of r = 2 cos θ

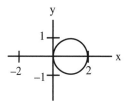

The graph forms a circle with radius 1. Each value of r falls on the circle.

• The family curves called ***Archimedes' spirals*** have the equation form r = aθ, where a is any constant. For example, r = θ is depicted as:

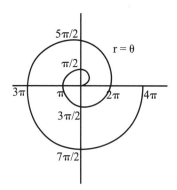

• A graph of a polar equation such as $r = a\theta$ in a polar coordinate system is obtained using the same method as in rectangular coordinates: by making a table of values that satisfy the equation and plotting the resulting values, or by entering the equation into a graphing utility.

• **Example:** Make a table for $r = 3\theta, 0 \leq \theta \leq 2\pi$ (θ is in radians), and graph. It is convenient to make calculations using multiples of $30°$ or $\pi/6$ radians.

θ	$r = 3\theta$	θ	$r = 3\theta$
0	0	$7\pi/6$	10.99
$\pi/6$	1.57	$4\pi/3$	12.57
$\pi/3$	3.14	$3\pi/2$	14.14
$\pi/2$	4.71	$5\pi/3$	15.71
$2\pi/3$	6.28	$11\pi/6$	17.28
$5\pi/6$	7.85	2π	18.85
π	9.42		

Graph of $r = 3\theta, 0 \leq \theta \leq 2\pi$

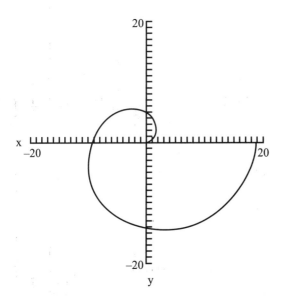

Each value of r for a given angle θ falls on the curve.

• Another important family of polar graphs is the **cardioids** with the equation forms $r = a + a \cos \theta$ and $r = a + a \sin \theta$.

• **Example:** Make a table for r = 1 + cos θ, 0 ≤ θ ≤ 360°, and graph in
a polar coordinate system. Begin making calculations using multiples of
30°. Remember that values between 0° and 360° represent one period of
the cosine function.

$$r = 1 + \cos \theta$$

θ	r	θ	r
0°	2.00	210°	0.13
30°	1.87	240°	0.50
60°	1.50	270°	1.00
90°	1.00	300°	1.50
120°	0.50	330°	1.87
150°	0.13	360°	2.00
180°	0.00		

Graph of r = 1 + cos θ, 0 ≤ θ ≤ 360°

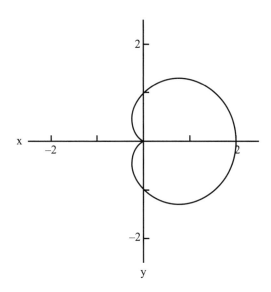

Each value of r for a given angle θ falls on the curve.

• A family of curves called *roses* is represented by the general equations
in the forms of r = a sin nθ and r = a cos nθ, having n petals if n is odd,
and 2n petals if n is even. For example, graphs of *three-leaved roses* have
equation forms r = a cos 3θ and r = a sin 3θ, graphs of *four-leaved roses*

have equation forms r = a cos 2θ and r = a sin 2θ, graphs of *five-leaved roses* have equation forms r = a cos 5θ and r = a sin 5θ, and graphs of *eight-leaved roses* have equation forms r = a cos 4θ and r = a sin 4θ. (Note that for n = 1, the equations for a rose become equations for circles r = a cos θ and r = a sin θ.)

• A graph of a rose equation is obtained using the same method as the other families of curves: by making a table of values that satisfy the equation and plotting the resulting values, or by using a graphing utility and entering the equation.

• An example of two *three-leaved rose* equations, r = cos 3θ and r = 3 cos 3θ, is depicted in the following graph, with r = cos 3θ represented by the smaller rose and r = 3 cos 3θ represented as the larger rose.

Graph of r = cos 3θ (smaller) and r = 3 cos 3θ (larger)

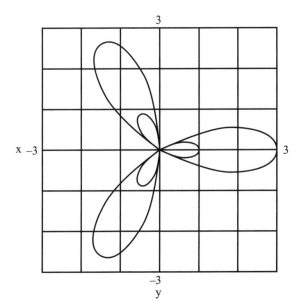

Each value of r for a given angle θ falls on the curve.

• An example of two *four-leaved rose* equations, r = cos 2θ and r = 2 sin 2θ, is depicted in the following graph, with r = cos 2θ represented by the smaller rose and r = 2 sin 2θ represented as the larger rose.

Graph of r = cos 2θ (smaller) and r = 2 sin 2θ (larger)

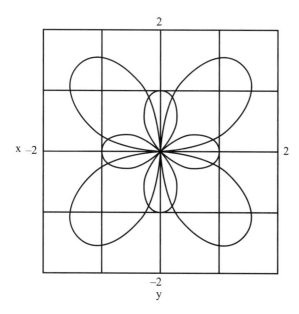

Each value of r for a given angle θ falls on the curve.

- **Lemniscates** have equation forms $r^2 = a^2 \cos 2\theta$ and $r^2 = a^2 \sin 2\theta$.

- An example of two *lemniscate* equations, $r = \pm\sqrt{(1/4) \cos 2\theta}$ and $r = \pm\sqrt{4 \sin 2\theta}$ is depicted in the following graph with $r = \pm\sqrt{(1/4) \cos 2\theta}$ represented by the smaller lemniscate and $r = \pm\sqrt{4 \sin 2\theta}$ represented as the larger lemniscate.

Both positive and negative roots were entered into the graphing program.

Graph of $r = \pm\sqrt{(1/4)\cos 2\theta}$ (smaller) and $r = \pm\sqrt{4\sin 2\theta}$ (larger)

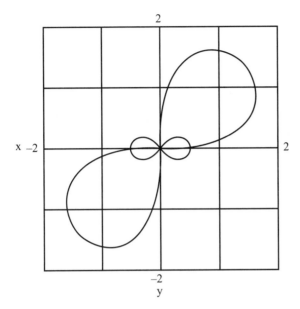

Each value of r for a given angle θ falls on the curve.

• The graphs of *straight lines* in rectangular coordinates are found by setting the variables x and y equal to constants, where x = a is a vertical line and y = b is a horizontal line. In polar coordinates, the equations for horizontal and vertical lines are more complicated. The equations for straight lines in polar equations include: *vertical lines* with equation form r = a / cos θ, *horizontal lines* with equation form r = a / sin θ, and *radial lines* with equation form θ = a.

• **Example:** The following graph is of r = 2 / cos θ, which is the vertical line at a distance of 2 from the origin, and of r = 2 / sin θ, which is the horizontal line at a distance of 2 from the origin.

Graph of r = 2/cos θ (vertical line) and of r = 2/sin θ (horizontal line)

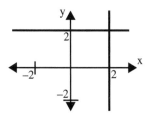

For 45°, r = 2/cos 45° ≈ 2.83 and r = 2/sin 45° ≈ 2.83, which is the r value where the lines cross.

• The coordinates of a *horizontal line* at a distance d away from the origin given by the equation r = d / sin θ have coordinates depicted by:

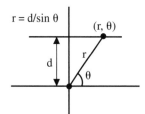

• A radial line given by the equation θ = a has the graph:

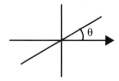

• As we have observed in the graphs in this section, polar coordinates are more appropriate for many curved shapes and rectangular coordinates are more appropriate for linear graphs.

• In three dimensions, *polar coordinates* become ***cylindrical coordinates*** and are given in terms of r, θ, and z, where:

$$x = r \cos θ$$
$$y = r \sin θ$$
$$z = z$$
$$r = \sqrt{x^2 + y^2}$$

When comparing the rectangular and *cylindrical coordinate systems*, the x- and y-components of the rectangular coordinate system are expressed in terms of polar coordinates, and the z-component is the same component as in the rectangular system. The r-component is measured from the Z-axis, the θ-component measures the distance around the Z-axis, and the z-component measures along the Z-axis.

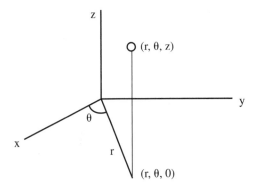

• An example of a shape represented in *cylindrical coordinates* is a *triangular wedge section*. In this figure the section $(\Delta\theta/2\pi)$ is a part of a whole area of the circle πr^2 that projects along the Z-axis to create a volume.

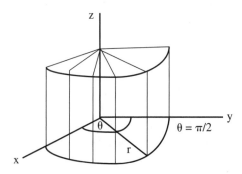

• Another coordinate system that is related to rectangular coordinates is the *spherical coordinates system*. In three dimensions, spherical coordinates are expressed in terms of ρ, θ, and ϕ, where ρ can range from 0 to ∞, θ can range from 0 to 2π, and ϕ can range from 0 to π. In spherical coordinates, the ρ component is measured from the origin, the θ component measures the distance around the Z-axis, and the ϕ component measures down from the Z-axis and is referred to as the *polar angle*. Note that ρ is measured from the origin rather than the Z-axis, as is the case with r in cylindrical coordinates. Also, θ and φ are similar to longitude and latitude on a globe. Spherical coordinates can be defined in terms of rectangular coordinates, x, y, and z as:

$$x = \rho \cos \theta \sin \phi$$
$$y = \rho \sin \theta \sin \phi$$
$$z = \rho \cos \phi$$
$$\rho = \sqrt{x^2 + y^2 + z^2}$$

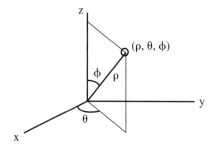

10.4 Parametric Equations

• Parametric equations are often the most efficient means to describe objects in *motion*, including airplanes, rockets, and baseballs. Projectiles can be described using parametric equations. For a *projectile*, the total time of travel, maximum height, time at which maximum height occurs, total horizontal distance traveled, and distance for a given t (time) value can be modeled using parametric equations. Parametric equations can also be used to model cycloids, which are the paths traced by a point on the circumference of a circle as it rolls along a line. Parametric equations may be used to represent the motion of a particle or object moving along a curve as well as to describe the curve itself.

• A set of points can be defined *parametrically* using the **parametric equations** x = f(t) and y = g(t), where t is called the **parameter**. It is possible to determine a set of ordered pairs in the plane using the equations x = f(t) and y = g(t), where parameter t is a real number defined in an interval. Each value of t corresponds with values of x and y, which results in (x, y) ordered pairs. It is important to note that more than one set of parametric equations may result in the same set of (x, y) ordered pairs or curve. In other words, parametric representation of a curve is not necessarily unique, and there may be more than one set of parametric equations for x and y that correspond to the rectangular form of the parametric equations.

- The parametric equations $x = f(t)$ and $y = g(t)$ can be calculated for various t values to produce ordered (x, y) pairs, and a graph can be plotted using the (x, y) values. Graphing calculators and graphing software are often able to graph parametric equations. A curve in a plane can be described parametrically using a set of points in a specified interval given by (x, y), where $x = f(t)$ and $y = g(t)$ with t as the parameter. For example, a unit circle can be graphed parametrically using $x = \cos t$ and $y = \sin t$. Both two- and three-dimensional geometric shapes can be described using parametric equations.

- **Example:** Consider a curve in a plane defined by the parametric equations $x = 2 \cos t$ and $y = 4 \sin t$ in the interval $0 < t \le 2\pi$. Write these equations in rectangular form and graph them using both the parametric and rectangular equations.

To transform parametric equations into rectangular form we would normally isolate t and write the equations in terms of x and y only. Because of the trigonometric functions it is difficult to isolate t; however, we can isolate cos t and sin t, then substitute the Pythagorean identity $\cos^2 t + \sin^2 t = 1$:

$$x = 2 \cos t \quad y = 4 \sin t$$
$$x/2 = \cos t \quad y/4 = \sin t$$

Substitute into $\cos^2 t + \sin^2 t = 1$:

$$(x/2)^2 + (y/4)^2 = 1, \text{ or } x^2/4 + y^2/16 = 1$$

which is the equation for $x = 2 \cos t$ and $y = 4 \sin t$ in rectangular form.

To graph first isolate y:

$$y^2/16 = 1 - x^2/4$$
$$y^2 = 16(1 - x^2/4)$$
$$y = \pm\sqrt{16(1 - x^2/4)} = \pm\sqrt{16 - 4x^2}$$

The following graph depicts curves for the equations $x = 2 \cos t$ and $y = 4 \sin t$, as well as both the positive and negative roots of $y = \pm\sqrt{16 - 4x^2}$. In the following graph, the rectangular and parametric equations overlap each other to form the ellipse. Note that the positive root of the rectangular equation forms the upper section of the ellipse above the X-axis, and the negative root forms the lower section of the ellipse below the X-axis. Also note that the ellipse has x-intercepts at ±2 and y-intercepts at ±4.

Graph of parametric equations x = 2 cos t and y = 4 sin t, and rectangular equation $y = \pm\sqrt{16 - 4x^2}$

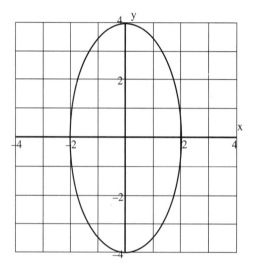

• A graph of parametric equations can be drawn by selecting values of the parameter, t, and calculating values of x and y or by entering the equations into a graphing calculator or graphing software.

• For example, consider the parametric equations x = t and y = t². We can create a graph by first calculating x and y values for selected t values and then plotting the results.

$x = t, y = t^2$

t	x	y
−3	−3	9
−2	−2	4
−1	−1	1
0	0	0
1	1	1
2	2	4
3	3	9

Graph of x = t and y = t²

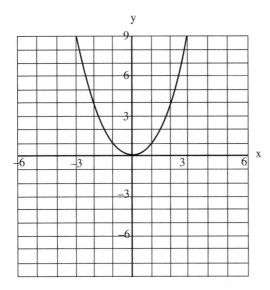

• A ***unit circle*** can be represented implicitly, explicitly, or parametrically as follows:

implicitly: $x^2 + y^2 = 1$

explicitly: $y = \sqrt{1-x^2}$ and $y = -\sqrt{1-x^2}$

parametrically: $x = \cos t, y = \sin t, 0 \leq t \leq 2\pi$

The equations $x = \cos t$, $y = \sin t$, $y = \sqrt{1-x^2}$, and $y = -\sqrt{1-x^2}$ can be plotted and will result in the same graph of a unit circle:

Graph of x = cos t, y = sin t, $y = \sqrt{1-x^2}$ and $y = -\sqrt{1-x^2}$

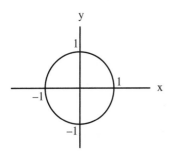

• In general, a *curve* can be represented implicitly, explicitly, or parametrically in an XY plane as:

1. *Implicitly* by an equation in x and y, or f(x,y);
2. *Explicitly* by equations for y in terms of x or x in terms of y, $y = f(x)$ or $x = g(y)$; or
3. *Parametrically* by a pair of equations for x and y in terms of a third variable or parameter t, $x = f(t)$ and $y = g(t)$.

• The **parameterization** of a half-circle from 0 to π is described by $x = \cos t, y = \sin t, 0 \leq t \leq \pi$, and depicted as:

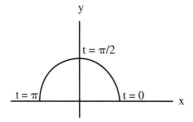

• *Parametric equations* can be used to describe the *motion of a particle* moving on a curve as well as to describe the *curve* itself. A curve is generally *parameterized* from one end to the other without retracing. To parameterize function $y = f(x)$, substitute the parameter t for x so that $x = t$ and $y = f(t)$, where parameter t may or may not represent time.

• *Motion of a particle in a plane* can be described using two *parametric equations*, $x = f(t)$ for horizontal motion along the x-coordinate and $y = g(t)$ for vertical motion along the y-coordinate. The parameter is t for time, such that at time t the particle is at point $(f(t), g(t))$.

• For example, the *motion of a particle on a circle* in a plane can be described using the parametric equations $x = \cos t$ and $y = \sin t$, where t represents time.

A circle with radius 1 can be expressed generally as $x^2 + y^2 = 1$, which can be written using the parameters $x = \cos t$ and $y = \sin t$ as:

$$\cos^2 t + \sin^2 t = 1$$

If a particle is moving at a uniform speed, it will travel around the circle in 2π units of time. As the particle travels around the circle, its motion can be reflected onto the X and Y axes while it goes from −1 through zero to +1 in both x and y directions (for a unit circle).

If this particle is moving in a counterclockwise direction uniformly at different t values, x and y are:

at t $= 0$: x $= 1$, y $= 0$

at t $= \pi/2$: x $= 0$, y $= 1$

at t $= \pi$: x $= -1$, y $= 0$

at t $= 3\pi/2$: x $= 0$, y $= -1$

at t $= 2\pi$: x $= 1$, y $= 0$

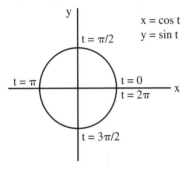

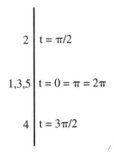

Y-axis reflection moving
through t values
of t $= 0$ to t $= 2\pi$

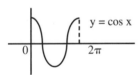

X-axis reflection
moving through points 1 to 5

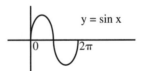

• Parametric equations can be used to describe the motion of a projectile. A *projectile* is defined as an object that experiences the force of gravity and no frictional forces and can be described using the x (horizontal) component, the y (vertical) component, and time t. The important quantities and key equations that are often required for projectile calculations are discussed in the following paragraphs and include:

1. *Horizontal distance traveled* at a given t value $x(t) = (v_0 \cos \theta)t$;
2. *Total horizontal distance traveled* when total time T occurs
   $x(T) = (v_0 \cos \theta)T = (v_0^2 \sin 2\theta)/g$;
3. *Vertical distance* at a given t value $y(t) = (v_0 \sin \theta)t - gt^2/2$;

4. *Maximum height* given by $y_{max} = (v_0 \sin \theta)^2/2g$, where y_{max} occurs at one-half of the time the projectile is in the air;
5. *Time at which maximum height* occurs, which is given by one-half of the total time $T/2$, where T is the total time; and
6. *Total time* T of travel given by $T = (2v_0 \sin \theta)/g$.

These equations can be developed as follows:

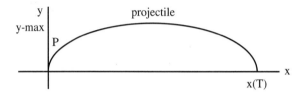

The initial position of an object is $x(t) = 0$ and $y(t) = 0$, or in some cases a starting height $y = h$ is specified. The initial velocity v_0 in the x direction is $(v_0 \cos \theta)$ and in the y direction is $(v_0 \sin \theta)$, where v_0 is the speed and θ is the angle the projectile makes with the horizontal axis. The force (or acceleration) of gravity g is in the y-direction and is given by the second derivative as $d^2y/dt^2 = -g$. Gravity affects the upward component of velocity so that it decreases by $(-gt)$. If there are no frictional forces, the horizontal component of velocity remains constant. Therefore, the x and y components of *velocity* are:

$$v_x = dx/dt = v_0 \cos \theta \text{ and } v_y = dy/dt = v_0 \sin \theta - gt.$$

(Note that derivatives are shown for thoroughness; however, knowledge of calculus is not required. See *Master Math: Calculus* for definitions and uses of the derivative.)

The distance along the X-axis $x(t)$ increases with time and the height along the Y-axis $y(t)$ increases, then decreases. The *distance traveled* or path of the projectile is obtained (from calculus) by velocity components with respect to time resulting in:

$$x(t) = (v_0 \cos \theta)t$$
$$y(t) = (v_0 \sin \theta)t - gt^2/2$$

The *maximum height* occurs where the rate of change (or derivative) in the y direction is zero, or $dy/dt = 0$. Therefore, the derivative:

$$dy(t)/dt = (d/dt)[(v_0 \sin \theta)t - gt^2/2] = v_0 \sin \theta - gt$$

when $dy/dt = 0$, then $v_0 \sin \theta = gt$

Solving $v_0 \sin \theta = gt$ for t results in:

$t = (v_0 \sin \theta)/g$

Substituting t into $y(t)$ we obtain y_{max}:

$y(t)$ at $y_{max} = (v_0 \sin \theta)(v_0 \sin \theta)/g - g((v_0 \sin \theta)/g)^2/2$

$= (v_0 \sin \theta)^2/g - (v_0 \sin \theta)^2/2g = (v_0 \sin \theta)^2/2g =$

$y_{max} = (v_0 \sin \theta)^2/2g$

where y_{max} occurs at one-half of the time the projectile is in the air.

The total *horizontal distance* $x(T)$ the projectile travels occurs when $y = 0$ and time $= T$, where T is the total time in the air. Therefore, at time $= T$:

$(v_0 \sin \theta)T = gT^2/2$

Solving for T gives *total time* T as:

$T = (2v_0 \sin \theta)/g$

The *total distance* $x(t)$ at $t = T$ is:

$x(T) = (v_0 \cos \theta)T = (v_0 \cos \theta)(2 v_0 \sin \theta)/g = (v_0{}^2 \sin 2\theta)/g$

Remember: $2 \sin x \cos x = \sin 2x$.

Because y_{max} occurs at one-half of the time the projectile is in the air, the *time when maximum height* occurs is $T/2$.

• **Example:** If a small rocket is projected upward from the ground with an initial velocity of 60 feet/second at an angle of 60°, how long will it be in the air, how high and far will it travel, and where is it at $t = 1$ second?

The total time is $T = (2 v_0 \sin \theta)/g$:

$T = ((2)(60 \text{ ft/s} \sin 60°) / (32 \text{ ft/s}^2) \approx 3.25 \text{ seconds}$

The total horizontal distance is $x(T) = (v_0 \cos \theta)T$:

$x(T) = (60 \text{ ft/s} \cos 60°) T = (30 \text{ ft/s})(3.25 \text{ s}) = 97.50 \text{ feet}$

The maximum vertical distance or height is $y_{max} = (v_0 \sin \theta)^2/2g$:

$y_{max} = (60 \text{ ft/s} \sin 60°)^2 / (2)(32 \text{ ft/s}^2) \approx 42.19 \text{ feet}$

At t = 1 second the location is given by x(t) = $(v_0 \cos \theta)t$ and

y(t) = $(v_0 \sin \theta)t - gt^2/2$. Therefore:

x(t) = (60 ft/s cos 60°)(1 s) = 30.00 ft

y(t) = (60 ft/s sin 60°)(1 s) − (32 ft/s²)(1 s)²/2 ≈ 35.96 ft

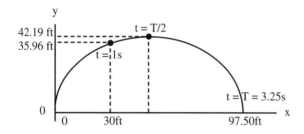

- Parametric equations can also be used to describe a cycloid. A **cycloid** represents the path of a point on the perimeter of a circle as it is rolled along a line or surface. An example of a cycloid may be a point on the wheel of a vehicle. If the circle has radius r, and point P begins at the bottom at x = 0, then if it is rolled along the X-axis, it makes a complete revolution at x = 2πr.

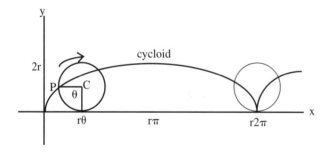

The parameter θ represents the angle through which the circle revolves. The circle rolls a distance of rθ along the X-axis and its center is at y = r and x = rθ. At θ = 0, the point is at x = 0, y = 0, and at θ = 2π, the point is at x = 2πr, y = 0. The segment between the center and the point is taken into account in measurements by subtracting (r sin θ) from x and (r cos θ) from y. Therefore:

x = rθ − r sin θ = r(θ − sin θ)

y = r − r cos θ = r(1 − cos θ)

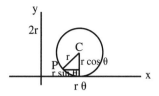

These equations for a cycloid can also be written using the parameter t as:

$$x = rt - r \sin t = r(t - \sin t)$$
$$y = r - r \cos t = r(1 - \cos t)$$

in the interval of all real numbers.

• **Example:** Graph the equations for a cycloid $x = rt - r \sin t$ and $y = r - r \cos t$ for $r = 2$ in the interval $-2\pi < t < 2\pi$.

When $r = 2$, the equations become:

$$x = 2t - 2 \sin t \text{ and } y = 2 - 2 \cos t$$

Note that the maximum y values are at $2r = 4$.

Graph of $x = 2t - 2 \sin t$ and $y = 2 - 2 \cos t$

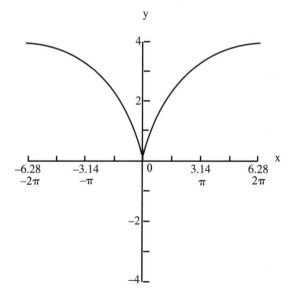

• We can also write the *parametric equations* using the *position vector*, which has its initial point at the origin of a rectangular coordinate system. A position vector is defined according to the location of its *terminal point*. A *position vector* can be used to locate the *position of a moving object* and can be written: $\mathbf{R}(t) = x(t)\mathbf{i} + y(t)\mathbf{j} + z(t)\mathbf{k}$

To *parameterize a surface* in three dimensions, such as a *cylinder*, first remember that a circle in two dimensions is described using $x = \cos t$, $y = \sin t$. If the circle is on an XY plane, then the z-dimension is zero, and the equations become $x = \cos t$, $y = \sin t$, and $z = 0$. If z and t are allowed to vary, then many circles along the Z-axis can exist to form a cylinder so that $x = \cos t$, $y = \sin t$, and $z = z$ describe many circles along z. The position vectors are:

$$\mathbf{R} = (x)\mathbf{i} + (y)\mathbf{j} + (z)\mathbf{k} = (\cos t)\mathbf{i} + (\sin t)\mathbf{j} + (z)\mathbf{k}$$

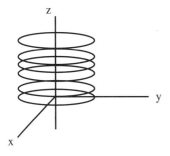

• A *sphere* can be parameterized using spherical coordinates. In three dimensions, spherical coordinates are expressed in terms of ρ rho, θ theta, and ϕ phi, where ρ can range from 0 to ∞, θ can range from 0 to 2π, and ϕ can range from 0 to π. In *spherical coordinates*, the ρ component is measured from the origin, the θ component measures the distance around the Z-axis, and the ϕ component measures down from the Z-axis and is referred to as the polar angle. The coordinates can be defined in terms of Cartesian (or rectangular) coordinates, x, y, and z:

$$x = \rho \cos \theta \sin \phi, y = \rho \sin \theta \sin \phi, z = \rho \cos \phi, \text{ and } \rho = \sqrt{x^2 + y^2 + z^2}$$

The parameters of a sphere centered at origin are:

$$x = \cos \theta \sin \phi, y = \sin \theta \sin \phi, \text{ and } z = \cos \phi$$

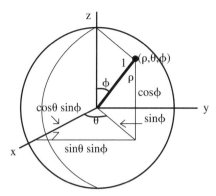

The equation can be written in the form:

$\mathbf{R}(\rho,\theta,\phi) = (\cos\theta \sin\phi)\mathbf{i} + (\sin\theta \sin\phi)\mathbf{j} + (\cos\phi)\mathbf{k}$

For example, if point P is centered at $(2, 2, 2)$ and the sphere has a radius of 2, then $\mathbf{Ro} = 2\mathbf{i} + 2\mathbf{j} + 2\mathbf{k}$ and $\mathbf{R}(\rho,\theta,\phi)$ is multiplied by 2 to expand the radius to 2. The equation becomes:

$\mathbf{R}(\rho,\theta,\phi) = 2\mathbf{i} + 2\mathbf{j} + 2\mathbf{k} + (2\cos\theta \sin\phi)\mathbf{i} + (2\sin\theta \sin\phi)\mathbf{j} + (2\cos\phi)\mathbf{k}$
$= (2 + 2\cos\theta \sin\phi)\mathbf{i} + (2 + 2\sin\theta \sin\phi)\mathbf{j} + (2 + 2\cos\phi)\mathbf{k}$

or

$x = 2 + 2\cos\theta \sin\phi$, $y = 2 + 2\sin\theta \sin\phi$, and $z = 2 + 2\cos\phi$

10.5 Chapter 10 Summary and Highlights

• *Polar coordinates* describe points in a plane or in space, similar to rectangular coordinates. The difference is that in polar coordinates, there is an r-coordinate that maps the distance of a point from the origin of the coordinate system, and there is a θ-coordinate that measures the angle the r-ray makes from the horizontal positive X-axis called the *polar axis*. A point P has coordinates (x, y) in the rectangular coordinate system, and point P designates the ordered pair (r, θ) in polar coordinates. Point P is located by giving the angle θ from the positive X-axis to ray OP and the distance r from the origin to point P. The angle θ can be measured in degrees or radians. Angle θ is positive if the rotation is counter-clockwise, and negative if the rotation is clockwise. The r-coordinate, which is the directed distance from the origin to the point P, is positive if measured from the origin along the terminal side of θ, and negative if measured along the terminal side extended through the origin. A point $(-r, \theta)$ with a negative r-value represents a point a distance r from the origin but in the opposite direction as point (r, θ). A given point in a polar coordinate system can be described by numerous polar coordinates.

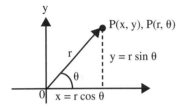

• Coordinates or equations can be converted between polar and rectangular forms. The relationships between point (x, y) in a two-dimensional rectangular coordinate system and point (r, θ) in a two-dimensional polar coordinate system are:

$$r^2 = x^2 + y^2, \text{ or } r = \sqrt{x^2 + y^2}, \tan \theta = y/x, \text{ where x is not 0}$$
$$x = r \cos \theta, \text{ or } \cos \theta = x/r, \text{ and } y = r \sin \theta, \text{ or } \sin \theta = y/r$$

• Polar equations can be graphed by calculating ordered pairs of r and θ values (r, θ) in a coordinate system, or by entering an equation into a graphing calculator or graphing software.

• *Parametric equations* are often the most efficient means to describe objects in *motion* including airplanes, projectiles, and cycloids. Parametric equations may be used to represent the motion of a particle or object moving along a curve as well as to describe the curve itself. A set of points can be defined *parametrically* using the *parametric equations* x = f(t) and y = g(t), where t is called the *parameter*. Each value of t corresponds with values of x and y, which results in (x, y) ordered pairs. The parametric equations x = f(t) and y = g(t) can be calculated for various t values to produce ordered (x, y) pairs, and a graph can be plotted using the (x, y) values. Graphing calculators and graphing software are often able to graph parametric equations.

Chapter 11

Complex Numbers and the Complex Plane

11.1 Complex Numbers Defined

• The concept of *complex numbers* came about because of early mathematicians encountering square roots of negative numbers. Complex numbers are the solution to roots of negative numbers. Complex numbers can be negative when they are squared, such that $x^2 = -1$. There are no *real number* solutions to this equation because for any real number x, $x^2 \geq 0$. Complex numbers are used in the design of electrical circuits, ships, airplane wings, and fractals (which are image-shapes that repeat themselves infinitely in continually decreasing dimensions).

How Complex Numbers Fit Into the Number Hierarchy

• Complex numbers encompass both real and imaginary numbers, real numbers encompass both rational and irrational numbers, rational numbers include integers, which include whole numbers, which include natural numbers. Remember that a *rational number* is a number that can be expressed in the form of a fraction, x/y, providing the denominator is not zero. The result of dividing two integers (with a nonzero divisor) is a rational number. Rational numbers can be represented in the form of decimals that either terminate or end, or in the form of a decimal that repeats one or more digits, such as 1/4 = 0.25 or 1/3 = 0.33333...

• *Irrational numbers* are numbers that cannot be expressed in the form of a fraction. Irrational numbers possess endless non-repeating digits to the right of the decimal point, such as, π = 3.14159..., and the roots $\sqrt{2} = 1.414213562...$, $\sqrt{3} = 1.732050807...$, $\sqrt{5} = 2.236067977...$, and other roots.

• *Real numbers* contain rational and irrational numbers and include natural numbers, whole numbers, integers, fractions, and decimals. All real numbers except zero are either positive or negative and correspond to points on the real number line, and all points on the number line correspond to real numbers. The real number line reaches from negative infinity ($-\infty$) to positive infinity ($+\infty$).

$$\overleftarrow{}\longrightarrow$$

| −4 | −3 | −2 | −1 | −.5 0 | 1 | √2 | 2 | 5/2 | 3 | π | 4 |

Real numbers include −0.5, −2, 5/2, and π (where π = 3.14159...).

• Every *real number* corresponds to a point on the real number line. There is, however, *no* real number equal to $\sqrt{-1}$ and no point on a real number line corresponding to $\sqrt{-1}$. This means that the equation $x^2 = -1$ has no *real* solutions.

• Because there is no number that when squared equals −1, the symbol i was introduced. The symbol i has the properties

$i = \sqrt{-1}$

$(i)^2 = -1$, because $(\sqrt{-1})(\sqrt{-1}) = -1$

$\sqrt{-x} = \sqrt{-1}\sqrt{x} = i\sqrt{x}$, where x is a positive number.

Numbers involving i or $\sqrt{-1}$ are called *complex numbers*.

For example,

$$(\sqrt{-4}\,)^2 = (\sqrt{-1}\,\sqrt{4}\,)(\sqrt{-1}\,\sqrt{4}\,) = (i\sqrt{4}\,)(i\sqrt{4}\,) = i^2\sqrt{(4)(4)} = -4.$$

The square root of -1 is often encountered when solving quadratic equations.

• Complex numbers involve i and are generally in the form $(x + iy)$ or $z = (x + iy)$, where x and y are *real numbers* and i is *imaginary*. Complex numbers are also written using a and b as $(a + ib)$, where a and b are *real numbers* and i is *imaginary*. In the expression $(x + iy)$ the x term is referred to as the *real part* and the iy term is referred to as the *imaginary part*. A real number multiplied by i forms an **imaginary number**, such that:

(real number) $\times$ i = (imaginary number)

A real number added to an imaginary number forms a **complex number**, such that:

(real number) + (real number)(i) = (complex number), or

(real number) + (imaginary number) = (complex number)

• **Complex numbers** include all real numbers and all imaginary numbers. Every real number can be written as a complex number:

x + 0i = x, where the imaginary part is zero.

• Two complex numbers a + bi and c + di are equal if a = c and b = d. Therefore, a + bi = c + di, if and only if a = c and b = d.

• The square root of a negative number such as $\sqrt{-1}$, $\sqrt{-3}$, $\sqrt{-5}$, or $\sqrt{-9}$ is an imaginary number. This is true because

$$\sqrt{-3} = (\sqrt{3})(\sqrt{-1}) = i\sqrt{3}$$
$$\sqrt{-5} = (\sqrt{5})(\sqrt{-1}) = i\sqrt{5}$$
$$\sqrt{-9} = (\sqrt{9})(\sqrt{-1}) = i\sqrt{9} = 3i$$

where $i = \sqrt{-1}$.

• Properties of i include:

$$(i)^2 = -1$$
$$(i)^3 = (i)^2(i) = (-1)i = -i$$
$$(i)^4 = (i^2)^2 = (-1)^2 = 1$$
$$(i)^5 = (i)^4(i) = (1)i = i$$

and so on.

• Using i in calculations helps to avoid errors that can occur when combining roots of negative numbers. For example,

$(\sqrt{-9})(\sqrt{-4})$ does *not* equal $\sqrt{36}$, but rather

$(\sqrt{-9})(\sqrt{-4}) = (\sqrt{9})(\sqrt{-1})(\sqrt{4})(\sqrt{-1}) = (3i)(2i) = 6i^2 = -6.$

11.2 The Complex Plane in Rectangular Form

• Complex numbers can correspond to points in a coordinate system, usually called the ***complex plane***. For example, $3 + 4i$ corresponds to $x = 3$ and $y = 4$, where the X-axis is real and the Y-axis is imaginary.

The coordinate $3 + 4i$ on a complex plane corresponds to point $(3, 4)$. Each complex number can be associated with a point in a ***rectangular coordinate system***.

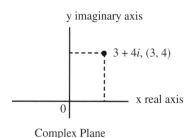

y imaginary axis

$3 + 4i, (3, 4)$

x real axis

0

Complex Plane

• At the origin the coordinates are $(0, 0)$ and represent the complex number $0 + 0i = 0$. All points on the *real* X-axis have coordinates in the form $(x, 0)$ and correspond to real numbers $x + 0i = x$. All points on the *imaginary* Y-axis have coordinates in the form $(0, y)$ and correspond to imaginary numbers $0 + yi = yi$.

• A complex number cannot only be represented in a *complex plane* that has a real axis and an imaginary axis, but a complex number can also be represented by a ***vector***. Therefore, each complex number can correspond to a ***position vector***. For example, *vector* $3\mathbf{i} + 4\mathbf{j}$ looks the same as *complex number* $3 + 4i$:

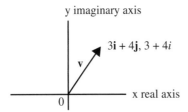

11.3 Addition and Subtraction of Complex Numbers in Rctangular Form

• In *rectangular form complex numbers are added or subtracted* by adding or subtracting the real terms and imaginary terms separately. The result is in the form (x + iy). (This is similar to combining like terms.) For example:

$$(1 + 2i) + (3 + 4i) = 1 + 3 + 2i + 4i = 4 + 6i$$
$$(5 + 4i) - (3 + 2i) = (5 - 3) + (4i - 2i) = 2 + 2i$$

• Note: To *add or subtract complex numbers in polar form (or trigonometric form)*, it is simplest to convert the polar form to rectangular form x + iy and add or subtract the real parts and imaginary parts separately. (See Sections 11.4 and 11.5 for polar and rectangular forms and converting between.)

• The *sum of two complex numbers* can be represented *graphically* as a *resultant vector* of the corresponding vectors represented by the two complex numbers. For example, represent graphically complex numbers: $(2 + i) + (1 + 3i) = 3 + 4i$

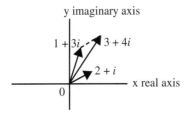

(See Section 9.4 for determining the sum and difference of two vectors.)

• **Example:** Calculate and represent graphically $(4 + 3i) - (2 - i)$.
$(4 + 3i) - (2 - i)$ can be rewritten $(4 + 3i) + (-2 + i)$ and added.
$$(4 + 3i) + (-2 + i) = 4 - 2 + 3i + i = 2 + 4i$$

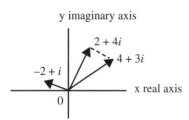

11.4 Complex Numbers in Polar Form and the Complex Plane

• We know from Chapter 10 that the relationship between polar and rectangular coordinates is:

$$x = r \cos \theta, y = r \sin \theta, r = \sqrt{x^2 + y^2}, \text{ and } \tan \theta = y/x \text{ or } \theta = \tan^{-1}(y/x)$$

Complex numbers are expressed in polar, or trigonometric, form by writing the real and imaginary parts using $r \cos \theta$ and $r \sin \theta$. Complex numbers can be expressed in polar coordinate form using two numbers, the absolute value r and the angle θ (depicted on the complex plane below).

• *Polar form of complex numbers* can be written using polar coordinates $x = r \cos \theta$ and $y = r \sin \theta$. Therefore, **complex numbers expressed in polar form** are written:

$$x + iy = r \cos \theta + i\, r \sin \theta = r(\cos \theta + i \sin \theta)$$

An abbreviated notation for complex numbers is written:

$$r(\cos \theta + i \sin \theta) = r \operatorname{cis} \theta$$

The **complex plane** using polar coordinates is drawn:

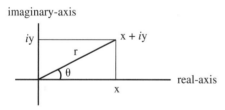

where $x + iy = r \cos \theta + i\, r \sin \theta$ and complex number $x + iy$ corresponds to the vector with magnitude r and direction θ.

• Note that $\cos \theta + i \sin \theta = e^{i\theta}$ is Euler's formula, where $e \approx 2.718$.

Therefore, another notation written in ***complex polar form*** is:

   $z = x + iy = re^{i\theta}$

where r cos θ + i r sin θ = r(cos θ + i sin θ) = $re^{i\theta}$.

• Sine and cosine are periodic functions with period 2π and can be represented as sin θ = sin(θ + n2π) and cos θ = cos(θ + n2π) where n is any integer. Using this periodic nature, we can write a more *general polar form* of a complex number:

   $z = x + iy = r(\cos(\theta + n2\pi) + i\sin(\theta + n2\pi)) = re^{i(\theta + n2\pi)}$

• The number r is called the ***modulus***, or ***absolute value***, of complex number x + iy. The absolute value of a complex number is the distance (or length r) from the origin to the point representing that number. Therefore, the absolute value of r is

   $mod(x + iy) = r = \sqrt{x^2 + y^2}$ and is a positive number.

In addition, the angle θ is the polar angle and is called the argument of x + iy. Angle θ is also sometimes referred to as the amplitude. (Remember: In a polar coordinate system, the polar axis lies on the positive side of the horizontal X-axis beginning at the origin so that the polar axis coincides with the positive X-axis.) The argument θ is generally chosen such that $-\pi < \theta \leq \pi$, or $-180° < \theta \leq 180°$. Remember that θ is not unique because there are many coterminal angles for θ. Angle θ is often chosen as the smallest positive angle for which tan θ = y/x, unless it is more convenient to select a coterminal angle of θ for a particular situation. In summary, a vector and the associated complex number may be described in terms of the length r of the vector and any positive angle θ, which the vector makes with the positive real X-axis.

• In this figure we can see that x = r cos θ and y = r sin θ. In the relationship x + iy = r cos θ + i r sin θ = r(cos θ + i sin θ), r(cos θ + i sin θ) is the polar or trigonometric form, and x + iy is the rectangular form of the complex number.

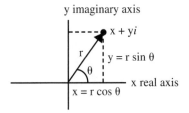

Complex Plane

• Note that for complex numbers written using a and b instead of x and y, the polar form of the complex number is written:

$a + bi = r(\cos \theta + i \sin \theta)$, where $a = r \cos \theta$ and $b = r \sin \theta$.

Also, $r = \sqrt{a^2 + b^2}$ and $\tan \theta = b/a$, or $\theta = \arctan(b/a)$.

• To **add or subtract complex numbers in polar form**, it is simplest to convert the polar form to rectangular form $x + iy$ and add or subtract the real parts and imaginary parts separately.

$(a + bi) + (c + di) = (a + c) + (bi + di)$
$(a + bi) - (c + di) = (a + bi) + (-c - di) = (a - c) + (bi - di)$

(See Section 11.5 for converting between rectangular and polar forms.)

11.5 Converting Between Rectangular and Polar Form

• To **convert from polar to rectangular form**, simply calculate x and y values for $r \cos \theta + i \, r \sin \theta$, where $x = r \cos \theta$ and $y = r \sin \theta$.

• **Example:** Convert $2 \cos 270° + i \, 2 \sin 270°$ in polar form to rectangular form.

$x + iy = 2 \cos 270° + i \, 2 \sin 270°$
$= (2)(0) + i \, (2)(-1) = 0 + i(-2) = -2i$

• To **convert from rectangular to polar form**, draw a graph of the number in a complex plane, then find r and θ using $r = \sqrt{x^2 + y^2}$ and $\tan \theta = y/x$, providing x is not equal to zero. If $x = 0$, determine θ by inspection. Also, be careful to note the quadrant of θ in the complex plane when making calculations.

• **Example:** Represent $2\sqrt{3}\mathbf{i} + 2\mathbf{j}$ in polar form, complex rectangular form, and complex polar form, and graph.

$2\sqrt{3}\,\mathbf{i} + 2\mathbf{j}$ is vector notation in rectangular form.

The rectangular complex form is $x + iy$, or $2\sqrt{3} + 2i$.

To convert to polar form, calculate r and θ:

$r = \sqrt{x^2 + y^2} = \sqrt{(2\sqrt{3})^2 + 2^2} = 4$
$\tan \theta = y/x = 2/2\sqrt{3}$, or $\theta = 30°$

Therefore, the (r, θ) polar coordinate is (4, 30°).

The complex polar form is r(cos θ + *i* sin θ) = 4(cos 30° + *i* sin 30°).

The graph of this complex number is depicted as:

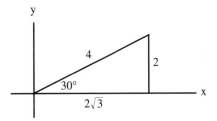

where the length of the hypotenuse of the graph of a complex number is called the absolute value of the complex number; in this case it is 4. Note that the calculations and graph are consistent with sin 30° = 2/4 = 1/2.

- **Example:** Convert complex number –*i* from rectangular form to polar form. Then convert back to rectangular form to check result. First draw graph of x + *i*y = 0 – *i*:

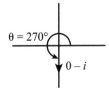

Calculate r and θ:

$$r = \sqrt{x^2 + y^2} = \sqrt{0^2 + (-1)^2} = \sqrt{1} = 1$$

tan θ = y/x

In this example, x = 0 so we find θ by inspection of the graph which indicates that θ = 270°.

Therefore, 0 – *i* in polar form is 1(cos 270° + *i* sin 270°).

Next convert back to rectangular form to check result.

1(cos 270° + *i* sin 270°) = 1(0 + *i*(–1)) = –*i*

which is the original complex number in rectangular form.

• **Example:** Convert complex number $-1 + i$ from rectangular to polar form. Then convert back to rectangular form to check result. First draw graph of $x + iy = -1 + i$:

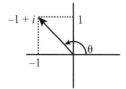

Calculate r and θ:

$$r = \sqrt{x^2 + y^2} = \sqrt{(-1)^2 + (1)^2} = \sqrt{2}$$

$\tan \theta = y/x = 1/(-1)$, or $\theta = -45°$. However, from the graph we can see that the angle represents a reference triangle. The angle θ is therefore $\theta = 180° - 45° = 135°$.

Therefore, $-1 + i$ in polar form is $\sqrt{2} (\cos 135° + i \sin 135°)$.

Next convert back to rectangular form to check result.

$$\sqrt{2} \cos 135° + i\sqrt{2} \sin 135° = -1 + i(1) = -1 + i$$

which is the original complex number in rectangular form.

11.6 Multiplication and Division of Complex Numbers in Rectangular and Polar Forms

• A primary advantage of converting complex numbers into polar (or trigonometric) form is that multiplication and division often become easier. Using polar form allows complex numbers to be multiplied and divided to give the product or quotient more directly in the same form.

Products of Complex Numbers in Rectangular and Polar Forms

• *Complex numbers in rectangular form are multiplied* as ordinary binomials, and $(i)^2$ is replaced by -1. Remember: To multiply binomials, each term in the first binomial is multiplied by each term in the second binomial, and like terms are combined (added).

- **Example:** Multiply $(1 + 2i)$ and $(3 + 4i)$ in rectangular form.

$$(1 + 2i)(3 + 4i) = (1)(3) + (1)(4i) + (2i)(3) + (2i)(4i)$$
$$= 3 + 4i + 6i + 8(i)^2 = 3 - 8 + 10i = -5 + 10i$$

- *Complex numbers in polar form are multiplied* using the *product theorem*. In *polar form two complex numbers are multiplied* as follows.

Given that $z = x + iy = r \cos \theta + i r \sin \theta = r(\cos \theta + i \sin \theta) = re^{i\theta}$, we can *multiply two complex numbers*

$$z_1 = r_1(\cos \theta_1 + i \sin \theta_1) \text{ and } z_2 = r_2(\cos \theta_2 + i \sin \theta_2),$$

using the *product theorem*, which is:

$$[r_1(\cos \theta_1 + i \sin \theta_1)][r_2(\cos \theta_2 + i \sin \theta_2)]$$
$$= r_1 r_2[\cos(\theta_1 + \theta_2) + i \sin(\theta_1 + \theta_2)]$$

The *product theorem* in abbreviated *cis* form (where cis $\theta = \cos \theta + i \sin \theta$) is:

$$[r_1 \text{ cis } \theta_1][r_2 \text{ cis } \theta_2] = r_1 r_2 \text{ cis}(\theta_1 + \theta_2)$$

The *product theorem* in *exponential polar form* is:

$$z_1 z_2 = r_1 e^{i\theta 1} r_2 e^{i\theta 2} = r_1 r_2 \, e^{i(\theta 1 + \theta 2)}$$

Note that this is consistent with the product rules for exponents:

$$x^a x^b = x^{a+b}$$

By using the product theorem to *multiply two complex numbers*, $r_1(\cos \theta_1 + i \sin \theta_1)$ and $r_2(\cos \theta_2 + i \sin \theta_2)$, we multiply r_1 and r_2 and add θ_1 and θ_2.

- The *product theorem* can be verified as follows.

Given that $x + iy = r \cos \theta + i r \sin \theta = r(\cos \theta + i \sin \theta) = re^{i\theta}$:

Multiply $r_1(\cos \theta_1 + i \sin \theta_1)$ and $r_2(\cos \theta_2 + i \sin \theta_2)$:

$$[r_1(\cos \theta_1 + i \sin \theta_1)][r_2(\cos \theta_2 + i \sin \theta_2)]$$
$$= r_1 r_2(\cos \theta_1 + i \sin \theta_1)(\cos \theta_2 + i \sin \theta_2)$$
$$= r_1 r_2(\cos \theta_1 \cos \theta_2 + i \cos \theta_1 \sin \theta_2 + i \sin \theta_1 \cos \theta_2 + i \sin \theta_1 \, i \sin \theta_2)$$
$$= r_1 r_2[\cos \theta_1 \cos \theta_2 - \sin \theta_1 \sin \theta_2 + i (\cos \theta_1 \sin \theta_2 + \sin \theta_1 \cos \theta_2)]$$

Substituting the sum identities for cosine and sine

$$\sin(x + y) = \cos x \sin y + \sin x \cos y$$
$$\cos(x + y) = \cos x \cos y - \sin x \sin y$$

where $x = \theta_1$ and $y = \theta_2$, results in:

$$r_1 r_2[\cos(\theta_1 + \theta_2) + i \sin(\theta_1 + \theta_2)]$$

which can be written in exponential polar form as: $r_1 r_2\, e^{i(\theta 1 + \theta 2)}$

• **Example:** Multiply $(1 + 2i)$ and $(3 + 4i)$ in polar form and compare with rectangular form.

To solve:

(a) Convert $(1 + 2i)$ and $(3 + 4i)$ into polar form.
(b) Calculate the product using the product theorem for polar form.
(c) Convert the product back into rectangular form and verify that the result from calculating in polar form is the same result found by calculating directly in rectangular form.

Solution:

(a) To *convert from rectangular to polar form*, draw a graph of the number in a complex plane, then find r and θ using $r = \sqrt{x^2 + y^2}$ and $\tan \theta = y/x$, providing x is not equal to zero.

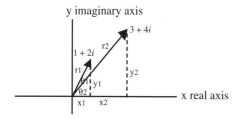

For $(1 + 2i)$: $r_1 = \sqrt{x_1^2 + y_1^2} = \sqrt{1^2 + 2^2} = \sqrt{5}$

$\tan \theta_1 = y_1/x_1 = 2/1 = 2$, or $\theta_1 = \arctan 2 \approx 63.43°$

Therefore, the polar form $x + iy = r(\cos \theta + i \sin \theta)$ is

$$(1 + 2i) = \sqrt{5}\,(\cos 63.43° + i \sin 63.43°)$$

For $(3 + 4i)$: $r_2 = \sqrt{x_2{}^2 + y_2{}^2} = \sqrt{3^2 + 4^2} = \sqrt{25} = 5$

$\tan \theta_2 = y_2/x_2 = 4/3$, or $\theta_2 = \arctan(4/3) \approx 53.13°$

Therefore, the polar form $x + iy = r(\cos \theta + i \sin \theta)$ is

$(3 + 4i) = 5(\cos 53.13° + i \sin 53.13°)$

(b) To calculate the product using the product theorem for polar form, use
$[r_1(\cos \theta_1 + i \sin \theta_1)][r_2(\cos \theta_2 + i \sin \theta_2)]$

$= r_1 r_2[\cos(\theta_1 + \theta_2) + i \sin(\theta_1 + \theta_2)]$:

$[\sqrt{5} \ (\cos 63.43° + i \sin 63.43°)][5(\cos 53.13° + i \sin 53.13°)]$

$= 5\sqrt{5} \ [\cos (63.43° + 53.13°) + i \sin (63.43° + 53.13°)]$

$= 5\sqrt{5} \ [\cos 116.56° + i \sin 116.56°]$.

(c) To convert the product back into rectangular form, use
$r(\cos \theta + i \sin \theta) = x + iy$, where $x = r \cos \theta$ and $y = r \sin \theta$:

$5\sqrt{5} \ [\cos 116.56° + i \sin 116.56°] = x + iy$

Using a calculator:

$5\sqrt{5} \ [\cos 116.56° + i \sin 116.56°] \approx 11.18[-0.447 + i\,0.894]$

$\approx -4.997 + i\,9.995 \approx -5 + 10i$

Verify that the result is the same as calculated directly in rectangular form, which is:

$(1 + 2i)(3 + 4i) = 3 + 4i + 6i + 8(i)^2 = 3 - 8 + 10i = -5 + 10i$

The results are the same for calculating the product in polar form and rectangular form.

• **Example:** Multiply $4(\cos 20° + i \sin 20°)$ and $2(\cos 40° + i \sin 40°)$ in polar form and compare with rectangular form.

Using the product theorem for complex numbers

$[r_1(\cos \theta_1 + i \sin \theta_1)][r_2(\cos \theta_2 + i \sin \theta_2)]$
$= r_1 r_2[\cos(\theta_1 + \theta_2) + i \sin(\theta_1 + \theta_2)]$

multiply the two numbers:

$[4(\cos 20° + i \sin 20°)][2(\cos 40° + i \sin 40°)]$
$= (4)(2)[\cos(20° + 40°) + i \sin(20° + 40°)]$
$= 8[\cos 60° + i \sin 60°]$

This can be written in abbreviated notation as 8 cis 60°.

We can check this result by converting each of the original complex numbers into rectangular form and multiplying.

The first number: $4 \cos 20° + 4i \sin 20° \approx 3.759 + i\,1.368$

The second number: $2 \cos 40° + 2i \sin 40° \approx 1.532 + i\,1.286$

Multiplying these in rectangular form:

$(3.759 + i\,1.368)(1.532 + i\,1.286)$
$= 5.759 + 4.834\,i + 2.096\,i - 1.759 = 4.00 + 6.93\,i$

Finally, compare with the result found using polar form by converting it to rectangular form:

$8 \cos 60° + 8\,i \sin 60°$ converted to rectangular form is: $4.00 + 6.93\,i$

Therefore, we find the same results multiplying in rectangular and polar forms.

• **Example:** Find the product of $2 \cos 300° + i\,2 \sin 300°$ and $2 \cos 210° + i\,2 \sin 210°$ in exponential polar form.

Write in exponential form:

$2 \cos 300° + i\,2 \sin 300° = 2e^{i300°}$
$2 \cos 210° + i\,2 \sin 210° = 2e^{i210°}$

Multiplying:

$2e^{i300°}\,2e^{i210°} = (2)(2)\,e^{i(300° + 210°)} = 4e^{i510°}$

Using the smallest coterminal angle for 510°

$510° - 360° = 150°$

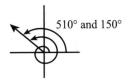

510° and 150°

Therefore, the product can be written: $4e^{i150°}$

Quotients of Complex Numbers in Rectangular and Polar Forms

• *Complex numbers are divided in rectangular form* by first multiplying the numerator and denominator by what is called the ***complex conjugate*** of the denominator. Then the numerator and denominator are divided and combined as with multiplication. For example, the complex conjugate of $(3 + 2i)$ is $(3 - 2i)$ and the complex conjugate of $(3 - 2i)$ is $(3 + 2i)$. The product of a complex number and its conjugate is a real number. Remember to replace $(i)^2$ by -1 during calculations.

• **Example:** Divide $(1 + 2i)$ by $(3 + 4i)$ in rectangular form.

$(1 + 2i) \div (3 + 4i) = (1 + 2i)(3 - 4i) \div (3 + 4i)(3 - 4i)$
$= (3 - 4i + 6i - 8i^2) \div (9 - 12i + 12i - 16i^2)$
$= (3 + 2i - 8(-1)) \div (9 - 16(-1)) = (11 + 2i) \div 25 = 11/25 + 2i/25$
$\approx 0.44 + i\,0.08$

• *Complex numbers in polar form can be divided using the quotient theorem.* In *polar form, two complex numbers are divided* as follows.

Given that $z = x + iy = r \cos \theta + i\,r \sin \theta = r(\cos \theta + i \sin \theta) = re^{i\theta}$, we can *divide two complex numbers*

$z_1 = r_1(\cos \theta_1 + i \sin \theta_1)$ and $z_2 = r_2(\cos \theta_2 + i \sin \theta_2)$

using the **quotient theorem**, which is:

$[r_1(\cos \theta_1 + i \sin \theta_1)] / [r_2(\cos \theta_2 + i \sin \theta_2)]$
$= (r_1 / r_2)[\cos(\theta_1 - \theta_2) + i \sin(\theta_1 - \theta_2)]$

The *quotient theorem* in abbreviated *cis* form is:

$[r_1 \text{ cis } \theta_1] / [r_2 \text{ cis } \theta_2] = (r_1 / r_2) \text{ cis}(\theta_1 - \theta_2)$

The *quotient theorem* in *exponential polar form* is:

$z_1/z_2 = (r_1 e^{i\theta_1})/(r_2 e^{i\theta_2}) = (r_1/r_2)e^{i(\theta_1 - \theta_2)}$

Note that this is consistent with the quotient rules for exponents:

$x^a/x^b = x^{a-b}$.

By using the quotient theorem to *divide two complex numbers* $r_1(\cos \theta_1 + i \sin \theta_1)$ and $r_2(\cos \theta_2 + i \sin \theta_2)$, we can divide r_1 by r_2 and subtract θ_1 and θ_2.

• **Example:** Divide $(1 + 2i)$ by $(3 + 4i)$ in polar form and compare with rectangular form. (This is similar to the example for multiplication earlier in this section.)

To solve:

(a) Convert $(1 + 2i)$ and $(3 + 4i)$ into polar form.
(b) Calculate the quotient using the quotient theorem for polar form.
(c) Convert the quotient back into rectangular form and verify that the result is the same as calculated in rectangular form.

Solution:

(a) To *convert from rectangular to polar form*, draw a graph of the number in a complex plane, then find r and θ using $r = \sqrt{5}$ and $\tan \theta = y/x$, providing x is not equal to zero.

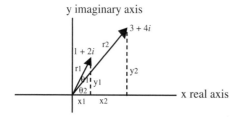

For $(1 + 2i)$: $r_1 = \sqrt{x_1^2 + y_1^2} = \sqrt{1^2 + 2^2} = \sqrt{5}$

$\tan \theta_1 = y_1/x_1 = 2$, or $\theta_1 = \arctan 2 \approx 63.43°$

Therefore, the polar form $x + iy = r(\cos \theta + i \sin \theta)$ is

$(1 + 2i) = \sqrt{5} (\cos 63.43° + i \sin 63.43°)$

For $(3 + 4i)$: $r_2 = \sqrt{x_2^2 + y_2^2} = \sqrt{3^2 + 4^2} = \sqrt{25} = 5$

$\tan \theta_2 = y_2/x_2 = 4/3$, or $\theta_2 = \arctan(4/3) \approx 53.13°$

Therefore, the polar form x + iy = r(cos θ + i sin θ) is

$(3 + 4i) = 5(\cos 53.13° + i \sin 53.13°)$

(b) To calculate the quotient using the quotient theorem for polar form, use $[r_1(\cos θ_1 + i \sin θ_1)] / [r_2(\cos θ_2 + i \sin θ_2)]$

$= (r_1 / r_2)[\cos(θ_1 - θ_2) + i \sin(θ_1 - θ_2)]$:

$[\sqrt{5}\ (\cos 63.43° + i \sin 63.43°)] / [5(\cos 53.13° + i \sin 53.13°)]$

$= (\sqrt{5}\ /5)[\cos (63.43° - 53.13°) + i \sin (63.43° - 53.13°)]$

$= (\sqrt{5}\ /5)[\cos 10.30° + i \sin 10.30°]$

(c) To convert the product back into rectangular form, use r(cos θ + i sin θ) = x + iy, where x = r cos θ and y = r sin θ:

$(\sqrt{5}\ /5)[\cos 10.30° + i \sin 10.30°] = x + iy$

Using a calculator:

$(\sqrt{5}\ /5)[\cos 10.30° + i \sin 10.30°] \approx 0.45[0.984 + i\ 0.179]$

$\approx 0.44 + i\ 0.08$

Then verify that the result is the same as calculated above in this section in rectangular form, which was:

$(1 + 2i) ÷ (3 + 4i) = (1 + 2i)(3 - 4i) ÷ (3 + 4i)(3 - 4i)$

$= (3 - 4i + 6i - 8i^2) ÷ (9 - 12i + 12i - 16i^2)$

$= (3 + 2i - 8(-1)) ÷ (9 - 16(-1)) = (11 + 2i) ÷ 25 = 11/25 + 2i/25$

$= 0.44 + i\ 0.08$

The results are the same for calculating the product in polar form and rectangular form.

• **Example:** Find the quotient of 2 cos 210° + i 2 sin 210° divided by cos 300° + i sin 300° in exponential polar form.

Convert to exponential form:

$2 \cos 210° + i\ 2 \sin 210° = 2e^{i210°}$

$\cos 300° + i \sin 300° = e^{i300°}$

Divide:

$2e^{i210°} / e^{i300°} = (2/1)e^{i(210°-300°)} = 2e^{i(-90°)}$

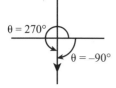

$\theta = 270°$

$\theta = -90°$

Use the smallest positive coterminal angle: $360° - 90° = 270°$

Therefore, the quotient is: $2e^{i270°}$

11.7 Powers and Roots of Complex Numbers

• *De Moivre's theorem* is used to find *powers* of *complex numbers in polar form*. This theorem also provides the foundation for the *nth-root theorem*, which is used to find all n *roots* of a complex number.

Powers of Complex Numbers-De Moivre's Theorem

• *De Moivre's theorem* is used to find the *power of a complex number in polar (or trigonometric) form*. Raising a number to a positive power is in essence repeatedly applying the product rule. By the *product rule*, we know that the product of two complex numbers in polar form can be found by multiplying the absolute values and adding the angles. Remember that the product theorem is:

$$[r_1(\cos \theta_1 + i \sin \theta_1)][r_2(\cos \theta_2 + i \sin \theta_2)]$$
$$= r_1 r_2[\cos(\theta_1 + \theta_2) + i \sin(\theta_1 + \theta_2)]$$

Raising a number to the power two is the same as taking the product of two numbers.

$$[r(\cos \theta + i \sin \theta)]^2 = [r(\cos \theta + i \sin \theta)][r(\cos \theta + i \sin \theta)]$$
$$= r^2(\cos 2\theta + i \sin 2\theta)$$

De Moivre's theorem expands this from the product of two complex numbers to the product of n complex numbers. For example,

$$[r(\cos \theta + i \sin \theta)]^3 = r^3(\cos 3\theta + i \sin 3\theta)$$
$$[r(\cos \theta + i \sin \theta)]^4 = r^4(\cos 4\theta + i \sin 4\theta)$$

and so on.

Therefore, the power of a complex number in polar form can be found using **De Moivre's theorem**, which is:

$$[r(\cos \theta + i \sin \theta)]^n = r^n(\cos n\theta + i \sin n\theta)$$

where n is a real number or integer.

De Moivre's theorem written in abbreviated cis form is:

$$[r(\text{cis } \theta)]^n = r^n(\text{cis } n\theta)$$

De Moivre's theorem in exponential polar form is:

$$(re^{i\theta})^n = r^n e^{in\theta}$$

(De Moivre's theorem can be derived using mathematical techniques.)

- **Example:** Find $(1 + i)^{13}$.

First convert $1 + i$ into polar form:

$$r^2 = x^2 + y^2 = 1 + 1 = 2, \text{ therefore}, r = \sqrt{2}$$
$$\tan \theta = y/x = 1/1, \text{ or } \theta = 45°, \text{ which lies in the first quadrant.}$$

The polar form is: $r (\cos \theta + i \sin \theta) = \sqrt{2} (\cos 45° + i \sin 45°)$

Raised to the power thirteen: $\sqrt{2}^{\;13} (\cos 45° + i \sin 45°)^{13}$

Because $\sqrt{2}^{\;13} = 2^{13/2}$, this becomes $2^{13/2}(\cos 45° + i \sin 45°)^{13}$

Apply *De Moivre's theorem*:

$$[r(\cos \theta + i \sin \theta)]^n = r^n (\cos n\theta + i \sin n\theta)$$
$$2^{13/2}(\cos 45° + i \sin 45°)^{13} = 2^{13/2}(\cos(45°(13)) + i \sin(45°(13)))$$
$$= 2^{13/2} (\cos 585° + i \sin 585°) = 2^{13/2} (\cos 225° + i \sin 225°)$$

(where 585° and 225° are coterminal)

$$= 2^6 2^{1/2} (\cos 225° + i \sin 225°) = 64\sqrt{2} (\cos 225° + i \sin 225°)$$
$$= 2^{6.5} (\cos 225° + i \sin 225°)$$

Therefore, $(1 + i)^{13} = 64\sqrt{2} (\cos 225° + i \sin 225°)$

We can also write this in rectangular form by taking sine and cosine of $225°$ and multiplying each by $64\sqrt{2}$:

$$(1 + i)^{13} = -64 - i\,64$$

- **Example:** Find $(\sqrt{3} - i)^{10}$.

First convert $\sqrt{3} - i$ into polar form:

$$r^2 = x^2 + y^2 = 3 + 1 = 4; \text{ therefore, } r = 2$$
$$\tan\theta = y/x = -1/\sqrt{3}, \text{ or } \theta = -30°$$

Find smallest positive angle, $360° - 30° = 330°$; therefore, $\theta = 330°$.

The polar form is: $r(\cos\theta + i\sin\theta) = 2(\cos 330° + i\sin 330°)$

Raised to the power ten: $2^{10}(\cos 330° + i\sin 330°)^{10}$

Apply *De Moivre's theorem*:

$$[r(\cos\theta + i\sin\theta)]^n = r^n(\cos n\theta + i\sin n\theta):$$
$$2^{10}(\cos 330° + i\sin 330°)^{10} = 2^{10}(\cos(10(330°)) + i\sin(10(330°)))$$
$$= 2^{10}(\cos(3300°) + i\sin(3300°)) = 2^{10}(\cos 60° + i\sin 60°)$$

(where $3300°$ and $60°$ are coterminal)

Therefore $(\sqrt{3} - i)^{10} = 1024(\cos 60° + i\sin 60°)$

In rectangular form: $1024\cos 60° + i1024\sin 60° = 512 + i1024(\sqrt{3}/2)$.

Therefore, $(\sqrt{3} - i)^{10} = 512 + i\,512\sqrt{3}$.

Roots of Complex Numbers-the nth-Root Theorem

- The *nth-root theorem* is used to find all n roots of a complex number. *De Moivre's theorem* provides the foundation for the nth-root theorem. Every complex number, $x + yi = r(\cos\theta + i\sin\theta)$, except zero has exactly n distinct nth roots. Therefore, a complex number has two square roots, three cube roots, four fourth roots, five fifth roots... n nth roots.

If $z = w^n$ ($n = 1, 2, 3, ...$), then for each value of w there corresponds one value of z. Therefore, it is clear that for a non-zero value of z there corresponds n distinct values of w, where each of these values is called an nth root of z, or, $w = \sqrt[n]{z} = z^{1/n}$

Similarly, if $w^2 = z$, then $w = \sqrt{z} = z^{1/2}$, or if $w^3 = z$, then $w = \sqrt[3]{z} = z^{1/3}$.

• For a complex number $z = re^{i\theta}$, by De Moivre's theorem, we know that $(re^{i\theta})^n = r^n e^{in\theta}$, or $z = [r(\cos \theta + i \sin \theta)]^n = r^n(\cos n\theta + i \sin n\theta)$.

For $\sqrt{z}$, it follows that:

$$z^{1/2} = \sqrt{z} = r^{1/2}e^{i\theta/2} = r^{1/2}(\cos \theta/2 + i \sin \theta/2)$$

When $-\pi < \theta \leq \pi$, this is called the principal root.

For n roots this becomes: $z^{1/n} = r^{1/n}e^{i\theta/n} = r^{1/n}(\cos \theta/n + i \sin \theta/n)$ with principal root $-\pi < \theta \leq \pi$, and for all other roots:

$$z^{1/n} = r^{1/n}e^{i(\theta+2\pi k)/n} = r^{1/n}[\cos((\theta+2\pi k)/n) + i \sin((\theta+2\pi k)/n)]$$

or in degrees:

$$z^{1/n} = r^{1/n}e^{i(\theta+360°k)/n} = r^{1/n}[\cos((\theta+360°k)/n) + i \sin((\theta+360°k)/n)]$$

for $k = 0, 1, 2, 3, ...n-1$

Therefore, the ***nth root theorem in exponential polar form*** states that for n as a positive integer:

$$z^{1/n} = r^{1/n}e^{i(\theta+2\pi k)/n} = r^{1/n}e^{i(\theta+360°k)/n}, \text{ for } k = 0, 1, 2, 3, ...n-1$$

there exist n distinct nth roots of $re^{i\theta}$.

The ***nth root theorem in polar form*** states that for n as a positive integer:

$$z^{1/n} = r^{1/n}[\cos((\theta+2\pi k)/n) + i \sin((\theta+2\pi k)/n)], k = 0, 1, 2, 3, ...n-1$$

or in degrees:

$$z^{1/n} = r^{1/n}[\cos((\theta+360°k)/n) + i \sin((\theta+360°k)/n)],$$

for $k = 0, 1, 2, 3, ...n-1$

The values of n lie on a *circle of radius* $r^{1/n}$, or $\sqrt[n]{r}$, with its center at the origin, and constitute the vertices of a *regular polygon of n sides*. The angles of a given complex number are equal and differ such that:

angles of third roots differ by 360°/3, or 120°,
angles of fourth roots differ by 360°/4, or 90°,
angles of fifth roots differ by 360°/5, or 72°, and
angles of nth roots differ by 360°/n.

• **Example:** If $z^n = 1$ (for a unit circle),
$z^{1/n} = r^{1/n}[\cos((\theta+2\pi k)/n) + i \sin((\theta+2\pi k)/n)]$,

then, for the third root, $\sqrt[3]{1} = 1$, or $(1)^{1/3} = 1$, where $r = 1$, find the three cube roots and graph.

For $z = 1 + i(0)$, $r = 1$ and $\theta = 0$.

$z^{1/n} = 1[\cos((\theta+2\pi k)/3) + i \sin((\theta+2\pi k)/3)]$ for $k = 0, 1$, and 2.
$k = 0$,
$\cos((0+2\pi 0)/3) + i \sin((0+2\pi 0)/3) = \cos 0 + i \sin 0 = 1 + i(0)$
$k = 1$,
$\cos((0+2\pi 1)/3) + i \sin((0+2\pi 1)/3) = -1/2 + i\sqrt{3}/2 \approx -0.50 + i0.866$
$k = 2$,
$\cos((0+2\pi 2)/3) + i \sin((0+2\pi 2)/3) = -1/2 - i\sqrt{3}/2 \approx -0.50 - i0.866$

Depicted on a unit circle are points $(1 + i(0))$, $(-0.50 + i0.866)$, and $(-0.50 - i0.866)$:

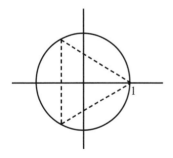

Note: This represents the three cube roots of $x^3 - 1 = 0$.

Normally, when we see $x^3 = 1$, we think of it having one root, 1. This is only true when we restrict solutions to real numbers. However, by including *complex numbers*, there are three roots: one real and two imaginary.

• **Example:** Find three third roots and six sixth roots of $(1 + i\sqrt{3}\,)$ in exponential polar form and graph each.

First convert $1 + i\sqrt{3}$ into polar form:

$$r^2 = x^2 + y^2 = 1 + 3 = 4, \text{ therefore}, r = 2$$
$$\tan\theta = y/x = \sqrt{3}/1, \text{ or } \theta = 60°$$

Therefore, polar form is:

$$r(\cos\theta + i\sin\theta) = 2(\cos 60° + i\sin 60°)$$

Exponential polar form is: $re^{i\theta} = 2e^{i60°}$

To find roots using polar form write:

$$r^{1/n}(\cos(\theta+360°k/n) + i\sin(\theta+360°k/n)), \text{ for } k = 0, 1, 2, 3, \ldots$$

or using exponential polar form: $r^{1/n}e^{i(\theta+360°k)/n}, k = 0, 1, 2, 3, \ldots$

Using exponential polar form to find n $= 3$ cube roots of $2e^{i60°}$ we write:

$2^{1/3}e^{i(60°+360°k)/3}$ for $k = 0, 1,$ and 2. Substituting k values:

$k = 0, 2^{1/3}e^{i(60°+360°(0))/3} = 2^{1/3}e^{i(20°)}$
$k = 1, 2^{1/3}e^{i(60°+360°(1))/3} = 2^{1/3}e^{i(140°)}$
$k = 2, 2^{1/3}e^{i(60°+360°(2))/3} = 2^{1/3}e^{i(260°)}$

We can also write these in polar complex and rectangular forms as:

$\sqrt[3]{2}\ (\cos 20° + i\sin 20°) \approx 1.18 + i\,0.43$
$\sqrt[3]{2}\ (\cos 140° + i\sin 140°) \approx -0.97 + i\,0.81$
$\sqrt[3]{2}\ (\cos 260° + i\sin 260°) \approx -0.22 - i\,1.24$

Graph these three points on a circle of radius $r^{1/n} = 2^{1/3} \approx 1.26$:

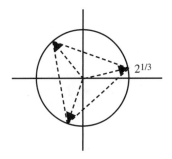

The points are spaced equally by 120°.

Use exponential polar form to calculate n = 6 sixth roots of $2e^{i60°}$ and give results in exponential polar and polar forms.

$2^{1/6}e^{i(60°+360°k)/6}$ for k = 0, 1, 3, 4 and 5. Substitute k values:

k = 0, $2^{1/6}e^{i(60°+360°(0))/6} = 2^{1/6}e^{i(10°)} = 2^{1/6}$ (cos 10° + i sin 10°)

k = 1, $2^{1/6}e^{i(60°+360°(1))/6} = 2^{1/6}e^{i(70°)} = 2^{1/6}$ (cos 70° + i sin 70°)

k = 2, $2^{1/6}e^{i(60°+360°(2))/6} = 2^{1/6}e^{i(130°)} = 2^{1/6}$ (cos 130° + i sin 130°)

k = 3, $2^{1/6}e^{i(60°+360°(3))/6} = 2^{1/6}e^{i(190°)} = 2^{1/6}$ (cos 190° + i sin 190°)

k = 4, $2^{1/6}e^{i(60°+360°(4))/6} = 2^{1/6}e^{i(250°)} = 2^{1/6}$ (cos 250° + i sin 250°)

k = 5, $2^{1/6}e^{i(60°+360°(5))/6} = 2^{1/6}e^{i(310°)} = 2^{1/6}$ (cos 310° + i sin 310°)

Depicted on a graph are these six points on a circle of radius $r^{1/n} = 2^{1/6} \approx 1.12$, at the six angle locations:

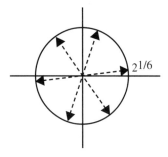

There is 60° between each point on this circle of radius $2^{1/6}$.

11.8 Chapter 11 Summary and Highlights

• Complex numbers are the solution to roots of negative numbers. Complex numbers can be negative when they are squared, such that $x^2 = -1$. Complex numbers are used in the design of electrical circuits, ships, and airplane wings. Complex numbers contain real numbers and imaginary numbers. Complex numbers involve i and are generally in the form (x + iy), or z = (x + iy), where x and y are *real numbers* and i is *imaginary*. In the expression (x + iy), the x term is referred to as the *real part* and the iy term is referred to as the *imaginary part*. A real number multiplied by i forms an *imaginary number*, such that:

(real number) × i = (imaginary number)

A real number added to an imaginary number forms a *complex number*:

(real number) + (real number)(*i*) = (complex number), or

(real number) + (imaginary number) = (complex number)

• Complex numbers correspond to points in a coordinate system called
the *complex plane*. For example, 3 + 4*i* corresponds to x = 3 and y = 4,
or point (3, 4), where the X-axis is real and the Y-axis is imaginary.
Each complex number can be associated with a point in a rectangular
coordinate system. A complex number can also be represented by a
vector. For example, *vector* 3**i** + 4**j** corresponds to complex number
3 + 4i. Complex numbers can also be written in polar form.

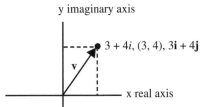

Complex Plane (rectangular and vector)

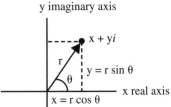

Complex Plane (polar and rectangular)

The *polar form of complex numbers* can be written using polar coordinates x = r cos θ and y = r sin θ. A complex number expressed in polar
form is: x + *i*y = r cos θ + *i* r sin θ = r(cos θ + *i* sin θ) = r$e^{i\theta}$

where r = $\sqrt{x^2 + y^2}$ and θ is the polar angle.

• To *convert from polar to rectangular form*, calculate x and y values for
r cos θ + *i* r sin θ, where x = r cos θ and y = r sin θ. To *convert from
rectangular to polar form*, draw a graph of the number in a complex
plane, then find r and θ using r = $\sqrt{x^2 + y^2}$ and tan θ = y/x, providing
x is not equal to zero. If x = 0, determine θ by inspection.

• To add or subtract complex numbers in rectangular form, add or subtract
the real terms and imaginary terms separately. To add or subtract complex
numbers in polar form, convert to rectangular form and add or subtract the
real parts and imaginary parts separately.

• Complex numbers in rectangular form can be multiplied as ordinary
binomials, where $(i)^2$ is replaced by –1. Complex numbers in polar form
are multiplied using the *product theorem,* which is:

$[r_1(\cos\theta_1 + i\sin\theta_1)][r_2(\cos\theta_2 + i\sin\theta_2)]$
$= r_1 r_2[\cos(\theta_1 + \theta_2) + i\sin(\theta_1 + \theta_2)] = r_1 e^{i\theta 1} r_2 e^{i\theta 2} = r_1 r_2 e^{i(\theta 1 + \theta 2)}$

By using the product theorem, we multiply r_1 and r_2 and add θ_1 and θ_2.

• Complex numbers can be divided in rectangular form by first multiplying the numerator and denominator by the complex conjugate of the denominator, then the numerator and denominator are divided and combined as with multiplication. Complex numbers in polar form can be divided using the *quotient theorem*, which is:

$[r_1(\cos\theta_1 + i\sin\theta_1)] / [r_2(\cos\theta_2 + i\sin\theta_2)]$
$= (r_1/r_2)[\cos(\theta_1 - \theta_2) + i\sin(\theta_1 - \theta_2)] = (r_1 e^{i\theta 1})/(r_2 e^{i\theta 2}) = (r_1/r_2)e^{i(\theta 1 - \theta 2)}$

By using the quotient theorem, we divide r_1 by r_2 and subtract θ_1 and θ_2.

• *De Moivre's theorem* is used to find *powers* of complex numbers in polar form. De Moivre's theorem is an expansion of the product rule and is: $[r(\cos\theta + i\sin\theta)]^n = r^n(\cos n\theta + i\sin n\theta) = (re^{i\theta})^n = r^n e^{in\theta}$

• The *nth-root theorem* is used to find all *n* roots of a complex number. A complex number has two square roots, three cube roots, four fourth roots, five fifth roots, or n nth roots.

The *nth root theorem in exponential and polar forms* is:

$r^{1/n}e^{i(\theta + 2\pi k)/n} = r^{1/n}e^{i(\theta + 360^\circ k)/n}$, for k = 0, 1, 2, 3, ...n–1
$= r^{1/n}[\cos((\theta + 2\pi k)/n) + i\sin((\theta + 2\pi k)/n)]$, k = 0, 1, 2, 3, ...n–1
$= r^{1/n}[\cos((\theta + 360^\circ k)/n) + i\sin((\theta + 360^\circ k)/n)]$, for k = 0, 1, 2, 3, ...n–1

The values of n lie on a *circle of radius* $r^{1/n}$, with its center at the origin, and constitute the vertices of a *regular polygon of n sides*. The angles of a given complex number are equal and differ such that:

angles of third roots differ by 360°/3, or 120°,

angles of fourth roots differ by 360°/4, or 90°,

angles of nth roots differ by 360°/n.

Chapter 12

Relationships Between Trigonometric Functions, Exponential Functions, Hyperbolic Functions, and Series Expansions

• *Trigonometric functions* are related to *exponential functions* and *hyperbolic functions* and can be written as *series*. Trigonometric functions can be expressed as exponential functions as well as series expansions or approximations. Exponential functions can also be expressed as a series. There is an inherent relationship between complex trigonometric functions and exponential functions, as we saw in Chapter 11 with Euler's identity, $e^{ix} = \cos x + i \sin x$. Other relationships exist between trigonometric functions and exponential functions and their expansions.

12.1 Relationships Between Trigonometric and Exponential Functions

• *Trigonometric functions* and *exponential functions* are related to each other. Following are identities defining the relationships between trigonometric functions and exponential functions:

$e^{iz} = \cos z + i \sin z$ (Euler's formula)

$e^z = e^{x+iy} = e^x(\cos y + i \sin y)$

$e^{-iz} = \cos z - i \sin z$

$e^{i(-z)} = \cos(-z) + i \sin(-z)$

$\cos z = (1/2)(e^{iz} + e^{-iz})$

$\sin z = (1/2i)(e^{iz} - e^{-iz})$

where $z = x + iy$ and $i = \sqrt{-1}$.

Expansions of Trigonometric and Exponential Functions

• Exponential functions and the trigonometric functions can be expanded into series. The expansions for e^x and the trigonometric functions are:

$e^z = 1 + z + z^2/2! + z^3/3! + z^4/4! + ... + z^n/n! + ...$

$\cos z = 1 - z^2/2! + z^4/4! - z^6/6! + z^8/8! - ... + (-1)^{n-1}z^{2n-2}/(2n-2)! + ...$

$\sin z = z - z^3/3! + z^5/5! - z^7/7! + z^9/9! - ... + (-1)^{n-1}z^{2n-1}/(2n-1)! + ...$

$\tan z = z + z^3/3 + 2z^5/15 + 17z^7/315 + ...$, $|z| < \pi/2$

$\cot z = 1/z - z/3 - z^3/45 - 2z^5/945 + ...$, $|z| < \pi$

$\csc z = 1/z + z/6 + 7z^3/360 + 31z^5/15120 + ...$, $|z| < \pi$

$\sec z = 1 + z^2/2 + 5z^4/24 + 61z^6/720 + ...$, $|z| < \pi/2$

Remember: !, or factorial, designates that a number is multiplied by each preceding whole number down to 1. For example,

$5! = (5)(4)(3)(2)(1) = 120$

• Writing functions in an expansion provides a means to calculate or estimate the function at a particular value. An expansion of sin x or cos x can be used to approximate values of these functions.

• **Example:** Use the series of degree 7 (or 7th power) for sine to estimate $\sin(\pi/3)$, and compare the result with that obtained using a calculator.

$\sin x = x - x^3/3! + x^5/5! - x^7/7!$

$\sin(\pi/3) \approx (\pi/3) - (\pi/3)^3/3! + (\pi/3)^5/5! - (\pi/3)^7/7!$

$\approx 1.047198 - 0.1913968 + 0.0104945 - 0.0002740 \approx 0.8660217$

Using a calculator, $\sin(\pi/3) \approx 0.8660254$ radians.
(Note there are rounding errors in these calculations.)

• **Example:** Use the series of degree 7 for sine to estimate sin 0.5, and compare the result with that obtained using a calculator.

$$\sin x = x - x^3/3! + x^5/5! - x^7/7!$$
$$\sin 0.5 \approx 0.5 - 0.5^3/3! + 0.5^5/5! - 0.5^7/7!$$
$$\approx 0.5 - 0.020833 + 0.000260 - 0.000002 \approx 0.479425$$

Using a calculator, sin 0.5 ≈ 0.479426.

• **Example:** Use the series of degree 7 to estimate e^x for x = 1, and compare the result with that obtained using a calculator.

$$e^x = 1 + x + x^2/2! + x^3/3! + x^4/4! + ... + x^n/n! + ...$$

For x = 1, this becomes:

$$e^x = 1 + 1 + 1/2! + 1/3! + 1/4! + 1/5! + 1/6! + 1/7!$$
$$\approx 1 + 1 + 0.5 + 0.166667 + 0.041667 + 0.008333$$
$$+ 0.001389 + 0.000198 \approx 2.718254$$

Using a calculator, $e^x = e^1 \approx 2.7182818$.

• Writing functions in an expansion provides a means to calculate or estimate the function at a particular value. The following summarizes how these expansions are derived.

12.2 Background: Summary of Sequences, Progressions, and Series, and Expanding a Function Into a Series

• The following is for background information and is a brief summary of sequences, arithmetic, and geometric progressions, and arithmetic and geometric series. It is important to understand that functions can be expressed as a series, and the following information is provided as a reference.

• A *sequence* is a set of numbers called *terms*, which are arranged in a succession in which there is a relationship or rule between each successive number. A sequence can be finite, having a last term, or infinite, having no last term. For example, the following is a finite sequence: {3, 6, 9, 12, 15, 18}. In this sequence, each number has a value of 3 more than the preceding number.

• An **arithmetic progression** is a sequence in which the difference between successive terms is a fixed number, and each term is obtained by adding a fixed amount to the term before it. This fixed amount is called the *common difference*. Arithmetic progressions can be represented by first-degree polynomial expressions. For example, the expression $(n + 1)$ can represent an arithmetic progression. Similarly, the sequence $\{3, 6, 9, 12, 15, 18\}$ is an arithmetic progression and can be represented by $(n + 3)$. A *finite arithmetic progression* can be expressed as:

$a, a + d, a + 2d, a + 3d, a + 4d, a + 5d,..., a + (n - 1)d$

where a is the first term, d is the fixed difference between each term, and $(a + (n - 1)d)$ is the last or "nth" term. Each term in this progression can be written as follows:

For $n = 1, a_1 = a + (1 - 1)d = a$
For $n = 2, a_2 = a + (2 - 1)d = a + d$
For $n = 3, a_3 = a + (3 - 1)d = a + 2d$, and so on.

In the arithmetic progression $\{3, 6, 9, 12, 15, 18\}$, $a = 3$ and $d = 3$. Therefore, for $n = 1, a_1 = 3$, for $n = 2, a_2 = 6$, for $n = 3, a_3 = 9$, and so on.

• A **geometric progression** is a sequence in which the *ratio* of successive terms is a fixed number, and each term is obtained by multiplying a fixed amount to the term before it. This fixed amount is called the *common ratio*. The terms in a geometric progression can be represented as:

$a, ar, ar^2, ar^3, ar^4, ar^5, ..., ar^{n-1}$,

where a is the first term, ar^{n-1} is the last term, and the ratio of successive terms is given by r such that:

$ar/a = r, ar^2/ar = r, ar^3/ar^2 = r$, etc.

An example of a geometric progression is: $\{2, 4, 8, 16, 32,...\}$, with $a = 2$ and $r = 2$, the geometric progression can be expressed as:

$2, 2(2), 2(2)^2, 2(2)^3, 2(2)^4, ..., 2(2)^{n-1}$

• A **series** is the **sum** of the terms in a progression or sequence. An *arithmetic series* is the sum of the terms in an arithmetic progression. A *geometric series* is the sum of the terms in a geometric progression. The *notation* used to express a series is *sigma notation*. The sigma notation that represents an *arithmetic series* is: $\sum_{n=1}^{m} a_n$, where a_n is the sequence function and the nth term, a_m is the last term, m is the index of the last term that is added, and n is the variable that changes between terms.

For example, in the *arithmetic progression* $\{3, 6, 9, 12,...\}$, the sum of the first three terms is the *arithmetic series*:

$$\sum_{n=1}^{3} a_n = 3 + 6 + 9 = 18$$

An *arithmetic series* can be calculated by determining the sum of the first and last terms in an arithmetic progression using the *formula* $(m/2)(a_1 + a_m)$. For example, applying this formula to the arithmetic progression $\{3, 6, 9\}$ results in: $(3/2)(3 + 9) = (3/2)(12) = 18$

- A *geometric series* can also be represented using sigma notation as:

$$\sum_{n=1}^{m} ar^{n-1} = a + ar + ar^2 + ar^3 + ar^4 + ar^5 + ... + ar^{n-1}$$

where a is the first term and $a \neq 0$, r is the ratio between successive terms, m is the index of the last term added, and n is the variable. For example, in the geometric progression $\{2, 4, 8, 16, 32, ...\}$, the sum of the first three terms is the *geometric series*:

$$\sum_{n=1}^{3} ar^{n-1} = 2 + 4 + 8 = 14$$

A geometric series can be calculated by determining the sum of the terms in the geometric progression using the *formula* $[(a)(1 - r^m)/(1 - r)]$, where r is the ratio. For example, applying this formula to the geometric progression $\{2, 4, 8\}$ results in:

$$2(1 - 2^3)/(1 - 2) = 14$$

In an *infinite geometric series*, m approaches infinity. As m approaches infinity, the formula for the series becomes: $\lim_{m \to \infty}[a(1 - r^m)/(1 - r)]$. For the case where $|r| < 1$ and $m \to \infty$, then r^m approaches zero and the sum of the infinite geometric series becomes $a/(1 - r)$.

- If a *series is infinite*, then there are an infinite number of terms in the progression or sequence that define the series. If the progression or sequence has an infinite number of terms, then the sum cannot be calculated exactly. However, under certain conditions the sum can be estimated. For example, if an infinite series has a *limit* so that as the terms are added the sum approaches a certain number, then the series will *converge* and the sum can be estimated. Conversely, if an infinite series has no limit so that as each additional term is added the sum approaches infinity, the series cannot be estimated. In general, to *estimate an infinite series*, it must be

determined whether the series has a limit and converges and what happens
to the sum as the number of terms approach infinity. To determine
whether an infinite series will converge, there are a variety of *tests for
convergence* that may be used including *the Comparison Test*, the *Ratio
Test*, tests for series with positive and negative terms, the *Integral Test*,
and the *Root Test*. (*See Master Math: Calculus* for a discussion on infinite
series and convergence.)

• A *series* can be differentiated, multiplied, added to, etc., and is sometimes written in terms of the variable x rather than r:

$$a + ax + ax^2 + ax^3 + ax^4 + ... + ax^{n-1} = a(1 - x^n)/(1 - x)$$

Expanding Functions Into Series

• Trigonometric and exponential functions can be estimated using series
expansions. When a *function* is written in the form of an *infinite series*,
it is said to be *"expanded" in an infinite series*. Two common series
representing expansions of functions are the **Maclaurin series** and the
Taylor series. Expanding functions into these series can be applied to
approximating functions including linear and quadratic approximations,
approximating solutions to differential equations, and estimating numerical values such as constructing tables of exponential, logarithmic, and
trigonometric functions.

• Representing a function in a Taylor series or a Maclaurin series involves
determining the *coefficients* $a_0, a_1, ... a_n$ of the series. The coefficients can
be found by *differentiation* providing the function has all its derivatives.
Obtaining all the derivatives of a function can be tedious, so other methods
including substitution and integration are employed.

The expansion of a function f(x) about x = a = 0 is known as the
Maclaurin series or the **Taylor series** *for f(x) expanded about the point
x = 0* and is given by:

$$f(x) = a_0 + a_1 x + a_2 x^2 + a_3 x^3 + a_4 x^4 + ... a_n x^n ... = \sum_{n=0}^{\infty} \frac{f^{(n)}(0)}{n!} x^n$$

$$= [f(0)] + [f\,'(0)]x + [f\,''(0)/2!]x^2 + [f\,'''(0)/3!]x^3 + ...[f^{(n)}(0)/n!]x^n...$$

• Using the *Taylor series* allows a function to be expanded about some
point, a, other than zero. For a function f(x):

$$f(x) = a_0 + a_1(x-a) + a_2(x-a)^2 + a_3(x-a)^3 + ...a_n(x-a)^n + ...$$

the coefficients a_n are computed by repeated differentiation as with the Maclaurin series. The resulting Taylor series for $f(x)$ is:

$$f(x) = [f(a)] + [f\,'(a)](x-a) + [f\,''(a)/2!](x-a)^2 + [f\,'''(a)/3!](x-a)^3$$

$$+...+ [f^{(n)}(a)/n!](x-a)^n... = \sum_{n=0}^{\infty} \frac{f^{(n)}(a)}{n!}(x-a)^n$$

This is the **Taylor series**, which is expanded about point $x = a$. If $a = 0$, the Taylor Series becomes the *Maclaurin Series*.

- The **trigonometric functions** *can be expanded* and computed for selected values. The **expansions of sine and cosine** are:

$$\sin x = x - x^3/3! + x^5/5! - x^7/7! +...+(-1)^{n-1}x^{2n-1}/(2n-1)! +...$$

$$\cos x = 1 - x^2/2! + x^4/4! - x^6/6! +...+(-1)^{n-1}x^{2n-2}/(2n-2)! +...$$

- To obtain the trigonometric series for $f(x) = \sin x$ for x near 0, begin with the Taylor series expansion at $x = 0$:

$$f(x) = a_0 + a_1x + a_2x^2 + a_3x^3 + a_4x^4 +...a_nx^n ...$$

$$= [f(0)] + [f\,'(0)]x + [f\,''(0)/2!]x^2 + [f\,'''(0)/3!]x^3 +...[f^{(n)}(0)/n!]x^n...$$

where $f\,'(0)$ represents the first derivative, $f\,''(0)$ represents the second derivative, and so on. We can determine the coefficients of the series by differentiation for a degree of, for example 7, and substitute into the above expansion. (Note that the derivative of sine is cosine, and the derivative of cosine is – sine.) Taking successive derivatives results in:

$$f(x) = \sin x \rightarrow f(0) = 0, \qquad f\,'(x) = \cos x \rightarrow f\,'(0) = 1,$$

$$f\,''(x) = -\sin x \rightarrow f\,''(0) = 0, \qquad f\,'''(x) = -\cos x \rightarrow f\,'''(0) = -1,$$

$$f^{(4)}(x) = \sin x \rightarrow f^{(4)}(0) = 0, \qquad f^{(5)}(x) = \cos x \rightarrow f^{(5)}(0) = 1,$$

$$f^{(6)}(x) = -\sin x \rightarrow f^{(6)}(0) = 0, \qquad f^{(7)}(x) = -\cos x \rightarrow f^{(6)}(0) = -1.$$

Then we can substitute each of these derivatives into the expansion resulting in $f(x) = \sin x$ about the point $x = a = 0$ which is:

$$\sin x = 0 + x + (0)x^2/2! - x^3/3! + (0)x^4/4! + x^5/5! + (0)x^6/6! - x^7/7! +...$$

$$\sin x = x - x^3/3! + x^5/5! - x^7/7! +...$$

We can check how good our approximation is by graphing $f(x) = \sin x$ and $\sin x = x - x^3/3! + x^5/5! - x^7/7!$ and comparing the curves. In the following graph the thin black curve represents $\sin x$ and the thicker gray curve represents $x - x^3/3! + x^5/5! - x^7/7!$.

Graph of y = sin x (thin) and sin x = x – x³/3! + x⁵/5! – x⁷/7! (thick)

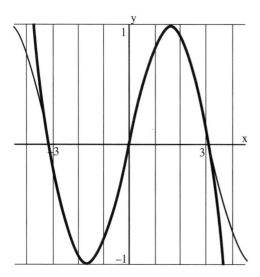

The series approximation overlaps well for values of x near zero.

The trigonometric series for $f(x) = \cos x$ for x near 0 can be found in the same manner as for $f(x) = \sin x$.

• An approximation of the ***exponential function*** e^x can also be computed using the Taylor or Maclaurin expansions:

The Maclaurin expansion of e^x for x near 0 is:

$$e^x = 1 + x + x^2/2! + x^3/3! + x^4/4! + ... + x^n/n! + ...$$

For x = 1 and n = 7, this becomes:

$$e^x = 1 + 1 + 1/2! + 1/3! + 1/4! + 1/5! + 1/6! + 1/7!$$
$$\approx 1 + 1 + 0.5 + 0.166667 + 0.041667 + 0.008333 + 0.001389$$
$$+ 0.000198 + ... \approx 2.718254$$

Using a calculator to compare the result, when x = 1: $e^x = e^1 \approx 2.71828$ Increasing the number of terms in the series will improve the accuracy.

• The series for e^x, sin x, and cos x all have $x^n/n!$ terms where the factorials lead to convergence for all x. In addition, term-by-term *differentiation* of series e^x yields e^x, and term-by-term differentiation of series sin x yields series cos x.

- In Chapter 7 we learned that $re^{i\theta} = r\cos\theta + ir\sin\theta = x + iy$. When $r = 1$, this becomes: $e^{i\theta} = \cos\theta + i\sin\theta$, which is known as *Euler's formula*. We can expand $e^{i\theta}$ into a series as:

$$e^{i\theta} = 1 + i\theta + (i\theta)^2/2! + (i\theta)^3/3! + (i\theta)^4/4! + \dots$$

It is possible to verify Euler's formula by showing that $e^{i\theta}$ is equal to $(\cos\theta + i\sin\theta)$ using series expansions:

$$\cos\theta + i\sin\theta = [1 - \theta^2/2! + \theta^4/4! - \theta^6/6!\dots] + i[\theta - \theta^3/3! + \theta^5/5! -\dots]$$
$$= 1 + i\theta - \theta^2/2! - i\theta^3/3! + \theta^4/4! + i\theta^5/5! - \theta^6/6!\dots$$

Substitute i^2 for -1, i^3 for $-i$, i^4 for 1, i^5 for i, etc:

$$= 1 + i\theta + (i\theta)^2/2! + (i\theta)^3/3! + (i\theta)^4/4! + (i\theta)^5/5! \dots = e^{i\theta}$$
$$= [1 - \theta^2/2! + \theta^4/4! - \theta^6/6!\dots] + i[\theta - \theta^3/3! + \theta^5/5! -\dots]$$
$$= e^{i\theta} = \cos\theta + i\sin\theta$$

which is *Euler's formula*, where the real part is $x = \cos\theta$ and the imaginary part is $y = \sin\theta$.

12.3 Hyperbolic Functions

- *Hyperbolic functions* are real, do not involve $i = \sqrt{-1}$, and are derived from the exponential functions e^z and e^{-z}, for $z = x + iy$. The hyperbolic functions include hyperbolic cosine (cosh), hyperbolic sine (sinh), hyperbolic tangent (tanh), hyperbolic cotangent (coth), hyperbolic cosecant (csch), and hyperbolic secant (sech).

- Definitions for the hyperbolic functions in terms of exponential functions include:

The *hyperbolic cosine*: $\cosh z = (1/2)(e^z + e^{-z})$

The *hyperbolic sine*: $\sinh z = (1/2)(e^z - e^{-z})$

The *hyperbolic tangent*: $\tanh z = (\sinh z / \cosh z) = (e^z - e^{-z}) / (e^z + e^{-z})$

The *hyperbolic cosecant*: $\operatorname{csch} z = 1 / \sinh z = 2/(e^z - e^{-z})$

The *hyperbolic secant*: $\operatorname{sech} z = 1 / \cosh z = 2/(e^z + e^{-z})$

The *hyperbolic cotangent*: $\coth z = \cosh z / \sinh z = 1/\tanh z$
$\qquad = (e^z + e^{-z})/(e^z - e^{-z})$

The graph of the hyperbolic cosine (cosh) and hyperbolic sine (sinh) together is depicted below. Note that as z gets large, cosh z and sinh z approach $(1/2)e^z$.

Graph of cosh z $= (1/2)(e^z + e^{-z})$ (thin) and sinh z $= (1/2)(e^z - e^{-z})$ (thick)

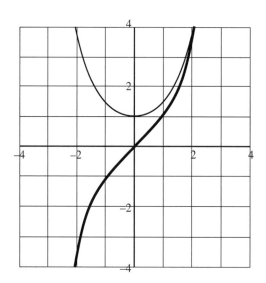

- The hyperbolic functions can be expressed in series form as:

  $$\cosh z = 1 + z^2/2! + z^4/4! + z^6/6! + z^8/8! + \ldots$$
  $$\sinh z = z + z^3/3! + z^5/5! + z^7/7! + z^9/9! + \ldots$$
  $$\tanh z = z - z^3/3 + 2z^5/15 - 17z^7/315 + \ldots, \ |z| < \pi/2$$
  $$\coth z = 1/z + z/3 - z^3/45 + 2z^5/945 + \ldots, \ |x| < \pi$$
  $$\operatorname{csch} z = 1/z - z/6 + 7z^3/360 - 31z^5/15120 + \ldots, \ |z| < \pi$$
  $$\operatorname{sech} z = 1 - z^2/2 + 5z^4/24 - 61z^6/720 + \ldots, \ |z| < \pi/2$$

- Hyperbolic functions are similar to trigonometric functions. For example, the exponential forms are similar:

  $$\cosh z = (e^z + e^{-z})/2 \text{ compares with } \cos z = (e^{iz} + e^{-iz})/2$$
  $$\sinh z = (e^z - e^{-z})/2 \text{ compares with } \sin z = (e^{iz} - e^{-iz})/2i$$

Like cosine, cosh is an *even function*:

  $$\cosh(-z) = \cosh z \text{ and } \cosh 0 = 1$$

Like sine, sinh is an *odd function*:

sinh(−z) = −sinh z and sinh 0 = 0

Properties that apply to cosh and sinh are similar to properties for cosine and sine, and they include the following:

$\cosh^2 z - \sinh^2 z = 1$

$e^z = \cosh z + \sinh z,$ $e^{-z} = \cosh z - \sinh z$

$\sinh^2 z = (1/2)(\cosh 2z - 1),$ $\cosh^2 z = (1/2)(\cosh 2z + 1)$

$\tanh^2(z) + \mathrm{sech}^2(z) = 1,$ $\coth^2(z) - \mathrm{csch}^2(z) = 1$

• Complex trigonometric functions and hyperbolic functions are related. Hyperbolic formulas can be derived from trigonometric identities by replacing z with iz. Relationships between trigonometric and hyperbolic functions include:

$\sinh z = -i \sin iz,$ $\cosh z = \cos iz,$ $\tanh z = -i \tan iz$

$\mathrm{csch}\, z = i \csc iz,$ $\mathrm{sech}\, z = \sec iz,$ $\coth z = i \cot iz$

$\sinh iz = \sin z,$ $\cosh iz = \cos z$

Formulas Involving Hyperbolic Functions

• For each formula involving the trigonometric functions, there is a similar (not necessarily identical) formula for the hyperbolic functions. Following are a number of the identities and formulas for hyperbolic functions:

Addition and subtraction formulas:

$\sinh(z_1 \pm z_2) = \sinh z_1 \cosh z_2 \pm \cosh z_1 \sinh z_2$

$\cosh(z_1 \pm z_2) = \cosh z_1 \cosh z_2 \pm \sinh z_1 \sinh z_2$

$\tanh(z_1 + z_2) = (\tanh z_1 + \tanh z_2) / (1 + \tanh z_1 \tanh z_2)$

$\coth(z_1 + z_2) = (1 + \coth z_1 \coth z_2) / (\coth z_1 + \coth z_2)$

Addition and subtraction of two hyperbolic functions:

$\sinh z_1 + \sinh z_2 = 2 \sinh[(z_1 + z_2)/2] \cosh[(z_1 - z_2)/2]$

$\sinh z_1 - \sinh z_2 = 2 \sinh[(z_1 - z_2)/2] \cosh[(z_1 + z_2)/2]$

$\cosh z_1 + \cosh z_2 = 2 \cosh[(z_1 + z_2)/2] \cosh[(z_1 - z_2)/2]$

$\cosh z_1 - \cosh z_2 = 2 \sinh[(z_1 + z_2)/2] \sinh[(z_1 - z_2)/2]$

$\tanh z_1 + \tanh z_2 = [\sinh(z_1 + z_2)] / [\cosh z_1 \cosh z_2]$

$\coth z_1 + \coth z_2 = [\sinh(z_1 + z_2)] / [\sinh z_1 \sinh z_2]$

Product formulas:

$$\sinh z_1 \cosh z_2 = (1/2)(\sinh(z_1 + z_2) + \sinh(z_1 - z_2))$$
$$\cosh z_1 \cosh z_2 = (1/2)(\cosh(z_1 + z_2) + \cosh(z_1 - z_2))$$
$$\sinh z_1 \sinh z_2 = (1/2)(\cosh(z_1 + z_2) - \cosh(z_1 - z_2))$$

Negative-angle formulas:

$$\cosh(-z) = \cosh z, \quad \sinh(-z) = -\sinh z, \quad \tanh(-z) = -\tanh z$$

Double-angle formulas:

$$\sinh 2z = 2 \sinh z \cosh z = [2 \tanh z] / [1 - \tan^2 z]$$
$$\cosh 2z = \cosh^2 z + \sinh^2 z = 1 + 2 \sinh^2 z = 2 \cosh^2 z - 1$$
$$\tanh 2z = 2 \tanh z / (1 + \tanh^2 z)$$

Half-angle formulas:

$$\sinh(z/2) = \pm [(\cosh z - 1)/2]^{1/2}$$
$$\cosh(z/2) = \pm [(\cosh z + 1)/2]^{1/2}$$
$$\tanh(z/2) = \pm [(\cosh z - 1)/(\cosh z + 1)]^{1/2}$$
$$= (\sinh z)/(\cosh z + 1) = (\cosh z - 1)/(\sinh z)$$

Hyperbolic Functions and Parameterization

• As we have seen, a unit circle can be parameterized by $(\cos t, \sin t)$, with t in interval $[0, 2\pi]$. Similarly, the hyperbolic functions $(\cosh t, \sinh t)$, with t in the interval from plus to minus infinity, can be used to parameterize the standard hyperbola $x^2 - y^2 = 1, x > 1$. In the figure a standard hyperbola is depicted with a point $(\cosh t, \sinh t)$, which has values of the parameter *t* along the curve. Consider the hyperbola for t = −1.5, the point is depicted on the curve at $\cosh(-1.5) \approx 2.35 = x$ and $\sinh(-1.5) \approx -2.13 = y$:

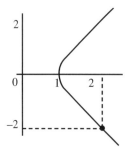

12.4 Chapter 12 Summary and Highlights

• *Trigonometric functions* are related to *exponential functions* and *hyperbolic functions* and can be written as *series*. Relationships between trigonometric functions and exponential functions include: $\cos z = (1/2)(e^{iz} + e^{-iz})$, $\sin z = (1/2i)(e^{iz} - e^{-iz})$, and $e^{iz} = \cos z + i \sin z$, which is Euler's formula, where $z = x + iy$.

The expansions for e^x, $\cos x$, and $\sin x$ are:

$$e^z = 1 + z + z^2/2! + z^3/3! + z^4/4! + \ldots + z^n/n! + \ldots$$

$$\cos z = 1 - z^2/2! + z^4/4! - z^6/6! + z^8/8! - \ldots + (-1)^{n-1}z^{2n-2}/(2n-2)! + \ldots$$

$$\sin z = z - z^3/3! + z^5/5! - z^7/7! + z^9/9! - \ldots + (-1)^{n-1}z^{2n-1}/(2n-1)! + \ldots$$

Writing functions in a series expansion provides a means to calculate or estimate the function at a particular value.

• *Hyperbolic functions* are real, do not involve $i = \sqrt{-1}$, and are derived from the exponential functions e^z and e^{-z}, for $z = x + iy$. Definitions for the hyperbolic functions include:

The *hyperbolic cosine*: $\cosh z = (1/2)(e^z + e^{-z})$

The *hyperbolic sine*: $\sinh z = (1/2)(e^z - e^{-z})$

The *hyperbolic tangent*: $\tanh z = (\sinh z / \cosh z) = (e^z - e^{-z}) / (e^z + e^{-z})$

The hyperbolic functions can also be expressed in series form.

Chapter 13

Spherical Trigonometry

13.1 Definitions and Properties

• Trigonometry and spherical trigonometry were primarily developed for and used in astronomy, geography, and navigation. Spherical trigonometry was developed to describe and understand applications involving triangles on spheres and spherical surfaces. The first trigonometric tables were created more than two thousand years ago for computations in astronomy. Trigonometry is currently used in numerous fields, including engineering, chemistry, surveying, navigation, physics, mathematics, astronomy, and architecture. An example in architecture is the triangular spherical shells of the famous Sidney Opera House in Sidney, Australia.

• *Spherical triangles* are triangles drawn on a spherical surface. Spherical triangles do not have straight-line sides; rather, they have sides that are *arcs of great circles* (which are the largest circles that can be drawn around a sphere's surface). *Great-circle arcs* (described below) form the sides of a spherical triangle, and where two arcs intersect, a *spherical angle* is formed. A spherical angle can be thought of as either an angle between the tangents of the two arcs at the point of intersection or as the angle between the planes of the two great circles where they intersect at the center of the sphere. Spherical angles are defined at the location where arcs of great circles meet. In other words, the arc lengths are a measure of the angle they subtend at the center of the sphere, and the spherical angles between the arcs are a measure of the angle at which the planes that form the arcs intersect. *Spherical trigonometry* involves relationships between the arc lengths (sides) and the spherical angles between the arcs.

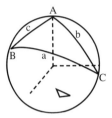

The large spherical triangle has angles A, B, and C, and has sides a, b, and c. If a spherical triangle is large in dimensions as compared to the size of the sphere, the spherical nature of the triangle must be considered. If a spherical triangle is small as compared to the size of the sphere, the triangle can be described using plane trigonometry. For this reason, when maps of cities are drawn, it is not necessary to use spherical triangles. However, if a map is drawn of a large area of the Earth, then the fact that the Earth is a sphere must be considered.

• *Sum of the angles in a triangle*: When a *planar triangle* is drawn on a flat sheet of paper, the sum of its three angles will always be 180 degrees. In *plane geometry*, the curvature is considered to be zero. In *hyperbolic geometry*, represented in two dimensions by a saddle-shaped surface, the angles of a triangle add up to less than 180 degrees. In hyperbolic geometry, there is a negative curvature. When a triangle is drawn on the surface of a sphere, the angles add up to more that 180 degrees, which is a characteristic of *spherical geometry*. The curvature in spherical geometry is therefore positive. Because the sum of the angles in a spherical triangle will always be greater than 180°, the arcs of great circles (sides) that make up the spherical triangle will each measure less than 180°. In addition, because the sides of a spherical triangle are arcs rather than straight lines, they are measured in radians or degrees.

Planar triangle   Hyperbolic triangle   Spherical triangle

• The need for spherical trigonometry becomes obvious when distance and angle measurements are made on a large area of the Earth. For example, suppose we know the distances and angle measurements of three cities that are far away from each other and form a large *right* triangle. If we attempt to use the *Pythagorean Theorem* to confirm the side and angle measurements, we will find a small disagreement between the measurements and the calculations of the sides and angles. In fact, the sum of the angles in this triangle will be greater than the 180 degrees that is always measured in planar triangles. This discrepancy occurs because the large triangle that was measured on the surface of the Earth is a spherical triangle. The discrepancy between the measurements and calculations (using planar equations) becomes greater the further apart the locations which are being measured are situated from each other. This occurs because the spherical triangle becomes larger relative to the size of the Earth's surface. Therefore, we can see that triangles on a sphere have different properties than triangles on a plane.

Great Circles, Small Circles, Latitude, and Longitude

• If a straight slice is made, or a plane is passed, through a sphere, a circle is formed on the surface of the sphere. If the diameter of the circle formed on the surface is less than the diameter of the sphere, the circle is called a *small circle*. Except for the equator, *latitude lines* on Earth form small circles. If a slice is made through the center of a sphere and the sphere is split into two equal hemispheres, the diameter of the circle formed on the surface will be the same as the diameter of the sphere. This is called a *great circle* and is the largest circle that can be drawn on the surface of a sphere. A great circle can be thought of as the circle formed on the surface if a plane is inserted through the center of a sphere. The *equator* on Earth is a *great circle* and is halfway between the North and South Poles. (The Earth is not a perfect sphere, but is often modeled as a sphere for practical purposes.)

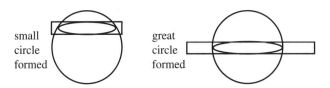

small circle formed

great circle formed

Longitude lines that form a circle around the Earth, crossing the poles, are great circles. The term *circle of longitude* is often used to refer to a *great circle* consisting of all points encircling a sphere along the given longitudes. Half of such a great circle, consisting of points with the same longitude from the North Pole to the South Pole, is called a *meridian*. The term *circle of latitude*, or *parallel of latitude*, is used for a circle consisting of all points with the same *latitude*. Parallels of latitude are everywhere equally distant as are concentric circles on a plane.

North Pole

South Pole

• A particular point on Earth can be located using its *longitude* and *latitude*. Latitude and longitude are usually expressed in degrees, although they can be measured in radians if it is more convenient. The equator is at 0°, and the poles are at plus and minus 90°. The starting point for longitude measurements is generally considered as the *prime meridian*, which passes through the Greenwich Observatory at London, England, and has zero longitude. Latitude and longitude are depicted on most globes.

• It is interesting to note that if you begin at the North Pole and travel due south, then travel due east or due west, then travel due north, you will end up back at the North Pole.

North Pole

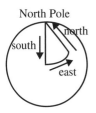

13.2 Measuring Spherical Triangles

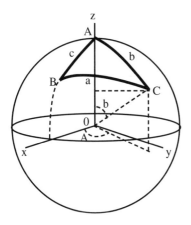

• When measuring distances or plotting a course on a sphere, it turns out that the *shortest distance or path between two points is the path that is part of a great circle* on which the two points are located. In other words, on a spherical surface, a great circle path, often called a *geodesic*, is always the shortest path between two points.

• A spherical triangle can be labeled with its sides represented using small letters a, b, and c, and the angles represented using capital letters such that angle A is opposite side a, angle B is opposite side b, and angle C is opposite side c.

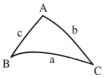

• Measuring spherical triangles can be confusing because both the sides and angles have angular measures. Therefore, it is helpful to measure the *angles* of spherical triangles in *degrees* and measure the *sides* in *radians*.

• The **length of a side of a spherical triangle** can be compared to the radian measure for arc length, in which *arc length* = (radius)(central angle measure in radians).

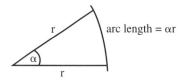

arc length = αr

In the ABC spherical triangle drawn above, a, b, and c represent the sides and are usually measured in radians. The *lengths of the sides* are ar, br, and cr. In the spherical triangle, a, b, and c also represent the angles subtended at the center of the sphere by the great circle arcs, such that:

a = (length of arc a) / (radius of sphere)

To obtain the *length of the arc* for a side, we can multiply the radian measure of a side by r. A side can be expressed in units such as kilometers, if r is the radius of the sphere and is measured in kilometers.

• *The angles between two curves:* Each curve (side) is a part of a great circle, which is formed by a plane intersecting the sphere and passing through its center. The angle between two curves (sides) is therefore the angle between the two planes. Angles may be measured or obtained using relations such as the Law of Sines and the Law of Cosines for spherical triangles.

13.3 The Law of Sines and the Law of Cosines for Spherical Triangles for Calculating Sides and Angles

• There are various formulas relating the sides and angles of a spherical triangle. Two that are of particular importance are the *sine rule* or **Law of Sines for spherical triangles** and the *cosine rule* or **Law of Cosines for spherical triangles**. The laws of sines and cosines for spherical triangles compared with the laws of sines and cosines for planar triangles are as follows:

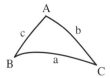

Spherical Triangles	Planar Triangles

Law of Sines

$$\frac{\sin a}{\sin A} = \frac{\sin b}{\sin B} = \frac{\sin c}{\sin C} \qquad\qquad \frac{a}{\sin A} = \frac{b}{\sin B} = \frac{c}{\sin C}$$

Law of Cosines

$\cos a = \cos b \cos c + \sin b \sin c \cos A \qquad a^2 = b^2 + c^2 - 2bc \cos A$

$\cos b = \cos c \cos a + \sin c \sin a \cos B \qquad b^2 = c^2 + a^2 - 2ca \cos B$

$\cos c = \cos a \cos b + \sin a \sin b \cos C \qquad c^2 = a^2 + b^2 - 2ab \cos C$

In a planar triangle, a, b, and c have units of length, and in the spherical triangle, a, b, and c are measured in radians or degrees and are the angles subtended at the center of the sphere by the great circle arcs, such that:

a = (length of arc a) / (radius of sphere)

To solve spherical triangles, we can use the laws of sines and cosines to determine required measurements of angles A, B, and C and sides a, b, and c, and then multiply a, b, and c by the radius of the sphere to obtain side lengths.

• The formulas for a spherical triangle can be reduced to those for a plane triangle when a, b, and c (in radians) are all considerably less than 1. To test this, suppose we have a spherical triangle with sides a and b equal to 0.1 radian and angle C equal to 30°. We can find side c using both the Law of Cosines for spherical triangles and the Law of Cosines for planar triangles. Then we can compare the results. Remember to make trigonometric calculations using the degree or radian mode in your calculator that is appropriate for each step.

Using the Law of Cosines for spherical triangles:

$\cos c = \cos a \cos b + \sin a \sin b \cos C$

$\cos c = \cos 0.1 \cos 0.1 + \sin 0.1 \sin 0.1 \cos 30°$

$\cos c \approx (0.99500)(0.99500) + (0.09983)(0.09983)(0.86602) \approx 0.99866$

$c \approx \arccos 0.99866 \approx 0.05177$

Using the Law of Cosines for planar triangles:

$$c^2 = a^2 + b^2 - 2ab \cos C$$
$$c^2 = 0.1^2 + 0.1^2 - 2(0.1)(0.1) \cos 30°$$
$$c^2 \approx 0.01 + 0.01 - 0.01732 = 0.00268$$
$$c \approx 0.05177$$

Therefore, in this small triangle, we can use either the planar or spherical Law of Cosines.

Solving Spherical Triangles

• The *Law of Cosines and Law of Sines for spherical triangles* can be used to solve most spherical triangles. Using either of these formulas can result in an ambiguous answer for side and angle calculations. For example, if the Law of Sines results in sin x = 1/2, then x may be 30° or 150°, because both sin 30° = 1/2 and sin 150° = 1/2. Similarly, if the Law of Cosines results in cos x = 1/2, then x may be 60° or 300° (–60°) because both cos 60° = 1/2 and cos 300° = 1/2. Therefore, when calculating values using either formula, verify that the answer makes sense.

• To solve for all angles and all sides of a spherical triangle, the measures of at least three of the parts must be known to find the other three values. In addition, unlike planar triangles, it is not possible to ascertain the measure of a third angle in a spherical triangle by subtracting the sum of the two known angles from 180°.

• To solve problems that involve spherical triangles, often the *distance between two points* and/or *a direction described by an angle* is desired. It is important to remember that the shortest distance between two points on the surface of a sphere is the great circle path between the two points. On the Earth, the equator and circles of longitude are natural great circles. In addition, any circular path around the Earth that cuts it into two equal hemispheres is a great circle. If you want to measure the shortest distance between two points that lie on the *equator*, which is a great circle, then the shortest path would be along the equator. If you want to measure the distance between two points that lie on the same longitude, where circles of longitude are great circles, then the shortest path would be directly north or south from the first point to the second point. If the two points are on the same latitude but not the equator, then moving along the latitude line would *not be the shortest path*, because circles of latitude are not great circles (with the exception of the equator).

• **Example:** Suppose you are an astronaut and you will be landing on a planet that is identical to the Earth, except for the locations, sizes, and shapes of the continents and islands. Latitude and longitude are designated as they are on the Earth. You are planning to land on a beach you name as Point 1, which is at latitude 20° and longitude 0°, and explore the area for some period of time. Then you plan to travel in your space ship/motorboat across the ocean to Point 2, which is at latitude 30° and longitude 50°, to continue exploration before returning to your mother ship. What is the shortest travel distance possible between Point 1 and Point 2, which will be the great circle path, and at what angle from Point 1 should you travel to arrive exactly at Point 2?

Begin by plotting your course on the spherical triangle, which has its vertices as Point 1, Point 2, and this planet's North Pole:

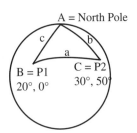

We want to find the length of side a, which is the great circle path between Point 1 and Point 2, and angle B so that we can travel in a north-east direction to Point 2.

What we know:

    Point 1 is at B and Point 2 is at C.

    Side a is the great circle path between Point 1 (B) and Point 2 (C).

    Point 1 (B) is located at latitude 20° and longitude 0°.

    Point 2 (C) is located at latitude 30° and longitude 50°.

    Side c and side b form the other two sides of the spherical triangle.

    Length of side c = 90° – latitude of B = 90° – lat 20°.

    Length of side b = 90° – latitude of C = 90° – lat 30°.

    (90° is the latitude at North Pole.)

Angle A measures the difference in longitude between Point 1 (B) and Point 2 (C), such that:

    A = longitude C – longitude B = longitude 50° – longitude 0°.

We can use this information to find side b, side c, and angle A. Then, we can use the Law of Cosines for spherical triangles to calculate side a, which is the shortest distance between Point 1 and Point 2. Finally, we can use the Law of Sines for spherical triangles to determine the angle of travel, B.

$c = 90° - \text{lat } 20° = 70° = (\pi/180°)70° \approx 1.2217$ radians

$b = 90° - \text{lat } 30° = 60° = (\pi/180°)60° \approx 1.0472$ radians

$A = \text{longitude C} - \text{longitude B} = \text{long } 50° - \text{long } 0° = 50°$

We can now use the Law of Cosines for spherical triangles to find the length of side a: $\cos a = \cos b \cos c + \sin b \sin c \cos A$.

Remember to have your calculator in the proper radian or degree mode for each calculation.

$\cos a \approx \cos 1.0472 \cos 1.2217 + \sin 1.0472 \sin 1.2217 \cos 50°$

$\cos a \approx (0.5000)(0.3420) + (0.8660)(0.9397)(0.6428) \approx 0.6941$

$a \approx \arccos 0.6941 \approx 0.8036$ radians

Therefore, the shortest (great circle) distance between Point 1 and Point 2 is:

distance = (a)(radius of Earth)

$\approx (0.8036)(6371 \text{ kilometers}) \approx 5120$ km

Calculated in miles this distance is:

distance = (a)(radius of Earth) = $(0.8036)(3959 \text{ miles}) \approx 3182$ mi

(Note that there are approximately 1.609 kilometers/mile. Also, values for the average radius of the Earth, which is an oblate spheroid, can vary depending on the source.)

Next, we can determine at what angle we must travel to get from Point 1 to Point 2. We know that Point 2 is at a latitude closer to the North Pole, so we will be angling up toward the pole. We also know that angle A is 50°. We can use the Law of Sines for spherical triangles to find angle B:

$\sin a / \sin A = \sin b / \sin B$

$\sin B = \sin A \sin b / \sin a$

$\sin B = \sin 50° \sin 1.0472 / \sin 0.8036$

$\approx (0.7660)(0.8660) / 0.7199 \approx 0.9215$

$B \approx \arcsin 0.9215 \approx 1.1719 \text{ rad} \times 180°/\pi \approx 67.15°$

Therefore, when you plot your course from Point 1 to Point 2, you will travel at an angle of 61.15° northeast for 5120 kilometers or 3182 miles.

• To solve problems involving spheres, it is a good idea to draw a triangle that reasonably represents the spherical triangle from which you plan to obtain values. This will reduce any errors that arise from angle calculations which may be ambiguous. Rechecking results by interchanging the use of the laws of sines and cosines for spherical triangles is also worthwhile when ambiguous angle measurements are calculated.

13.4. Celestial Sphere

• An application of spherical trigonometry is astronomy and the so-called *celestial sphere*. The celestial sphere is what we see when we look into the sky at night and observe the stars, planets, galaxies, etc., which appear to be located on the inside of a sphere. The *North Celestial Pole* is located above the Earth's North Pole and is the place in the sky to which the axis of Earth points in the north direction. The *South Celestial Pole* is located above the Earth's South Pole and is the place in the sky to which the axis of Earth points in the south direction. The *Celestial Equator* is above the Earth's equator.

• Locations can be identified and measurements made on the celestial sphere in the same manner as measures are made on the surface of the Earth. However, instead of using latitude and longitude, celestial measurements are made using *declination* and *right ascension.*

Declination (Dec) is the celestial equivalent of *latitude*. The declination is the position of a star measured in the north-south direction on the sky. Like latitude, declination describes the angular distance between a star and the Celestial Equator. Stars on the equator have a declination of 0°, and the North Star has a declination of 90°. As Earth turns, the declination does not change. Declination is measured in degrees, minutes, and seconds of an arc. Each degree can be divided into 60 minutes of an arc. Calculations are sometimes expressed in decimal degrees.

Right ascension (RA) is the celestial equivalent to *longitude*. The right ascension is the position of a star measured in the east-west direction on the sky. Like longitude, right ascension describes the distance along the Celestial Equator between a star and the right ascension starting point. Right ascension increasing from west to east begins with RA $0^h\ 00^m\ 00^s$, which is a semicircle centered on the center of Earth from the North Celestial Pole to the South Celestial Pole. For example, a star at RA 3^h 00^m 00^s is three hours east of a star at RA 0^h 00^m 00^s regardless of each stars' declination. An hour of RA is equal to an arc of 15° on the Celestial Equator, given that 24 hours x 15° = 360°, which is one complete circle around the Earth. A minute of RA is a measure of an angle on the sky that is 1/60th of an hour of RA. A second of RA is 1/60th of a minute of RA. Calculations are sometimes expressed in decimal degrees.

• To locate a star with, for example, right ascension 36° and declination 25°, we can move a telescope 36° from the declination starting point along the Celestial Equator, and then aim it 25° north from there.

• In the celestial sphere, a spherical triangle can be observed between the North Star, a star of interest, and the zenith, which is directly over the observer.

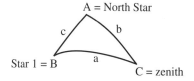

A = North Star

c   b

Star 1 = B   a

C = zenith

In this spherical triangle:

Side a represents the angular distance between the Star 1 and the zenith;

Side b represents the angular distance between the North Star and the zenith; and

Side c represents the angular distance between the North Star and the Star 1, such that: c = 90° − (declination of the star).

Because the North Star and the zenith are stationary, and the star changes position with respect to the zenith as the Earth turns, side c will be constant. However, side a and angle A will change as the Earth rotates. Angle A is called the *hour angle* of the star. Angle A depends on the right ascension of the star as well as the time, such that: Hour angle = local *sidereal time* − right ascension. Local sidereal time measures how much the celestial sphere has turned. The sidereal time is the time with respect to stars, and a sidereal day is 23 hours and 56 minutes and 4 seconds with respect to the stars. The difference between this day and a 24 hour day, is 3 minutes and 56 seconds, which is approximately 1/365 of a day. During a day the Earth moves through 1/365 of its orbit around the Sun.

• The Law of Cosines and Law of Sines for spherical triangles can be used to determine values for spherical triangles on the celestial sphere.

13.5. Chapter 13 Summary and Highlights

• Spherical trigonometry was developed to describe and understand applications involving triangles on spheres and spherical surfaces called *spherical triangles*. Spherical triangles do not have straight-line sides; rather, they have sides that are arcs of great circles (which are the largest circles that can be drawn around a sphere's surface). Great-circle arcs form the sides of a spherical triangle, and where two arcs intersect, a spherical angle is formed. A spherical angle can be thought of as either an angle between the tangents of the two arcs at the point of intersection or as the angle between the planes of the two great circles where they intersect at the center of the sphere. *Spherical trigonometry* involves relationships between the arc lengths (sides) and the spherical angles between the arcs. Because the sides of a spherical triangle are arcs rather than straight lines, they are measured in radians or degrees. In addition, unlike a planar triangle, in which the sum of the angles will always be 180°, in a spherical triangle, the angles add up to more than 180°. When measuring distances or plotting a course on a sphere, it turns out that the shortest distance or path between two points is the path that is part of a great circle on which the two points are located.

• The *length of a side of a spherical triangle* can be compared to the radian measure for arc length, in which *arc length* = (radius)(central angle measure in radians). In a spherical triangle with angles A, B, and C and sides a, b, and c, the *lengths of the sides* are ar, br, and cr, where r is the radius of the sphere. In the spherical triangle, a, b, and c are also the angles subtended at the center of the sphere by the great circle arcs (sides) such that: a = (length of arc a) / (radius of sphere). To obtain the *length of the arc* for a side, multiply the radian measure of the side by r. Because both the sides and angles have angular measures, measure the *angles* of spherical triangles in *degrees* and measure the *sides* in *radians*.

• The *Law of Cosines for spherical triangles* and *Law of Sines for spherical triangles* can be used to solve most spherical triangles, and are:

Law of Sines for spherical triangles

$$\frac{\sin a}{\sin A} = \frac{\sin b}{\sin B} = \frac{\sin c}{\sin C}$$

Law of Cosines for spherical triangles

$\cos a = \cos b \cos c + \sin b \sin c \cos A$

$\cos b = \cos c \cos a + \sin c \sin a \cos B$

$\cos c = \cos a \cos b + \sin a \sin b \cos C$

Index

of variables, 102
of vectors, 293–295

U

unit circles, 120–125
  trigonometric equations, solving,
    276–286
unit vectors, 295, 298–300

V

values
  absolute, 7
  absolute, functions, 100
  absolute, vectors, 295
  angles, 124
  of cross products, 316
  of decimals, comparing, 9
  domains, 187–188
  graphs of functions, 178
  multiple, 188
  negative, 190
  positive, 190
  principal, 186, 197
  ranges, 187–188
variables
  functions with multiple, 102–103
  linear two-variable functions, 103
vectors, 318, 358, 379
  addition, 300–304
  coordinate systems, 295–298
  directions of, 303
  examples of simple problems, 304–309
  overview of, 293–295
  positions, 352
  products, 314–317, 319
  radius, 104
  scalars, multiplication with, 309–310
  subtraction, 300–304
  unit, 298–300
  zeros, 300
velocity
  angular, 125–127, 131
  linear, 125–127
  tangents, 133
  vectors, 294
verifying
  identities, 225–227
  product theorem, 365
vertical
  angles, 22
  axis, 40

distances, 347
lines, 332, 339
lines, tests, 94, 178
shifting, 145, 174
shifting, cosines, 151–153
shifting, sines, 151–153
sides, 137
stretching, 145
vertices, 20, 40
  hyperbolas, 44
  points of parabolas, 41
voltage, 159
volume of cubes, 45

W

water waves, 163
waveforms, 132
wave patterns, 134
whole numbers, 4

X

X-axis, 11–12
  absolute value functions, 100
  complex numbers, 358
  functions, 93
  hyperbolas, 44
  polar coordinate system, 322
  references, 129
  reflection, 347
  roots, 266
  standard position angles, 108
  vectors, 303
  vertical sides, 138
x-intercept method, 270, 272, 287

Y

Y-axis, 11
  circular functions, 130
  complex numbers, 358
  functions, 93
  hyperbolas, 44
  inverse trigonometric functions, 186
  reflection, 347

Z

Z-axis, 11, 13, 341
  zeros, 4–5, 175
  vectors, 294, 300